Managing Engineering and Technology

Seventh Edition

Managing Engineering and Technology

Seventh Edition

Lucy C. Morse

Associate Professor—Emerita

University of Central Florida

William J. Schell IV

Associate Professor

*Norm Asbjornson College of Engineering,
Montana State University*

Daniel L. Babcock

Professor—Emeritus

Missouri University of Science and Technology

Senior Vice President Courseware Portfolio Management: Marcia J. Horton
Director, Portfolio Management: Engineering, Computer Science & Global Editions: Julian Partridge
Executive Portfolio Manager: Holly Stark
Portfolio Management Assistant: Emily Egan
Managing Content Producer: Scott Disanno
Content Producer: Carole Snyder
Rights and Permissions Manager: Ben Ferrini
Manufacturing Buyer, Higher Ed, Lake Side Communications, Inc. (LSC): Maura Zaldivar-Garcia and Deidra Smith

Inventory Manager: Bruce Boundy
Product Marketing Manager: Yvonne Vannatta
Field Marketing Manager: Demetrius Hall
Marketing Assistant: Jon Bryant
Cover Designer: Black Horse Designs, Inc.
Cover Photo: Ioana Davies/123RF
Printer/Binder: LSC Communications, Inc.
Full-Service Project Management: Integra Publishing Services, Inc.

Contents

Preface   **xi**
Acknowledgments   **xiii**

Part I   Introduction to Engineering Management   **1**

Chapter 1   Engineering and Management   **3**

Preview   3
Learning Objectives   3
Engineering   4
Management   7
Engineering Management: A Synthesis   13
Discussion Questions   17
Sources   18
Statistical Sourcebook   19

Chapter 2   Historical Development of Engineering Management   **20**

Preview   20
Learning Objectives   20
Origins   21
The Industrial Revolution   23
Management Philosophies   27
Scientific Management   27
Administrative Management   33
Behavioral Management   35
Contemporary Contributions   39
Discussion Questions   42
Sources   43

Part II Functions of Technology Management 45

Chapter 3 Leading Technical People 47

Preview 47
Learning Objectives 48
Leadership 48
Motivation 58
Motivating and Leading Technical Professionals 66
Discussion Questions 70
Sources 72
Statistical Sourcebook 73

Chapter 4 Planning and Forecasting 74

Preview 74
Learning Objectives 75
Nature of Planning 75
The Foundation for Planning 76
Some Planning Concepts 81
Forecasting 84
Strategies for Managing Technology 91
Discussion Questions 94
Problems 94
Sources 95
Statistical Sourcebook 95

Chapter 5 Decision Making 96

Preview 96
Learning Objectives 97
Nature of Decision Making 97
Management Science 99
Tools for Decision Making 102
Computer-Based Information Systems 114
Implementation 115
Discussion Questions 115
Problems 115
Sources 117

Chapter 6 Organizing 118

Preview 118
Learning Objectives 119

Nature of Organizing   119
Traditional Organization Theory   121
Technology and Modern Organization Structures   128
Teams   130
Discussion Questions   134
Sources   135

Chapter 7   Some Human Aspects of Organizing   **136**

Preview   136
Learning Objectives   137
Staffing Technical Organizations   137
Authority and Power   146
Delegation   148
Committees   150
Teams   151
Discussion Questions   152
Sources   152
Statistical Sourcebook   153

Chapter 8   Controlling   **154**

Preview   154
Learning Objectives   155
The Process of Control   155
Financial Controls   158
Human Resource Controls   166
Discussion Questions   168
Problems   169
Sources   170

Part III   Managing Technology   **171**

Chapter 9   Managing Research and Development   **173**

Preview   173
Learning Objectives   174
Product and Technology Life Cycles   174
Nature of Research and Development   176
Research Strategy and Organization   177
Selecting R&D Projects   178
Making R&D Organizations Successful   181
Creativity, Innovation, Entrepreneurship   190
Discussion Questions   195

Problems   196
Sources   196

Chapter 10   Managing Engineering Design   198

Preview   198
Learning Objectives   199
Nature of Engineering Design   199
Systems Engineering/New Product Development   200
Concurrent Engineering   203
Control Systems in Design   205
Design Criteria   209
Other Criteria in Design   217
Discussion Questions   221
Problems   221
Sources   222

Chapter 11   Planning Production Activity   223

Preview   223
Learning Objectives   224
Introduction   224
Planning Manufacturing Facilities   227
Quantitative Tools in Production Planning   232
Production Planning and Control   238
Manufacturing Systems   244
Discussion Questions   248
Problems   249
Sources   249

Chapter 12   Managing Quality and Production Operations   251

Preview   251
Learning Objectives   252
Assuring Product Quality   252
Total Quality Management   258
Productivity   264
Maintenance and Facilities (Plant) Engineering   268
Other Manufacturing Functions   271
Discussion Questions   273

Problems   273
Sources   274

Chapter 13   Engineers in Marketing and Service Activities   **276**

Preview   276
Learning Objectives   277
Marketing and the Engineer   277
The Process of Marketing   277
The 4Ps of the Marketing Mix   279
Marketing and Engineers—Partnerships, R&D, and Technical Sales   279
Engineers in the Service Economy   281
Discussion Questions   285
Sources   286

Part IV   Managing Projects   287

Chapter 14   Project Planning and Acquisition   **289**

Preview   289
Learning Objectives   290
Characteristics of a Project   290
The Project Proposal Process   291
Project Planning Tools   293
Monitoring and Controlling   305
Discussion Questions   310
Problems   311
Sources   313

Chapter 15   Project Organization, Leadership, and Control   **314**

Preview   314
Learning Objectives   315
Project Organization   315
The Project Manager   322
Motivating Project Performance   324
Types of Contracts   330
Discussion Questions   331
Sources   332

Part V Managing Your Engineering Career

333

Chapter 16 Engineering Ethics

335

Preview 335
Learning Objectives 335
Professional Ethics and Conduct 336
Engineering Codes of Ethics 339
Corporate Codes of Ethics 341
Ethical Problems in Consulting and Construction 342
Ethical Problems in Industrial Practice 347
Summary: Making Ethical Decisions 355
Discussion Questions 358
Sources 358
Case Study Websites 359

Chapter 17 Achieving Effectiveness as an Engineer

360

Preview 360
Learning Objectives 361
Getting off to the Right Start 361
Charting Your Career 365
Communicating Your Ideas 367
Staying Technically Competent 373
Professional Activity 376
Diversity in Engineering and Management 379
Management and the Engineer 381
Managing Your Time 386
Discussion Questions 389
Sources 390

Chapter 18 Globalization and Challenges for the Future

393

Preview 393
Learning Objectives 394
Globalization 394
Engineering Grand Challenges 403
Future Considerations in Engineering and Management 405
Discussion Questions 407
Sources 408
Global Websites 408

Index

409

Preface

The seventh edition of *Managing Engineering and Technology* maintains the focus of prior editions on supporting the growth of engineers into engineering managers, while considering the changing professions of engineering and engineering management. Engineers are talented problem solvers who often lack the training or expertise to solve problems through others. To build that expertise, *Managing Engineering and Technology* provides readers with the foundations of engineering management in five parts. In Part 1, we introduce the concept of engineering management, its relationship to engineering, and its historical underpinnings. In Part 2, we provide the core of management thought, including the four traditional roles of management—planning, organizing, leading, and controlling—with a particular focus on leadership and how leadership fits into an engineering management context. In addition, we provide tools to understand how human motivation and leadership are used to promote effectively working with and managing technical professionals. In Part 3, we explore both traditional (e.g., managing design) and non-traditional (e.g., managing marketing) roles of the engineering manager when managing technology. In Part 4, we provide an overview of the Project Management process. And in Part 5, we provide tools and discuss key topics needed to be successful in an engineering management career, including an exploration of engineering ethics, tools for career management, and key concepts of the forces changing the worlds of engineering and engineering management, including globalization.

WHAT'S NEW IN THIS EDITION?

This edition welcomes a new author with substantial experience as a practicing engineering manager and engineering management educator. That change brings a number of new and improved content to assist in the development of students' engineering management skills. The text is updated throughout, with new and revised content in each chapter. Key enhancements include:

- **New vignettes** in each chapter that explore modern developments in the management of engineering and technology with an application or example tied to the material from that chapter. These include discussions of highly successful engineering managers and those with headline grabbing ethical failings.
- **An entirely rewritten Chapter 13 on marketing**, with a focus on the movement toward digital marketing and how an engineering toolset can be used in this data-driven world.

- **Extensive new material in Chapter 16 on ethics**, with new ethical models incorporated that are typically easier for undergraduate students to relate to and utilize.
- **A substantially rewritten Chapter 18**, updated to reflect the current state of globalization and aspects of political unrest around the world.
- **Refreshed and updated content** in each chapter that highlights current trends and topics, such as the many roles of engineering in Amazon, Inc., updated leadership models, and examinations of leadership and management from beyond the Western world. In addition, changes to modernize the language and make it more welcoming and inclusive were made throughout the text. These updates included substantial streamlining of several chapters, reducing the overall text length by 10% while maintaining all key concepts and content.

FOR THE INSTRUCTOR

All of these considerable enhancements were made while retaining the same organization and topical flow from the sixth edition to allow for a smoother adoption for instructors. At the same time, this new material has led to considerable changes in the exercises for each chapter. Most chapters have 25%+ new questions from the prior edition. An updated instructor's guide and chapter slides are available at www.pearson.com/engineering-resources.

Acknowledgments

Together we'd like to thank Dan Babcock for his initial vision for this text and all of our colleagues at the American Society for Engineering Management and the American Society for Engineering Education who have helped clarify our thinking and writing about Engineering Management and Engineering Management education over the years. Those acknowledged in prior editions whose work continues to contribute to this edition include Henry Metzner (Missouri S&T), Jean Babcock, Ted Eschenbach (U. Alaska-Anchorage), Thomas A. Crosby (Pal's Sudden Service), Charles W. Keller (U. Kansas), Brian Goldiez (U. Central Florida), Nabeel Yousef (Daytona State College), and C. Steven Griffin (CSR).

Notable supporters for the thinking that went into this edition include Craig Downing (Rose-Hulman), Ted Eschenbach (U. Alaska-Anchorage), and Paul Kauffmann (East Carolina University). In addition, we thank our students who, over the years, have both intentionally and unwittingly helped us to identify opportunities to improve the text and areas of new knowledge that needed to be incorporated. For this edition we are extremely grateful to Norm Asbjornson of AAON, Inc. and Doug Melton of the Kern Entrepreneurial Engineering Network for use of their materials when developing the vignettes for Chapters 4 and 9, and to the team at Pearson for their guidance and support.

Most importantly, we thank our families for their continued support and encouragement. Without the love and patience of Jack Selter and Melanie, Ana, Megan, and William Griffin Schell the journey to create this text would not have been possible.

Part I

Introduction to Engineering Management

1

Engineering and Management

PREVIEW

Today's technological society is constantly changing, and with this change comes a need for the engineer to be able to address society's technological challenges as well as the opportunities for the future. Engineers play a key role in maintaining technological leadership and a sound economy as the world becomes flatter in today's global economy. To do this, the engineer needs to remain alert to changing products, processes, technologies, and opportunities. To make the transition from successful engineer to successful engineering manager, engineers must learn and apply a new set of tools.

To assist the engineer prepare for a productive life and position of leadership, this chapter begins with a discussion of the origins of engineering practice and education, the nature of the engineering profession, and the types of engineers, their work, and their employers. Next, management is defined and managerial jobs and functions are characterized. Finally, these topics are synthesized by defining engineering management and a discussion of the expectation of managerial responsibilities in an engineering career.

LEARNING OBJECTIVES

When you have finished studying this chapter, you should be able to do the following:

- Describe the origins of engineering practice.
- Identify the functions of management.
- Explain what engineering management is.
- Explain the need for engineers in management.

ENGINEERING

Origins of Engineering

The words *engineer* and *ingenious* both stem from the Latin *ingenium,* which means a talent, natural capacity, or clever invention. Early applications of *clever inventions* often were military ones, and *ingeniarius* became one of several words applied to builders of such *ingenious* military machines.

Heritage of the Engineer.   By whatever name, the roots of engineering lie much earlier than the time of the Romans, and the engineer today stands on the shoulders of giants. William Wickenden said this well in 1947:

> Engineering was an art for long centuries before it became a science. Its origins go back to utmost an-
> tiquity. The young engineer can say with truth and pride, "I am the heir of the ages. Tubal Cain, whom
> Genesis places seven generations after Adam and describes as the instructor of every artificer in brass and
> iron, is the legendary father of my technical skills. The primitive smelters of iron and copper; the ancient
> workers in bronze and forgers of steel; the discoverers of the lever, the wheel, and the screw; the daring
> builders who first used the column, the arch, the beam, the dome, and the truss; the military pioneers who
> contrived the battering ram and the catapult; the early Egyptians who channeled water to irrigate the land;
> the Romans who built great roads, bridges, and aqueducts; the craftsmen who reared the Gothic cathe-
> drals; all these are my forbears. Nor are they all nameless. There are: Hero of Alexandria; Archimedes of
> Syracuse; Roger Bacon, the monk of Oxford; Leonardo da Vinci, a many-sided genius; Galileo, the father
> of mechanics; Volta, the physician; the versatile Franklin. Also, there are the self-taught geniuses of the
> industrial revolution: Newcomen, the ironmonger; Smeaton and Watt, the instrument makers; Telford, the
> stone mason; and Stephenson, the mine foreman; Faraday and Gramme; Perronet, Baker, and Roebling;
> Siemens and Bessemer; Lenoir and Lavassor; Otto and Diesel; Edison, Westinghouse, and Steinmetz; the
> Wright brothers, and Ford. These are representative of the trail blazers in whose footsteps I follow."

Beginnings of Engineering Education.   Florman contrasts the French and British traditions of engineering education in his *Engineering and the Concept of the Elite,* and the following stems both from that
and from Daniel Babcock's writings. In 1716 the French government, under Louis XV, formed a civilian
engineering corps, the *Corps des Ponts et Chausées*, to oversee the design and construction of roads and
bridges, and in 1747 founded the *Ecole des Ponts et Chausées* to train members of the corps. This was the
first engineering school in which the study of mathematics and physics was applied not only to roads and
bridges, but also to canals, water supply, mines, fortifications, and manufacturing. The French followed by
opening other technical schools, most notably the renowned *Ecole Polytechnique* under the revolutionary
government in 1794. In England, on the other hand, gentlemen studied the classics, and it was not until
1890 that Cambridge added a program in *mechanical science,* and 1909 when Oxford established a chair
in *engineering science.* True, the Industrial Revolution began in England, but *[k]nowledge was gained
pragmatically, in the workshop and on construction sites, and engineers learned their craft—and such
science as seemed useful, by apprenticeship.*
   America is heir to both traditions. Harvard and other early colleges followed the British classical
tradition, and during the Revolutionary War, we borrowed engineers from France and elsewhere to
help build (and destroy) military roads, bridges, and fortifications. "In the early days of the United

States, there were so few engineers—less than 30 in the entire nation when the Erie Canal was begun in 1817—that America had no choice but to adopt the British apprenticeship model. The canals and shops—and later the railroads and factories—were the 'schools' where surveyors and mechanics were developed into engineers. As late as the time of World War I, half of America's engineers were receiving their training 'on the job.'"

The U.S. Military Academy was established in 1802, at the urging of Thomas Jefferson and others, as a school for engineer officers, but they did not distinguish themselves in the War of 1812. Sylvanus Thayer, who taught mathematics at the Academy, was sent to Europe to study at the *Ecole Polytechnique* and other European schools; on his return in 1817 as superintendent of the Academy, he introduced a four-year course in civil engineering, and hired the best instructors he could find. As other engineering schools opened, they followed this curriculum and employed Academy graduates to teach from textbooks authored by Academy faculty. Florman continues:

> Perhaps the most crucial event in the social history of American engineering was the passage by Congress of the Morrill Act—the so-called "land grants" act—in 1862. This law authorized federal aid to the states for establishing colleges of agriculture and the so-called "mechanic arts." The founding legislation mentioned "education of the industrial classes in their several pursuits and professions in life." With engineering linked to the "mechanic arts," and with engineers expected to come from the "industrial classes," the die was cast. American engineers would not be elite polytechnicians. They would not be gentlemen attending professional school after graduation from college [as law and medicine became]....Engineering was to be studied in a four-year undergraduate curriculum.

Engineering as a Profession

The first issue (1866) of the English journal *Engineering* began with a description of

> the profession of the engineer as defined in the charter that Telford obtained [in 1818 for the Institute of Civil Engineers] for himself and his associates from [King] George the Fourth—"the art of directing the great sources of power in nature, for the use and convenience of man."

A more modern definition was created in 1979 by American engineering societies, acting together through the Engineers' Council for Professional Development (ECPD), the precursor to ABET (previously the Accrediting Board for Engineering and Technology). ECPD's definition focused on the application of math and science knowledge to develop novel solutions for the benefit of mankind.

This definition was modernized again in 2013 by the International Engineering Alliance (whose members include ABET). This update expands the definition to acknowledge the potential adverse consequences of engineering activity and note the ethical responsibility of engineers to to manage these risks and safeguard society and the environment.

Certainly, engineering meets all the criteria of a proud profession. Engineering undergraduates recognize the need for "intensive preparation" to master the specialized knowledge of their chosen profession, and practicing engineers understand the need for lifelong learning to keep up with the march of technology. In Part V of this book, we look at engineering societies and their ethical responsibilities in maintaining standards of conduct. Finally, engineers provide a public service not only in the goods and services they create for the betterment of society, but also by placing the safety of the public high on their list of design criteria. Each generation of engineers has the opportunity and obligation to preserve and enhance by its actions the reputation established for this profession by its earlier members.

What Engineers Do

Engineering.   Before a description of engineers can be made, the term *engineering* must be defined. We can define engineering as follows:

> En-gi-neer-ing *n*: a branch of science and technology concerned with the invention, design, building, maintenance, and improvement of structures, machines, devices, systems, materials, and processes.

In other words, engineering is the means by which people make possible the realization of human dreams by extending our reach in the real world. Engineers are the practitioners of the art of managing the application of science and mathematics, a practice that is generally accomplished through projects. By this description, engineering has a limitless variety of possible disciplines.

Engineers.   Engineering has been differentiated from other academic paths by the need for people to logically apply quantifiable principles. Academic knowledge, practical training, experience, and work-study are all avenues to becoming an engineer. The key attribute for engineers is the direct application of that knowledge and experience. The most up-to-date information on opportunities available for engineers can be found at various websites on the internet, industry publications, professional associations, and personal contacts within industry. Like other fields of endeavor, engineering no longer represents a static career choice. The basic idea is to be adept, adaptable, and aware.

Types of Engineers.   The rigid classification of engineers into specific specialties and careers has been eroding swiftly. Many engineering applications require cross-pollination or integration of multiple disciplines. Aerospace engineers require knowledge of metallurgy, electronic control systems, computers, production limitations and possibilities, finance, life cycle logistic planning, and customer service. These are all required to produce a viable commercial product such as an airliner or a fighter. The previous focus on a speciality is no longer as important as being able to communicate and team with others. These teams are composed of various specialists knowledgeable in several primary fields. The primary specialization allows the engineer to contribute in a core area. This knowledge is required to properly integrate and implement the ideas of others. Along those lines, the list of core technologies is expanding and mutating rapidly. During the early age of computers, the late 1950s, software engineers were electrical engineers. The computer operating systems were custom tailored to the internal logic design. As advances in design created the need for software specialists, the electrical engineers evolved into software engineers. Today, software engineers are split among the various types of applications. Desktop, internet, server, Internet of Things (IoT), and mobile operating system gurus are eagerly sought in a wide variety of industries. A similar process can be observed in construction, mechanical systems, chemical engineering, and industrial engineering. Another indicator of the change in engineering has been the development of the field of engineering technology. Engineering technology emerged in direct response to industry needs for a person having a practical applications education. Experience and training will increasingly determine an engineer's actual specialty. Adding to the confusion is the expectation that a person will change careers five or more times in their life, a trend that accelerates with each new generation. Flexibility and interpersonal skills will be the hallmark of the new generation of engineering disciplines.

Engineering Employment.   Traditional paths for a career in engineering have mirrored other fields of employment. Rarely will a person work for the same employer for their entire working lifetime. The simple fact is that the corporations and firms of the past no longer exist. Those currently in existence will have to change to meet the needs of customers. Employment opportunities lie with companies of all sizes. Greater size can mean greater work stability, albeit usually limited flexibility. This limitation is accompanied by the fact that larger firms have greater resources to implement change. A smaller firm may be less stable, but can rapidly adapt to changing circumstances. Unfortunately, smaller firms have fewer resources to respond to the changing circumstances. This means that engineers of the future should expect to be constantly improving their skills and marketability. Continuing education, flexibility, and a willingness to shift employment will be required of successful engineers.

Government employment traditionally meant steady employment with a relatively secure career path. This situation changed as government embraced business-based practices to reduce costs by outsourcing and contracting. A greater reliance on information technologies also reduced the workforce requirements through better communications. Although a large number of engineers remain employed by various governmental agencies, their main focus is evolving into oversight managers and controllers. Seniority currently guides progression in government service. However, the same forces found in the civilian market will generate a similar need in government employment for flexibility, continuing education, and willingness to switch jobs.

Engineering Jobs in an Organization.   Organizations of all types, from manufacturing to retail and financial services to government offer many types of jobs for engineers. Figure 1-1 displays a depiction of a basic organization chart for Amazon.com, along with some of the types of engineering positions available within the company. Amazon is a large and complex business, with engineers in roles throughout, including many in technology, research and design (R&D), and operations. The role of engineering positions in research and design is discussed in Chapters 9 and 10. Engineering functions in operations are discussed in Chapters 11 and 12. The more technically complex the product, the more engineers will be involved in technical sales, field service engineering, and logistics support, as discussed in Chapter 13. Finally, we discuss how in today's age of technical complexity, many general management positions are held by engineers.

MANAGEMENT

Management Defined

The Australian Edmund Young, in supplementary notes used in teaching from the original edition of this chapter, wrote that

> "[m]anagement" has been one of the most ubiquitous and misused words in the 20th century English language. It has been a "fad" word as well. Civil engineers discuss river basin management and coastal management, doctors discuss disease management and AIDS management, and garbage collectors are now waste management experts.

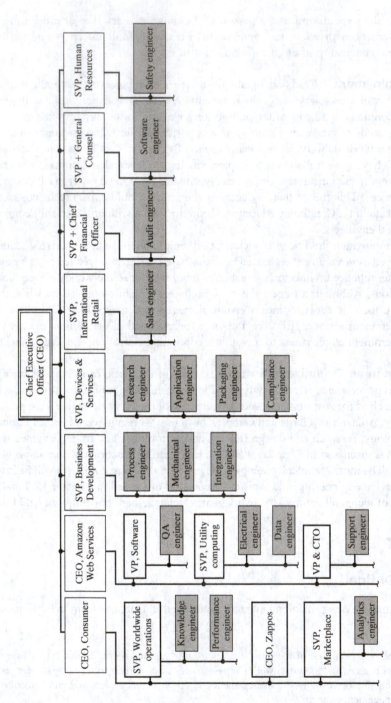

Figure 1-1 Selection of engineering roles in the organization of Amazon.com.

8

McFarland traces the meaning of the words *manage* and *management* as follows:

> The word *manage* seems to have come into English usage directly from the Italian *maneggiare*, meaning "to handle," especially to handle or train horses. It traces back to the Latin word *manus*, "hand." In the early sixteenth century *manage* was gradually extended to the operations of war and used in the general sense of taking control, taking charge, or directing....*Management* was originally a noun used to indicate the process for managing, training, or directing. It was first applied to sports, then to housekeeping, and only later to government and business.

McFarland continues by identifying "four important uses of the word *management*, as (1) an organizational or administrative process; (2) a science, discipline, or art; (3) the group of people running an organization; and (4) an occupational career." Sentences illustrating each of these in turn might be (1) "He practices good management"; (2) "She is a management student"; (3) "Management *doesn't really believe* in quality"; and (4) (heard from innumerable college freshmen) "I want to get into management." Of these four, most authors of management textbooks are referring to the first meaning (the *process*) when they define "management." According to some of these authors, management is defined in the following ways:

- The work of creating and maintaining environments in which people can accomplish goals efficiently and effectively (Albanese)
- The process of achieving desired results through efficient utilization of human and material resources (Bedeian)
- The process of reaching organizational goals by working with and through people and other organizational resources (Certo)
- A set of activities (including planning and decision making, organizing, leading, and controlling) directed at an organization's resources (human, financial, physical, and information) with the aim of achieving organizational goals in an efficient and effective manner (Griffin)
- The process by which managers create, direct, maintain, and operate purposive organizations through coordinated, cooperative human effort (McFarland)
- The process of acquiring and combining human, financial, informational, and physical resources to attain the organization's primary goal of producing a product or service desired by some segment of society (Pringle, Jennings, and Longnecker)

Albanese provides a set of definitions of the word *management* suggested by a sample of business executives:

- Being a respected and responsible representative of the company to your subordinates
- The ability to achieve willing and effective accomplishments from others toward a common business objective
- Organizing and coordinating a profitable effort through good decision making and people motivation
- Getting things done through people
- The means by which an organization grows or dies
- The overall planning, evaluating, and enforcement that goes into bringing about "the name of the game"—profit
- Keeping your customers happy by delivering a quality product at a reasonable cost
- Directing the actions of a group to accomplish a desired goal or objective in the most efficient manner

Management Levels

Ensign or admiral, college president or department chair, maintenance foreman, plant manager, or company president—all are managers. What skills must they have, what roles do they play, what functions do they carry out, and how are these affected by the level at which they operate? Let us look at each of these questions in order.

Management is normally classified into three levels: first-line, middle, and top. Managers at these three levels need many of the same skills, but they use them in different proportions. The higher the management level is, the further into the future a manager's decisions reach, and more resources placed at risk.

First-line managers directly supervise nonmanagers. They hold titles such as foreman, supervisor, or section chief. Generally, they are responsible for carrying out the plans and objectives of higher management, using the personnel and other resources assigned to them. They make short-range operating plans governing what will be done tomorrow or next week, assign tasks to their workers, supervise the work that is done, and evaluate the performance of individual workers. First-line managers may only recently have been appointed from among the ranks of people they are now supervising. They may feel caught in the middle between their former coworkers and upper management, each of which feels the supervisor should be representing them. Indeed, they must provide the *linking pin* between upper management and the working level, representing the needs and goals of each to the other.

Many engineers who go into a production or construction environment quickly find themselves assigned as a foreman or supervisor. The engineer may find such an assignment a satisfying chance to make things happen through their own actions and decisions. Doing so effectively, while according the workers the courtesy and respect merited by their years of experience, requires tact and judgment. If the engineer can achieve this balance, they may be surprised to find that the team members are respectful in return and are helpful to the engineer in learning *their* job.

Middle managers carry titles such as plant manager, division head, chief engineer, or operations manager. Although there are more first-line managers than any other in most organizations, most of the *levels* in any large organization are those of middle management. Even the lowest middle manager (the second-line manager, who directly supervises first-line managers) is an *indirect manager* and has the fundamentally different job of managing the efforts of employees through other managers. Middle managers make intermediate range plans to achieve the long-range goals set by top management, establish departmental policies, and evaluate the performance of the units and people in their organization. Middle managers also integrate and coordinate the short-range decisions and activities of first-line supervisory groups to achieve the long-range goals of the enterprise. Over the last two generations, middle management positions have decreased as organizations became "flatter" in an effort to become more competitive and get closer to their customers.

Top managers bear titles such as chairman of the board, president, or executive vice president; the top one of these will normally be designated *chief executive officer* (CEO). In government, the top manager may be the administrator (of NASA), secretary (of state or commerce), governor, or mayor. While top managers may report to some policymaking group (the board of directors, legislature, or council), they have no full-time manager above them.

Top managers are responsible for defining the character, mission, and objectives of the enterprise. They must establish criteria for and review long-range plans. They evaluate the performance of major departments, and evaluate leading management personnel to gauge their readiness for promotion to key executive positions. Bedeian paints a picture of the typical top manager: a college graduate (85 percent), probably with some postgraduate work (58 percent) and often a graduate degree (40 percent); usually from a middle-class background, often born to parents in business or a profession; age 50 to 65, with work experience concentrated in one, two, or three companies; and with a work week of 55 to 65 hours. Often, an organization will look for a top manager with particular strength in the functional area in which the enterprise is currently facing a challenge.

Managerial Skills

Katz suggests that managers need three types of skills: technical, interpersonal, and conceptual. *Technical skills* are skills (such as engineering, accounting, machining, or word processing) practiced by the group supervised. Figure 1-2 shows that the lowest level managers have the greatest need for technical skills, since they are directly supervising the people who are doing the technical work, but even top managers must understand the underlying technology on which their industry is based. *Interpersonal skills*, on the other hand, are important at every management level, since every manager achieves results through the efforts of other people. *Conceptual skills* represent the ability to "see the forest for the trees"—to discern the critical factors that will determine an organization's success or failure. This ability is essential to the top manager's responsibility for setting long-term objectives for the enterprise, although it is necessary at every level.

Managerial Roles—What Managers Do

Henry Mintzberg gives us another way to view the manager's job by examining the varied *roles* a manager plays in the enterprise. He divides them into three types: *interpersonal*, *informational*, and *decisional* roles, further described as follows:

- *Interpersonal* roles are primarily concerned with the manager's interactions with other people. This role can be as figurehead, focused on appearances and outward relationships; leader, focused on people below them in the organization; and liaison, focused on horizontal relationships and networking.

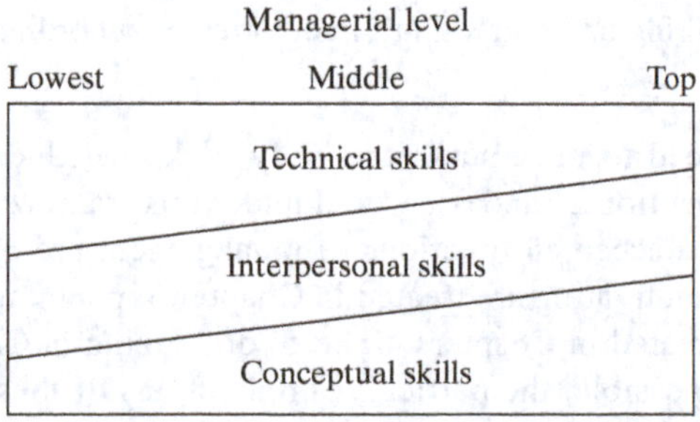

Figure 1-2   Blend of skills required at different management levels.

- *Informational* roles encompass how a manager exchanges and processes information. This role can be viewed as monitor, who collects information from inside and outside the organization; disseminator, who provides information to others within the organization; and spokesperson, who provides information to those outside the organization.
- *Decisional* roles describe how a manager uses information to make decisions. This role includes the entrepreneur, who initiates change and assumes risk; disturbance handler, who works to resolve problems or crises; resource allocator, who determines how the organization's resources of time and money are distributed; and negotiator, who handles bargaining and agreements inside and outside the organization.

Functions of Managers

Henri Fayol, the famous French mining engineer and executive, divided managerial activities into five elements: planning, organizing, command, coordination, and control. These elements, now called **functions of managers**, have proven remarkably useful and durable over the decades. Although each management author has their favored set of functions, almost all include planning, organizing, and controlling on their list. Command has become too authoritative a word in today's participative society and has been replaced by leading, motivating, or actuating. Few authors treat coordinating as a separate function. Nonetheless, as the late management author Harold Koontz concluded, "There have been no new ideas, research findings, or techniques that cannot readily be placed in these classifications." Koontz chose and (with coauthor Heinz Weihrich) defined his favored list of the functions of managers as follows:

- **Planning** involves selecting missions and objectives and the actions to achieve them; it requires decision making—choosing future courses of action from among alternatives.
- **Organizing** is that part of managing that involves establishing an intentional structure of roles for people to fill in an enterprise.
- **Staffing** [included with *organizing* by most authors] involves filling, and keeping filled, the positions in the organizational structure.
- **Leading** is influencing people to strive willingly and enthusiastically toward the achievement of organization and group goals. It has to do predominantly with the interpersonal aspect of managing.
- **Controlling** is the measuring and correcting of activities of subordinates to ensure that events conform to plans.

Engineering managers need to understand the body of knowledge that has been developed by management theorists and practitioners and organized under this framework, and this is the purpose of Part II of this book. Today the accepted functions of management are planning, organizing, leading, and controlling. Leading and motivating are treated in Chapter 3, planning and the associated subfunction of decision making are treated in Chapters 4 and 5, organizing in Chapters 6 and 7, and controlling in Chapter 8. Wherever possible, the particular implications of these functions for the technical employee and the technology-affected organization are emphasized.

The engineering manager also needs to understand the particular problems involved in managing research, development, design, production/operations, projects, and related technical environments. Parts III and IV treat the application of these management functions to the specific environments in which most engineers and engineering managers will work.

Management: Art or Science?

Earlier in this chapter the characteristics of a profession were discussed, and engineering was shown to meet all the criteria of a profession. Management also has a body of *specialized knowledge*, which is introduced in Part II. Many managers will have first completed bachelor's or master's degree programs in business administration, public administration, or engineering management, but the following applies, as Babcock has observed elsewhere:

> The knowledge need not be obtained only in such formal programs. It may be acquired by personal study, in-house employee education programs, seminars by all kinds of consultant entrepreneurs, or programs of many professional societies. Sometimes this formal or informal education is obtained before promotion [into] the management hierarchy, but often it occurs after promotion.

> A very small proportion of the broad range of managers belong to management-specific organizations such as the American Management Association, the Academy of Management, or (for engineers) the American Society for Engineering Management. They are more likely (especially in technical areas) to belong to management divisions or institutes within discipline-oriented professional societies. Considerations of standards, ethics, certification, and the like become those of the parent societies, not the management subset.

ENGINEERING MANAGEMENT: A SYNTHESIS

What Is Engineering Management?

Some writers would use a narrow definition of "engineering management," confining it to the direct supervision of engineers or of engineering functions. This would include, for example, supervision of engineering research or design activities. Others would add an activity we might consider the *engineering of management*—the application of quantitative methods and techniques to the practice of management (often called *management science*). However, these narrow definitions fail to include many of the management activities engineers actually perform in modern enterprises.

If engineering management is broadly defined to include the general management responsibilities engineers can grow into, one might well ask how it differs from *ordinary* management.

> The engineering manager is distinguished from other managers because they possesses both an ability to apply engineering principles and a skill in organizing and directing people and projects. They are uniquely qualified for two types of jobs: the management of *technical functions* (such as design or production) in almost any enterprise, or the management of broader functions (such as marketing or top management) in a *high-technology enterprise*.

Other Engineering Management Definitions

Engineering management is the art and science of planning, organizing, allocating resources, and directing and controlling activities that have a technological component.	American Society for Engineering Management
Engineering management is designing, operating, and continuously improving purposeful systems of people, machines, money, time, information, and energy by integrating engineering and management knowledge, techniques, and skills to achieve desired goals in techno- logical enterprise through concern for the environment, quality, and ethics.	Omurtag (1988)
Engineering management is the discipline addressed to making and implementing decisions for strategic and operational leadership in current and emerging technologies and their impacts on interrelated systems.	IEEE (1990) and Kocaoglu (1991)

Source: Timothy Kotnour and John V. Farr, "Engineering Management: Past, Present, and Future," *Engineering Management Journal*, vol. 17, no. 1, March 2005.

Need for Engineers in Management

Herbert Hoover, a very successful mining engineer and manager, recognized the importance of the American engineering manager in an address to engineers the year he was elected president of the United States:

> Three great forces contributed to the development of the engineering profession. The first was the era of intense development of minerals, metallurgy, and transportation in our great West....Moreover, the skill of our engineers of that period owes a great debt to American educators. The leaders of our universities were the first of all the educators of the world to recognize that upon them rested the responsibility to provide fundamental training in the application of science to engineering under the broadening influence and cultivation of university life. They were the first to realize that engineering must be transformed into a practice in the highest sense, not only in the training and character but that the essential quality of a profession is the installation of ethics....A third distinction that grew in American engineering was the transformation from solely a technical profession to a profession of administrators—the business manager with technical training.

There are several reasons engineers can be especially effective in the general management, especially in technically oriented organizations. High-technology enterprises make a business of doing things that have never been done before. Therefore, extensive planning is needed to make sure that every- thing is done right the first time—there may not be a second chance. Planning must emphasize recognizing and resolving the uncertainties that determine whether the desired product or outcome is feasible. Since these critical factors are often technical, the engineer is best capable of recognizing them and managing their resolution. In staffing a technically based enterprise, engineering managers can best evaluate the capability of technical personnel when they apply for positions and rate their later performance. Further, they will better understand the nature and motivation of the technical specialist and can more easily gain their respect, confidence, and loyalty. George H. Heilmeier, president and CEO of Bellcore (and an electrical engineer), makes clear the advantages of an understanding of technology in top management:

Competition is global, and the ability to compete successfully on this scale is fostered by corporate leaders who can do the following:

- Understand the business at a deep level.
- Understand both the technology that is driving the business today and the technology that will change the business in the future.
- Treat research and development as an investment to be nurtured, rather than an expense to be minimized.
- Spend more time on strategic thinking about the future as they rise higher in the corporation.
- Be dedicated to solving a customer's problem or satisfying a need.
- Place a premium on innovation.

Management and the Engineering Career

It is common for engineers to move into a management role or pursue management-related advanced degrees. A 2006 study by the National Science Foundation found over 15 percent of those employed 10 years after earning an engineering degree hold a management role. That percentage grows to over 20 percent about 10 years later. The Bureau of Labor Statistics currently records the total employment for engineering managers to be over 180,000, with a 5.5 percent increase expected by 2026. Despite this, undergraduate engineering education offers little preparation for such a possibility. To meet this need, many engineering schools now provide degree programs and courses in engineering management. These courses and programs blend business and engineering, as shown in Figure 1-3. Professional societies are an additional way engineers may improve their managerial skills with many providing a variety of educational opportunities. Many engineering related professional socities (e.g. the American Society of Mechanical Engineers, the Institute of Industrial and Systems Engineers, etc.) have subgroups for engineering management. In addition, the American Society for Engineering Management (ASEM) is solely dedicated to development of Engineering Management professionals.

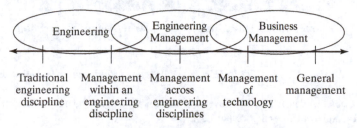

Figure 1-3   The field of engineering management.

Tim Cook: Picture of the Successful Engineering Manager

As CEO of Apple, Tim Cook is one of the most famous business leaders in the world and almost every profile of him makes prominent mention of his background in engineering. Cook graduated from Auburn University in 1982 with a Bachelor of Science in Industrial Engineering and later earned an MBA from Duke University. The skills developed in his industrial engineering education and how he used those skills at IBM and Compaq are prominent reasons that Steve Jobs recruited Cook to join Apple in 1998 where his first position was Senior Vice President of Worldwide Operations.[1] Tasked with improving Apple's complex supply chain, at that time a severe drag on the company's performance—Apple lost nearly 33 percent of its market value in the two years prior to him joining—Cook developed a system that continues to enable the company's product innovation. The results were almost immediate, and two years after Cook's arrival, Apple's value had increased over 460 percent and today it is commonly the most valuable company in the world and the first to reach $1 trillion in valuation. Cook recognizes the importance of blending both engineering and management skills and as an advisor at his alma mater has pushed for both excellence in technical education and well-rounded engineers.[2] He has noted that as "an engineer, you want to analyze things a lot. But if you believe that the most important data points are people, then you have to make conclusions in relatively short order."[3] As CEO, Cook has exemplified the tenants of the engineering profession to serve society and engineering ethics (see Chapter 16), dramatically increasing the company's charitable giving, social outreach, and recently stating "Whatever you do in your life, and whatever we do at Apple, we must infuse it with the humanity that each of us is born with."[4]

Source: Tobias Hase/dpa picture alliance/Alamy Stock Photo

Sources

1. Weinberger, M. *The rise of Apple CEO Tim Cook, the most powerful business leader in the world.* Business Insider, 2016.
2. Davis, K., *Tim Cook, Apple CEO and Auburn alum, discusses his time spent at Auburn*, in *The Auburn Plainsman*. 2014, Auburn University: Auburn, AL.
3. Lashinsky, A., *Apple's Tim Cook leads different*, in *Fortune*. 2015.
4. Cook, T. *Full transcript: Tim Cook delivers MIT'S 2017 commencement speech.* Quartz, 2017.

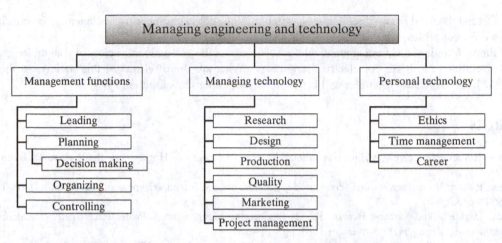

Figure 1-4   Managing engineering and technology text organization.

This book provides some insight into the nature of management and the environments in which the engineer is most likely to encounter the need for an understanding of management as their career progresses. Chapters 3 through 8 examine the functions of technology management. Chapters 9 through 13 examine the management of technology through the product life cycle. In the last three chapters, the career implications for the engineer moving to management are considered. The organization of these concepts within the book is shown in Figure 1-4.

DISCUSSION QUESTIONS

1-1. The precursors of today's engineers listed in the quotation from Wickenden had no classes and few or no books from which to learn scientific principles. How can you explain their success?

1-2. Create your argument for why *engineering management* is different than *management*. Why is this field needed?

1-3. Why is it so difficult to answer the simple question "How many engineers are there in the United States?" Is the question "How many physicians are there in the United States?" any easier? Why or why not?

1-4. Compare and contrast the role of the engineer with the role of the manager. How are they similar and how are they different?

1-5. What are the similarities in the definitions of *management* quoted from authors of management textbooks? How do you define *management*?

1-6. How does the job of supervisor or first-line manager differ from that of a middle manager?

1-7. Engineers often move into management of their organizations. Explain the ways that an engineering degree prepares an individual for this transition? What are the problems with this path?

1-8. Identify the three types of skills needed by an effective manager, as conceived by Robert L. Katz, and describe how the relative need for them might vary with the level of management.

1-9. Defend the need for engineering management. Why should engineering management be considered a different profession than simply "management"?

1-10. Find the engineering management related sub group for the professional society of your undergraduate discipline (e.g. IEEE for electrical engineers). What are the offerings of this society vs. those of ASEM? Which do you think will better serve your career development? Why?

SOURCES

Albanese, Robert, *Management: Toward Accountability for Performance* (Homewood, IL: Richard D. Irwin, Inc., 1975), p. 28.

Albanese, Robert, *Managing: Toward Accountability for Performance,* 3rd ed. (Homewood, IL: Richard D. Irwin, Inc., 1981), p. 5.

Babcock, Daniel L. and Sarchet, Bernard R., "Is Engineering Management a Profession?" *IEEE Transactions on Engineering Management,* November 1981, pp. 107–109.

Babcock, Daniel L., "Is the Engineering Manager Different?" *Machine Design,* March 9, 1978, pp. 82–85.

Bedeian, Arthur G., *Management,* 2nd ed. (New York: Holt, Rinehart and Winston, 1989), p. 6.

"Breaking Ground," *Engineering* [London], 1:1, January 5, 1866, p. 1.

Certo, Samuel C., *Modern Management: Diversity, Quality, Ethics, and the Global Environment,* 6th ed. (Needham Heights, MA: Allyn and Bacon, 1994), p. 6.

The Engineering Team (New York: Engineers' Council for Professional Development (now Accreditation Board for Engineering and Technology), 1979).

Fayol, Henri, *Administration Industrielle et Générale,* Constance Storrs, trans. (London: Sir Isaac Pitman & Sons Ltd., 1949).

Florman, Samuel C., "Engineering and the Concept of the Elite," *The Bridge,* National Academy of Engineering, Fall 1991.

Griffin, Ricky W., *Management,* 4th ed. (Boston: Houghton Mifflin Company, 1993), pp. 5–6.

Heilmeier, George H., "Room for Whom at the Top?: Promoting Technical Literacy in the Executive Suite," *THE BENT of Tau Beta Pi,* Spring 1994.

Hoover, Herbert C., "The Engineer's Contribution to Modern Life," an address to the American Institute of Mining and Metallurgical Engineers on receiving their Saunders Mining Medal at their 1928 annual meeting, reprinted in Dugald C. Jackson, Jr. and W. Paul Jones, eds., *The Profession of Engineering* (New York: John Wiley & Sons, Inc., 1929), pp. 119–120.

Katz, Robert L., "Skills of an Effective Administrator," *Harvard Business Review,* 52:5, September–October 1974, pp. 90–112.

McFarland, Dalton E., *Management: Foundations and Practices,* 5th ed. (New York: Macmillan Publishing Company, 1979), pp. 4–5.

Mintzberg, Henry, selected excerpts from *The Nature of Managerial Work,* Chapter 4. Copyright © 1973 by the author. Reprinted by permission of Harper Collins Publishers, Inc.

Pringle, Charles D., Jennings, Daniel F., and Longnecker, Justin G., *Managing Organizations: Functions and Behaviors* (Columbus, OH: Merrill Publishing Company, 1988), p. 4.

Weihrich, Heinz and Koontz, Harold, *Management: A Global Perspective,* 10th ed. (New York: McGraw-Hill Book Company, 1993), pp. 20–21.

Wickenden, William E., drafted before his 1947 death, later edited and collated by G. Ross Henninger as *A Professional Guide for Young Engineers,* rev. ed. (New York: Accreditation Board for Engineering and Technology, 1981), p. 7.

Young, Edmund J., personal communication, August 1988.

STATISTICAL SOURCEBOOK

The following is a useful source website. (May 2019)

http://www.bls.gov/ooh/Management/home.htm, *Occupational Outlook Handbook,* U.S. Bureau of Labor Statistics | Office of Occupational Statistics and Employment Projections, Washington, DC Management Occupations

2

Historical Development of Engineering Management

PREVIEW

The story of the development of management thought and of our ability to organize and control complex activities has already been documented. Two excellent books on this subject are by Claude George and Daniel Wren. In this chapter, only a small part of this history is introduced, concentrating on the people and situations of greatest significance to the engineer in management. First, the great construction projects of ancient civilizations are considered, and then the medieval production facility that was the Arsenal of Venice is discussed. Discussion of the Industrial Revolution examines its changes to manufacturing and society, first in England, and then in America.

As the nineteenth century ended and the twentieth century began, the United States led the world in finding better, more efficient ways to do things, in a movement that became known as **scientific management**, while Europeans such as Max Weber and Henri Fayol were developing philosophies of management at the top level. Around 1930, a series of experiments at the Hawthorne Works near Chicago led to studies on the impact of individual and group behavior on the effectiveness of managing. Engineering management continues to evolve, with the development in the second half of the twentieth century of methods for managing large projects such as the Apollo program, customer-centered organizations, globalization, and the information revolution.

LEARNING OBJECTIVES

When you have finished studying this chapter, you should be able to do the following:

- Describe the origins of engineering management.
- Identify the different basic management philosophies.
- Discuss the future issues that will affect the continued development of engineering management.

ORIGINS

Ancient Civilizations

Even the earliest civilizations required management skills wherever groups of people shared a common purpose: tribal activities, estates of the rich, military ventures, governments, or organized religion. Indeed, the prototypes of civil engineering and construction management became necessary as soon as "plants and animals were domesticated and people began living in communities. By 6000 B.C. these communities sometimes contained over 1,000 people, and Jericho is known to have had a wall and defensive towers." according to Davey, by 4500 B.C. the first canals diverted water from a river in eastern Iraq for crop irrigation. As canals proliferated, it became possible to store crops for commerce, and written records as well as management organization became necessary. According to Wren:

> In ancient Mesopotamia, lying just north and west of Babylon, the temples developed an early concept of a "corporation," or a group of temples under a common body of management. Flourishing as early as 3000 B.C., temple management operated under a dual control system: one high priest was responsible for ceremonial and religious activities, while an administrative high priest coordinated the secular activities of the organization. Records were kept on clay tablets, plans made, labor divided, and work supervised by a hierarchy of officials.

Many ancient civilizations left behind great stone structures that leave us wondering how they could have been created with the few tools then available. Examples include the Great Wall of China, the monoliths on Easter Island, Mayan temples in South America, and Stonehenge in England. Especially impressive are the pyramids of Egypt. The great pyramid of Cheops, built about 4,500 years ago, covers 13 acres and contains 2,300,000 stone blocks weighing an average of 5,000 pounds apiece. Estimates are that it took 100,000 men and 20 to 30 years to complete the pyramid—about the same effort in worker-years as it later took the United States to put a man on the moon. The only construction tools available were levers, rollers, and immense earthen ramps. Yet the difference in height of opposite corners of the base is only $\frac{1}{2}$ inch!

Hammurabi (2123–2081 B.C.) of Babylon "issued a unique code of 282 laws which governed business dealings....and a host of other societal matters." One law that should interest the civil engineer is the following:

> If a builder builds a house for a man and does not make its construction firm, and the house which he has built collapses, and causes the death of the owner of the house, that builder shall be put to death.

Today's engineer should be thankful that, while penalties for faulty design can be expensive and damaging to one's career, they are not terminal!

Problems of controlling military operations and dispersed empires have made necessary the development of new management methods since ancient times. Alexander the Great (356–323 B.C.) is generally credited with the first documented use of the staff system in the Western world. He developed an informal council whose members were each entrusted with a specific function (supply, provost marshal, and engineer).

Imperial Rome governed an estimated 50 million people spread from England to Syria and from Europe to North Africa by dividing the empire in turn into four major regions, 13 dioceses, and 110 provinces for civil government, with a separate structure for the military forces garrisoned throughout the provinces to maintain control. The great Roman roads that made it possible to move messages and Roman legions quickly from place to place were an impressive engineering achievement that enabled the empire's growth and helped it survive as long as it did.

It should not be inferred that early management skills were confined to Western civilization as it developed around the Mediterranean Sea (see Management Philosophies Outside Western Culture). George describes the consistent use of advisory staff by Chinese emperors as early as 2350 B.C., and "ancient records of Mencius and Chow (1100 to about 500 B.C.) indicate that the Chinese were utilizing principles of organizing, planning, directing, and controlling." In India, one Brahman Kautilya described in *Arthashastra* in 321 B.C. a wide range of topics on government, commerce, and customs. Because he analyzed objectively rather than morally the political practices that brought success in the past, his name *has become synonymous with sinister and unscrupulous management* in India (just as has Niccolo Machiavelli's name for his similar analysis in *The Prince* in the early seventeenth century in Italy).

The Arsenal of Venice

George abstracts from Lane a fascinating story of "what was perhaps the largest industrial plant of the [medieval] world." As Venice's maritime power grew, the city needed an armed fleet to protect her trade, and by 1436 it was operating its own government shipyard, the Arsenal. The Arsenal "had a threefold task: (1) the *manufacture* of galleys, arms, and equipment; (2) the *storage* of the equipment until needed; and (3) the *assembly and refitting* of the ships on reserve."

Most impressive was the assembly line used to outfit ships. A Spanish traveler, Pero Tafur, wrote in 1436:

> And as one enters the gate there is a great street on either hand with the sea in the middle, and on the one side are windows opening out of the houses of the Arsenal, and the same on the other side, and out came the galley towed by a boat, and from the windows they handed out to them from one the cordage, from another the bread, from another the arms, and from another the balistas and mortars, and so from all sides everything which was required, and when the galley had reached the end of the street, all the men required were on board, together with the complement of oars, and she was equipped from end to end. In this manner there came out ten galleys, fully armed, between the hours of three and nine.

George identifies several other industrial management practices of the Arsenal that were ahead of their time:

1. Systematic warehousing and inventory control of the hundreds of masts, spars, and rudders, and thousands of benches, footbraces, and oars needed to make the assembly line work
2. Well-developed personnel policies, including piecework pay for some work (making oars) and day wages for both menial labor and artisans (the latter with semiannual merit reviews and raises)
3. Standardization, so that any rudder would meet any sternpost, and all ships were handled the same way

4. Meticulous accounting in two journals and one ledger, with annual auditing
5. Cost control. As an example, one accountant discovered that lumber was stored casually in piles, and the process of searching through the piles to find a suitable log was costing three times as much as it did to buy the log in the first place; as a result of this early industrial engineering study an orderly lumberyard was established, which not only saved time and money but also permitted accurate inventory of lumber on hand.

An important innovation developed in Venice during this period was *double-entry bookkeeping*. Luca Pacioli published an instruction manual (*Summa de arithmetica, geometria, proportioni et proportionalia*) in 1494 describing the system then in use and recommending it. His discussion of the use of memorandum, journal, and ledger, supporting documents, and internal checks through periodic audits were so modern that *[m]any excerpts from Pacioli's writing could be inserted into our current accounting textbooks with virtually no change in wording.* Pacioli's work was translated into English about 50 years later and was in widespread use by the early eighteenth century.

THE INDUSTRIAL REVOLUTION

End of Cottage Industry

Before the late eighteenth century, farm families would spin cotton, wool, or flax to yarn or thread on a spinning wheel, weave it on a hand loom, wet the goods with mild alkali, and spread them on the ground for months to bleach in the sun before selling these *gray goods* at a local fair for whatever price they could get. Even when under the "putting out" system, where merchants at the fairs would provide the family with materials and buy their output at a negotiated rate, the work could be done in the farm cottage.

In the last third of the eighteenth century, a series of eight inventions (six British and two French) changed society irretrievably. Summarized from Amrine et al., they are the following:

1. The *spinning jenny*, invented by James Hargreaves in 1764, which could spin eight threads of yarn (later, 80) at once instead of one
2. The *water frame*, a spinning machine driven by water power, patented by Richard Arkwright and incorporated by him in 1771 in the first of many successful mills
3. The *mule*, a combination of the spinning jenny and water frame invented by Samuel Crompton in 1779, which enormously increased productivity and eliminated hand spinning
4. The *power loom*, a weaving machine patented in 1785 by Edmund Cartwright, which with time and improvements ended the ancient system of making cloth in the home
5. *Chlorine bleach*, discovered in 1785 by the French chemist Claude Louis Berthollet (and bleaching powder in 1798 by Charles Tennant), which provided quick bleaching without the need for large open areas or constant sunlight
6. The *steam engine*, patented by James Watt in 1769 and used in place of water power in factories beginning about 1785
7. The *screw-cutting lathe*, developed in 1797 by Henry Maudslay, which made possible more durable metal (rather than wood) machines
8. *Interchangeable manufacture*, commonly attributed to the American Eli Whitney in carrying out a 1798 contract for 10,000 muskets, but perhaps adopted by him as a result of a letter dated May 30,

1785, from Thomas Jefferson (while in France) to John Jay, describing the approach of Leblanc at the *manufacture de Versailles*:

> An improvement is made here in the Construction of muskets, which it may be interesting to Congress to know....It consists in the making of every part of them so exactly alike, that what belongs to any one, may be used for every other musket in the magazine.... [Leblanc] presented me the parts of fifty locks, taken to pieces, and arranged in compartments. I put several together myself, taking pieces at hazard as they came to hand, and they fitted in the most perfect manner. The advantage of this when arms are out of repair [is] evident.

Problems of the Factory System

The innovations of the late eighteenth century just described caused major upheavals in England's society as well as its economy. Cottage industry could not compete with factories powered first by water and then by steam. Underground coal mines provided fuel for the steam engine, and the steam engine powered the pumps that removed water seepage from the mines. A mass movement of workers from the farms and villages to the new industrial centers was required.

The new factory managers had problems of recruiting workers, training the largely illiterate workforce, and providing discipline and motivation to workers who had never developed the habits of industry. Wren quotes Powell: "If a person can get sufficient [income] in four days to support himself for seven days, he will keep holiday for the other three." Wren adds, "Some workers took a weekly holiday they called 'Saint Monday' which meant either not working or working very slowly at the beginning of the week." Today, plant managers who hire the chronically unemployed can face exactly this same problem with workers who have neither personal experience nor family tradition with the *habits of industry,* such as regular attendance and punctuality.

Explosive growth of the English mill towns led to filthy, overcrowded living conditions, widespread child labor, crime, and brutality. Falling wages, rampant unemployment, and rising food prices led to a rash of smashing of textile machinery by the Luddites, peaking between 1811 and 1812. This movement soon died for lack of leadership by dint of hanging Luddites in at least four cities.

England's agrarian history provided no source of professional managers. Supervisors often were illiterate workers who rose from the ranks and were paid little more than the workers they supervised, and there was no common body of knowledge about how to manage. Upper management often consisted of the sons and relatives of the founders, a condition that persists today in many areas. Gradually, the forerunners of modern factory management began to develop. One early firm was Boulton and Watt, founded by Matthew Boulton and James Watt to manufacture Watt's steam engine. By 1800, their sons inherited the firm and instituted innovations at their Soho Engineering Foundry such as factory layout planning, inventory control, production planning, work-flow study, sophisticated analysis of piecework rates, and paid overtime.

Another pioneer was Robert Owen, part owner of a mill complex in New Lanark, Scotland. Owen was ahead of his time in proposing that as much attention be paid to vital "human machines" as to inanimate ones. He told a group of factory owners the following:

> Your living machines may be easily trained and directed to procure a large increase of pecuniary gain. Money spent on employees might give a 50 to 100 percent return as opposed to a 15 percent return on machinery. The economy of living machinery is to keep it neat and clean, treat it with kindness that its mental movements might not experience too much irritating friction.

Owen was ahead of his time in adopting practices that treated workers well. He reshaped the whole village of New Lanark, improving housing, streets, sanitation, and education. Although he continued to employ children, he lobbied for legislation that ultimately forbade employing children under the age of nine, limited the workday to $10\frac{1}{2}$ hours, and forbade night work for children.

Industrial Development in America

England regarded her colonies as markets for English factories; as early as 1663, all manufactured goods were required to be purchased in England (even if made elsewhere in Europe), and the 1750 English Iron Act made it illegal to set up in the Colonies mills and furnaces for the manufacture of finished products. Although emigration of skilled labor to America was prohibited after the American Revolution, an experienced textile machinery builder and mechanic named Samuel Slater emigrated from England as a *farmer* and joined with three prosperous Rhode Island merchants to build the first technically advanced American textile mill at Pawtucket, Rhode Island, in 1790; by 1810 the census listed 269 mills in operation. Although growth of American industry was accelerated by the War of 1812 with England, most American firms before 1835 were small, family owned, and water powered. Only 36 firms employed more than 250 workers: 31 textile firms, three in iron, and two in nails and axes. The greatest sophistication in manufacturing was at the government-owned Springfield, Massachusetts, Armory, and this knowledge provided the basis for the later manufacture of axes, shovels, sewing machines, clocks, locks, watches, steam engines, reapers, and other products in the 1840s and 1850s.

Canals provided the first construction challenge for the new nation. Although the Middlesex Canal Company obtained the rights to build a canal from Boston to Lowell, Massachusetts, in 1793, they experienced great difficulty until they called in an immigrant engineer, William Weston, who had worked under a canal builder in England. Weston went on to provide know-how for all the major projects of that period in New England. This experience was available when the Erie Canal was built, 363 miles from Albany to Buffalo, New York between 1816 and 1825, a project that provided the training for many of early American civil engineers.

Railroads and steel were the high-technology growth industries of the nineteenth century. Colonel John Stevens, dubbed the *father of American engineering,* built the first rail line—the 23-mile Camden and Amboy Railroad—in 1830; by 1850 there were 9,000 miles of track extending west to Ohio. Morse's first experimental telegraph line was built in 1844; by 1860 there were 50,000 miles of telegraph line—much of it along railroad right of way and used in part to facilitate rail shipment.

The railroads presented management problems of a dimension not seen before, and the men who mastered these challenges became the leaders of the American industrial explosion. One such person was Andrew Carnegie (1835–1919), who at age 24 became superintendent of the largest division of the nation's largest railroad (the Pennsylvania). In 1872, attracted by Sir Henry Bessemer's new process, he moved into steelmaking—integrating operations, increasing volume, and selling aggressively. In 1868 the United States produced 8,500 tons of steel while England produced 110,000 tons; by 1902, thanks to Carnegie and others, America produced 9,100,000 tons to England's 1,800,000 tons.

The large industrial firms of the nineteenth century were precursors of the industrial giants of the twentieth century, headquartered primarily in Europe, the United States, or Japan, but manufacturing and selling all over the world. The nature of these *multinational corporations* and the opportunities they offer engineers are discussed in Chapters 17 and 18.

Development of Engineering Education

Most engineering skills through the eighteenth century were gained through apprenticeship to a practitioner. This is described by John Mihalasky:

> The first engineering school was probably established in France in 1747 when [Jean Rodolphe] Perronet, engineer to King Louis XV, set up his staff as a school. This group was later chartered in 1775 under the official name Ecole des Ponts et Chaussées [School of Bridges and Roads]. Other early schools were the Bergakademie at Freiburg in Saxony (1766), Ecole Polytechnic in Paris (1794), Polytechnic Institute in Vienna (1815), Royal Polytechnic of Berlin (1821), and University College of London (1840).

When the American colonies revolted in 1776, they did not have the engineering resources needed to build (or destroy) fortifications, roads, and bridges, and they had to rely on French, Prussian, and Polish assistance. At the urging of Thomas Jefferson and others, the new nation quickly established the U.S. Military Academy at West Point, New York, in 1802 to provide training in, among other things, practical science. Graduates did not acquit themselves as well as hoped in the War of 1812 with England, so Sylvanus Thayer, assistant professor of mathematics at the Academy (1810–1812), and Lt. Colonel William McRee were sent to Europe to examine the curricula at Ecole Polytechnic, the most famous scientific military school in the world. On their return in 1817, Thayer was appointed Superintendent at West Point. He collected the best teachers of physics, engineering, and mathematics available and set up a four-year civil engineering program. Ross emphasizes the importance of this program:

> The influence of the Academy extended far beyond the institution's cadets. "Every engineering school in the United States founded during the nineteenth century copied West Point, and most found their first professors and president among academy graduates." Many of the great canals, railroads, and bridges constructed during the nineteenth century were built by West Point graduates. The faculty, recruited by Thayer, wrote textbooks that dominated the subjects of mathematics, chemistry, and engineering during the 1800s.

For example, Mihalasky reports that Captain Partridge, an early Academy superintendent, founded the first civilian engineering school in the country in 1819, which later became known as Norwich University, followed by Rensselear (New York) Polytechnic Institute in 1823 with *a practical school of science,* and 12 years later a school of civil engineering. Other early engineering schools were Union College (1845), Harvard, Yale, and Michigan (1847). Mihalasky reports that only these six engineering schools existed in the United States when the Civil War opened, although Reynolds and Seely *have identified at least 50 institutions that at one time or another offered instruction in engineering before 1860* (although not necessarily as full curricula).

As reported elsewhere, "the event that had the greatest influence on engineering education was passage of the Morrill Land Grant Act in 1862. This act gave federal land (ultimately [totaling] 13,000,000 acres, an area 46 percent greater than Taiwan) to each state to support 'at least one college where the leading object shall be...scientific and classical studies...agriculture and mechanic arts.'" This made education in the "mechanic arts" (which became engineering) available and affordable throughout the country. By 1928, president-elect Herbert Hoover (himself a distinguished engineer and manager) could say the following:

> The leaders of our universities were the first of all the educators of the world to ... provide fundamental training in the application of science to engineering under the broadening influence and cultivation of

university life.... [Another] dimension that grew in American engineering was the transformation from solely a technical profession to a profession of administrators—the business manager with technical training.

An International Engineering Congress, with one division of the meetings on engineering education, was held as part of the 1893 Columbian Exposition in Chicago. Since there were then more than 100 engineering schools in the country, the engineering education sessions were well attended. Interest there led to the 1893 formation of the Society for the Promotion of Engineering Education, which became the American Society for Engineering Education (ASEE). In the century since, the meetings, journals, and studies of the ASEE have represented another major factor improving the quality of American engineering education.

MANAGEMENT PHILOSOPHIES

Over time, the number of management philosophies has been numerous. All have had, as their goal, to obtain optimal organizational performance, with the overall business environment guiding the selection of a particular style of management. Some theories have been fads that have not influenced a company's performance in the long term, while others have enhanced quality and productivity. Each theory has had its merits and drawbacks. These philosophies may be grouped into general categories of scientific, administrative, and behavioral.

SCIENTIFIC MANAGEMENT

Charles Babbage

Babbage (1792–1871) lived in England during the Industrial Revolution. His work was far ahead of its time: Wren calls him both *patron saint of operations research and management science* and *grandfather of scientific management*; and so, he will be discussed here, out of chronological sequence. Wren continues by describing the work for which Babbage is popularly known:

> He demonstrated the world's first practical mechanical calculator, his difference engine, in 1822. Ninety-one years later its basic principles were being employed in Burroughs' accounting machines. Babbage had governmental support in his work on the difference engine, but his irascibility cost him the support of government bureaucrats for his analytical engine, a versatile computer that would follow instructions automatically. In concept, Babbage's computer had all the elements of a more modern version. It had a store or memory device, a mill or arithmetic unit, a punch card input system, an external memory store, and conditional transfer [the modern "if statement"].

Babbage's engines never became a commercial reality, largely because of the difficulty in producing parts to the necessary precision and reliability. This frustration led him to visit a wide variety of English factories, and his fascination with what he observed there led to the publication of his very successful book *On the Economy of Machinery and Manufactures*, in 1832. In this he described at length his ideas on division of labor, his *method of observing manufactures,* and methods of optimizing factory size and location, and he proposed a profit-sharing scheme. He showed a sophisticated understanding of effective time-study methods:

> If the observer stands with his watch in his hand before a person heading a pin, the workman will almost certainly increase his speed, and the estimate will be too large. A much better average will result from inquiring what quantity is considered a fair day's work. When this cannot be ascertained, the number of operations performed in a given time may frequently be counted when the workman is quite unconscious that any person is observing him. Thus, the sound made by...a loom may enable the observer to count the number of strokes per minute...though he is outside the building.

Henry Towne and the American Society of Mechanical Engineers

From the first large-scale human endeavors, the science of management made little progress over the centuries, largely because almost no one considered management as a legitimate subject for study and discussion. Although engineers frequently became enterprise managers, the first American engineering societies (the American Society of Civil Engineers, founded in 1852, and the American Institute of Mining Engineers, founded in 1871) were not interested in machine shop operation and management. Wren believes that the first American forum for those interested in factory management was the *American Machinist*, an illustrated journal of practical mechanics and engineering founded in 1877, which soon began including a series of letters to the editor from James Waring See on machine shop management. The *Machinist* was instrumental in the formation of the American Society of Mechanical Engineers (ASME), which elected its first officers on April 7, 1880, at the Stevens Institute of Technology in Hoboken, New Jersey; ASME was formed to address itself to *those issues of factory operation and management that the other groups had neglected.* Speaking in this vein before the May 1886 ASME meeting in Chicago was an engineer named Henry R. Towne, who was cofounder of Yale Lock Company and president of Yale & Towne Manufacturing Company. Towne began his famous paper *The Engineer as Economist* with the following passage:

> The monogram of our national initials, which is the symbol of our monetary unit, the dollar, is almost as frequently conjoined to the figures of an engineer's calculations as are the symbols indicating feet, minutes, pounds, or gallons. The final issue of his work, in probably a majority of cases, resolves itself into an issue of dollars and cents, of relative or absolute values.

Towne then observed that

> although engineering had become a well-defined science, with a large and growing literature of its own,...the matter of shop management is of equal importance with that of engineering, as affecting the successful conduct of most, if not all, of our great industrial establishments, and that the *management of works* has become a matter of such great and far reaching importance as perhaps to justify its classification also as one of the modern arts.

Towne cited the need for a medium for the interchange of management experience *by the publication of papers and reports, and by meetings for the discussion of papers and interchange of opinions and* called for a new section of the ASME to carry this out. Although such a management section was not organized until 1920, consideration of matters of shop management became part of ASME meetings, and the ASME Management Division dates its official history from Towne's 1886 paper.

Frederick W. Taylor

Frederick Winslow Taylor (1856–1915), called the *father of scientific management,* was born in 1856 to a well-to-do family in Germantown (Philadelphia) and completed a four-year apprenticeship as a machinist. In 1878, he joined Midvale Steel Company as a laborer, and was promoted to time clerk and then foreman of a machine shop.

As foreman, he was frustrated because his machinists were producing only about a third of what Taylor (as a machinist himself) knew and demonstrated they should be producing. Even on piecework pay, production did not improve because the workers knew that as soon as they increased production, the rate paid per piece would be decreased, and they would be no better off. With permission of the president of Midvale Steel, Taylor began a series of experiments in which work was broken down into its "elements" and the elements timed to establish what represented a fair day's work.

During this period, he was a mechanical engineering student at Stevens Institute, where the ASME held its first meeting, graduating in 1883. The next year, at the age of 28, Taylor became chief engineer at Midvale Steel; a year later he joined the ASME, and in May 1886 attended its meeting in Chicago. Biographers report that Taylor was encouraged there to continue his studies of work methods and shop management by Henry Towne's paper (described previously). Another paper at that meeting was by Captain Henry Metcalf, describing a *shop-order system of accounts* he established at the Frankford (Pennsylvania) and Watervliet (New York) Arsenals in 1881 that helped management determine direct and indirect costs of work activity. In the extensive discussion that followed, Taylor reported on a similar system Midvale had been using for 10 years. For the first time in recorded history, engineers now had a medium for sharing their management problems and solutions. Taylor contributed further to this interchange with papers presented to the ASME in 1895 (*A Piece Rate System*) and in 1903 (*Shop Management*), and became president of ASME in 1906. Today, many of the engineering societies have active management divisions, with the American Society for Engineering Management totally devoted to such concerns.

Taylor's *piece rate system* involved breaking a job into elementary motions, discarding unnecessary motions, examining the remaining motions (usually through stopwatch studies) to find the most efficient method and sequence of elements, and teaching the resulting method to workers. The workers would be paid according to the quantity of work produced. Taylor went further in his *differential piecework* method by establishing one piece rate if the worker produced the standard number of pieces, and a higher rate for all work if the worker produced more. For example, if three pieces were deemed a standard day's work and the two rates were 50 and 60 cents per piece, the worker would earn $1.50 for making three pieces a day, but $2.40 for four.

The best-known examples of Taylor's studies occurred after he became a consultant to Bethlehem Iron (later Bethlehem Steel) Company in 1898. One was a study of a crew of pig-iron handlers: workers who picked up 92-pound *pigs* of iron, carried them up an inclined plank, and loaded them onto railroad flat cars. By developing a method that involved frequent rest periods to combat the cumulative fatigue resulting from such drudgery, Taylor was able to increase the number of long tons loaded by a worker in a day from 12.5 to 47.5.

In another example, Taylor examined the work of shoveling at Bethlehem:

> Operation of the three blast furnaces and seven large open-hearth furnaces required a steady intake of raw materials—sand, limestone, coke, rice coal, iron ore, and so forth. Depending on the season, 400 to 600 men were employed as shovelers in the 2-mile-long and a half-mile wide Bethlehem yard. Taylor noted that the shovelers were organized into work gangs of 50 to 60 men under the direction of a single foreman. Each owned his own shovel and used it to shovel whatever he was assigned....Taylor's analysis revealed that a shovel-load (depending on the shovel and the substance shoveled) varied in weight from 3.5 to 38.0 pounds, and that a shovel-load of 21.5 pounds yielded the maximum day's work. As a result, instead of permitting workers to use the same shovel regardless of the material they were handling, Taylor designed new shovels so that for each substance being shoveled the load would equal 21.5 pounds.

In the latter example, the average amount shoveled per day increased from 16 to 59 tons. In both of these cases the savings produced were shared. Workers' earnings increased from $1.15 to about $1.85 a day, while management's cost per ton handled was reduced by 55 percent or more.

Taylor summarized his methods in his 1911 book *Principles of Scientific Management* as a combination of four principles:

> *First.* Develop a science for each element of a man's work, which replaces the old rule-of-thumb method.
> *Second.* Scientifically select, then train, teach, and develop the workmen, whereas in the past he chose his own work and trained himself as best he could.

> *Third.* Heartily cooperate with the men so as to insure all of the work being done in accordance with the principles of the science which has been developed.

> *Fourth.* There is an almost equal division of the work and the responsibility between the management and the workmen. The management take over all work for which they are better fitted than the workmen [defining *how* work is to be done], while in the past almost all of the work and the greater part of the responsibility were thrown upon the men.

The Gilbreths

Frank B. Gilbreth (1868–1924) passed the entrance exams for the Massachusetts Institute of Technology, but he chose instead to apprentice himself as a bricklayer. Trying to learn the trade, he found that bricklayers used three sets of motions: one when working deliberately but slowly, another when working rapidly, and a third when trying to teach their helpers. Gilbreth resolved to find the *one best way*. He described it later in a testimony before the U.S. Interstate Commerce Commission:

> Bricks have been laid the same way for 4,000 years. The first thing a man does is to bend down and pick up a brick. Taylor pointed out that the average brick weighs ten pounds, the average weight of a man above his waist is 100 pounds. Instead of bending down and raising this double load, the bricklayer could have an adjustable shelf built so that the bricks would be ready to his hand. A boy could keep these shelves at the right height. When the man gets the brick in his hand, he tests it with his trowel. If anything, this is more stupid than stooping to pick up his material. If the brick is bad he discards it, but in the process, it has been carried up perhaps six stories, and must be carted down again. Moreover, it consumes

the time of a $5-a-day man when a $6-a-week boy could do the testing on the ground. The next thing the bricklayer does is to turn it over to get its face. More waste: more work for the $6 boy. Next what does the bricklayer do? He puts the brick down on the mortar and begins to tap it with his trowel. What does his tapping do? It gives the brick a little additional weight, so it will sink into the mortar. If anything, this is more stupid than any of the others. For we know the weight of the brick and it would be a simple matter in industrial physics to have the mortar mixed so that just that weight will press it down into the right layer. And the result? Instead of having eighteen motions in the laying of a brick, we have only six. And the men put to work to try it lay 2,700 with no more effort than they laid a thousand before.

By 1895 Gilbreth had his own construction firm based on *speed work*. He analyzed each job to eliminate unnecessary motions, devising a system of classifying hand motions into 17 basic divisions (which he called *therbligs* from his last name) such as *search*, *select*, *transport loaded*, *position*, and *hold*. He soon became one of the best-known building contractors in the world, but by 1912 had given up the construction business and was devoting full time to management consulting.

Lillian Moller Gilbreth (1878–1972) earned a bachelor's and master's degree in English at the University of California (and qualified for Phi Beta Kappa, although as a woman she was not included on the official list of recipients). She interrupted her Ph.D. studies for a trip to Europe by way of the port of Boston, where she met Frank Gilbreth on the outgoing leg and married him on her return. She later completed her Ph.D. at Brown University.

Lilian's contributions to both the fields of management and industrial psychology are significant. However, what is often missed in discussions of Lillian is how she helped others in her era achieve impact. In a presentation on the contributions of scientific management pioneers, Lyndall Urwick, a noted management scholar discussed later in this chapter, stated that the marriage of Frank and Lillian was providential. For without Lillian's deep understanding of human beings, the three engineers who led the scientific management revolution – Taylor, Gannt, and Gilbreth – might not have overcome the struggle to achieve wide-spread adoption of their techniques. It was Lillian's ability to bring the human element into the field that led to recognition that management could be learned and was not available only to those born with the ability. These are some of the reasons Wren called Lillian the "First Lady of Management."

Lillian quickly became interested in Frank's work and assisted him in preparation of six books published between 1908 and 1917 (*Field System*, *Concrete System*, *Bricklaying System*, *Motion Study*, *Fatigue Study*, and *Applied Motion Study*). Meanwhile, she continued work on her Ph.D. thesis, *The Psychology of Management*, one of the earliest contributions to understanding the human factor in industry, and submitted it in 1912. The work was serialized in *Industrial Engineering Magazine* and finally published as a book (the latter with the proviso that the author be listed as *L. M. Gilbreth* without identifying her as a woman, so that it might have some credibility).

Frank prepared an invited paper for the 1925 International Management Conference in Prague, but he died suddenly of a heart attack on June 14, 1924. Lillian presented the paper in his place, then continued Frank's work and established a strong reputation of her own as one of the creators of industrial psychology. She was the first woman admitted to the Society of Industrial Engineers (now the Institute of Industrial and Systems Engineers) and the ASME, the first woman professor of management at an engineering school (Purdue University and later the Newark College of Engineering), and the only woman to date to be awarded the Gilbreth Medal, the Gantt Gold Medal, or the CIOS Gold Medal. She has understandably been called the *first lady of management*. Many scholars have noted the signs of Lillian in Frank's earlier work, noting that her contributions to management are probably even greater than she was credited. Her life was so long (she outlived Frank by 48 years) and distinctive that many women engineers throughout the 20th century spoke of her as an inspiration that led them into engineering.

Growth and Implications of Scientific Management

Taylor's work attracted many disciples who propagated the scientific management method. Carl Barth, a mathematics teacher, was recruited by Taylor to help Henry Laurence Gantt solve the speed and feed problems in metal-cutting studies conducted at Bethlehem. Barth later helped Taylor apply scientific management to the problems of Link Belt, Fairbanks Scale, and Yale & Towne companies, and then helped George D. Babcock install scientific management at the Franklin Motor Car Company (1908–1912).

Gantt (1861–1919) earned degrees from Johns Hopkins University and Stevens Institute of Technology (in mechanical engineering in 1884, a year after Taylor). He joined Taylor at Midvale Steel in 1887, followed him to Simond's Rolling Machine Company and then to Bethlehem Steel, and became an independent consulting industrial engineer in 1901. Gantt modified Taylor's *differential piece rate* by providing a standard day rate regardless of performance, which provided security to workers during training and work delays due to materials not being available; workers who accomplished the specified daily production received an additional bonus, as did their foremen. Gantt is also noted for his work in developing charts that graphed function of performance against time; their application to project management is discussed in Chapter 14.

Another protégé of Taylor was Morris L. Cooke (1872–1960), a mechanical engineer (Lehigh University, 1895) who began *applying a questioning method to the wastes of industry long before he met or heard of Taylor*, then started reading Taylor's writings, and met him. Taylor funded Cooke to study the administrative effectiveness of ASME, sent him to perform an "economic study" of administration in educational organizations for the Carnegie Foundation for the Advancement of Teaching, and then sent him to help the newly elected reform mayor of Philadelphia improve the efficiency and effectiveness of municipal government (1911–1915). Cooke later advised the president of the American Federation of Labor (Samuel Gompers) and coauthored a book with the president of the Congress of Industrial Organizations (Phillip Murray), emphasizing that labor was as important for production as management. Later, he headed the Rural Electrification Administration, which brought inexpensive electric power to rural America.

Taylor's system received extensive publicity in the 1911 Eastern Rate case. The Eastern-railroads petitioned the Interstate Commerce Commission for an increase in rates, but Boston lawyer (and later Supreme Court justice) Louis D. Brandeis took up the cause of shippers with the theme that no increase would be necessary if railroads would only apply *scientific management* (the name adopted instead of the *Taylor system* in a meeting of Brandeis, Gilbreth, Gantt, and others in preparation for this case). The parade of witnesses supporting this view included Taylor, Gilbreth (as quoted above), Henry Towne, and others. Harrington Emerson (1853–1931), who had been very successful as a troubleshooter on the Burlington Railroad and then a consultant to the Santa Fe Railroad, testified that preventable labor and material waste was costing the railroad industry a million dollars a day. Scientific management spread rapidly because media and institutions for the sharing of knowledge and experience were becoming available in an unprecedented way. Many of the practitioners were active in ASME; they presented and critiqued papers at their meetings. Industrial and popular journals were increasing in number, and they reported on progress in scientific management and even serialized books by Taylor and Lillian Gilbreth. Most of the major participants authored several books each, many of which were widely read. Universities increasingly decided management was, after all, worthy of study. Taylor was persuaded to lecture at what would become the Harvard Business School, Lillian Gilbreth and Carl Barth each lectured at two universities, and Henry Gantt lectured at four. Bachelor's degree programs that combined engineering and business were founded at Stevens in 1902, Yale in 1911, and MIT in 1913. The discipline of industrial engineering (and the Institute of Industrial and Systems Engineers) originated from the work of scientific management, and the newer discipline of engineering management owes a great debt to it as well.

Scientific management did, however, have some negative impacts, which still affect us today. Taylor divided work into planning and training (a management responsibility) and rote execution (by the uneducated laborer of the day). Only in the last four decades have executives in mass production industries such as General Motors realized how much they were losing by only hiring workers for their physical labor rather than encouraging them to participate in improving work methods. This realization has led many modern companies to seek to eliminate what Konosuke Matsushita, founder of Matsushita Electric Industrial Company, termed "Taylorized heads"—situations where workers are not contributing their thoughts, just their labor. Efforts to eliminate the idea that "managers manage and workers work" are widespread in most Western service and technology companies today. In Chapter 2 some of the theories of human motivation are examined, and Chapter 12 looks at their application to production operations using techniques of total quality management and empowered teams.

ADMINISTRATIVE MANAGEMENT

As we have seen, initial American management study emphasized management at the production-shop level. In the meantime, two Europeans, Henri Fayol and Max Weber, were making significant contributions to general management theory.

Henri Fayol

Fayol (1841–1925) was an 1860 graduate of the National School of Mines at St. Etienne, France. His distinguished career is described in Urwick's foreword to the 1949 English translation of his most noted work, *Administration Industrielle et Générale* (General and Industrial Management). He believed that the activities of industrial undertakings could be divided into six groups: technical (production), commercial (marketing), financial, security, accounting, and administrative activities. The first five he considered well known, but the last, administrative (French has no exact equivalent of the word *management*), he considered most important above the first two levels of management, yet least understood. Fayol divided administration into planning/forecasting (*prevoyance*), organization, command, coordination, and control. He decried the absence of management teaching in technical schools, but stated that without a body of theory, no teaching is possible. He then proceeded to develop a set of 14 "general principles of administration," most of which have meaning to this day.

Today's critics of engineering education would agree with Fayol that:

> [o]ur young engineers are, for the most part, incapable of turning the technical knowledge received to good account because of their inability to set forth their ideas in clear, well-written reports, so compiled as to permit a clear grasp of the results of their research or the conclusions to which their observations have led them.

Engineering educators today would be less comfortable with his observation that "[l]ong personal experience has taught me that the use of higher mathematics counts for nothing in managing businesses and that engineers, mining or metallurgical, scarcely ever refer to them." However, every engineering student should consider his advice to future engineers:

> You are not ready to take over the management of a business, even a small one. College has given you no conceptions of management, nor of commerce, nor of accounting, which are requisite for a manager. Even if it had given you them, you would still be lacking in what is known as practical experience, and which is acquired only by contact with [people] and with things....Your future will rest much on your technical ability, but much more on your managerial ability. Even for a beginner, knowledge of how to plan, organize, and control is the indispensable complement of technical knowledge. You will be judged not on what you know but on what you do, and the engineer accomplishes but little without other people's assistance, even when [they] start out. To know how to handle [people] is a pressing necessity.

Max Weber and Bureaucracy

A contemporary of Fayol, the German sociologist Max Weber (1864–1920) influenced classical organization theory more than any other person. Weber developed a model for a rational and efficient large organization, which he termed a bureaucracy. Weber described the following as characteristics of *legal authority with a bureaucratic administrative staff*:

- The basic organizational unit is the *office* or position, which is designated a specific set of functions (based on division of labor), with clearly defined authority and responsibility.
- Members of the organization owe loyalty to the office, not (as with traditional authority or charismatic authority) to the individual.
- Candidates for offices are selected and appointed (not elected) based on their technical capability.
- Offices are organized in a clearly defined hierarchy: each lower office is under the control and supervision of a higher office.
- Officials (office holders) are *subject to strict and systematic discipline and control in the conduct of the office*, and subordinates have a right of appeal.
- Administrative acts, decisions, and rules must be reduced to writing.
- The office is the primary occupation of the incumbent, who is reimbursed by a fixed salary.
- Promotion is based on the judgment of superiors.
- Officials are not the owners of the organization.

The term *bureaucracy* need not imply an organization that is mired in red tape, delay, and inefficiency, with no concern for the human dimension. Most of Weber's elements are necessary in any large organization to assure consistent and reasonably efficient operation. The U.S. Postal Service or Internal

Revenue Service must have the same rules of operation at every local office; an army must have common procedures so that replacement officers and men can function quickly on assuming new positions; General Motors, or a large university or hospital, or the Boy Scouts of America must have fairly uniform structures and rules among their divisions to function smoothly. The challenge of a large organization is to incorporate into this necessary structure some flexibility to handle exceptions and an ability to recognize and reward individual contributions.

Russell Robb

Robb (1864–1927) was an American electrical engineer and manager whose original contributions on organization theory have not received the attention they deserve. After graduating from MIT, Robb spent most of his career as an executive in the Stone and Webster Engineering Corporation. He expressed his views on organization in three lectures presented to the Harvard University Graduate School of Business Administration in 1909 and later published. Young summarizes their import:

> These three lectures...contain more practical observations on organizations and concepts of organization theory than Weber. He was a practising engineer manager, whereas Weber was a sociologist....His penetrating observation of organizations as "only a means to ends—it provides a method" and analysis of principles and concepts make him more a "pioneer of organization theory" than Weber.

Lyndall Urwick

Urwick was an Englishman who majored in history at Oxford. His contribution lay not in creating concepts of management, but in being the first to try to develop a unified body of knowledge. Using Fayol's management functions as a framework, he analyzed the writings of Fayol, Taylor, Mary Parker Follett, James Mooney, and others, and attempted to correlate them with some of his own views into a consistent system of management thought. His 1943 book, *The Elements of Administration*, can therefore be viewed as the first general textbook on, as opposed to personal observations about, management. Toward the end of his long career he summarized his observations on the contribution of engineers to management:

> The study of management, as we all know, started with engineers. It was the sciences underlying engineering practice—mathematics, physics, mechanics, and so on—which were first applied by Frederick Winslow Taylor to analyzing and measuring the tasks assigned to individuals. That is where the science of management started.

BEHAVIORAL MANAGEMENT

Hawthorne Studies

What was arguably the single biggest direction change in early management thinking grew out of a series of studies conducted in the 1920s and early 1930s at the Hawthorne Works (near Cicero, Illinois) of the Western Electric Company (eventually becoming Lucent Technologies before being sold as pieces to other companies including Avaya and Nokia). The first phase of the studies, known as the Illumination Experiments, was conducted between 1924 and 1927 under the direction of Vannevar Bush, an electrical engineer from MIT who later developed systems that made the modern computer possible. The original intent was to find the level of illumination that made the work of female coil winders, relay assemblers, and small parts inspectors most efficient. Workers were divided into test and control groups, and lighting for the test group was increased from 24 to 46 to 70 foot-candles. Production of the test group increased as expected, *but production of the control group increased roughly the same amount.* Again, when lighting for the test group was decreased to 10, and then three foot-candles, their output *increased*, as did that of the control group. Production did not drop appreciably until illumination was lowered to that of moonlight (0.06 foot-candle).

To try to understand these unexpected results, Australian-born Harvard professor Elton Mayo and his colleague Fritz Roethlisberger conducted a second phase (1927–1932), known as the Relay Assembly Test Room Experiments. A large number of women were employed in assembling about 40 parts into the mechanical relays that were needed for telephone switching in the days before solid-state electronics. Six women whose prior production rates were known were moved from the large assembly room to a special test room to test the effects of changes in length and frequency of rest periods and hours worked. The women were given regular physical examinations (with free ice cream), their sleep each night and food eaten were carefully recorded, and room temperature and humidity were controlled. The room had an observer who recorded events as they happened and maintained a friendly atmosphere. The women had no supervisor, but they increasingly assumed responsibility for their own work and were allowed to share in decisions about changes in their work (a precursor of today's emphasis on *empowered teams*). Birthdays were regularly celebrated at work, and the women became fast friends after hours as well. Incentive pay had been used in the main workroom based on overall production of a large number of workers, but in the test room incentive pay was based just on production of the group of six.

After production rates had been stabilized in the new room, rest periods were added and maintained for periods of four or five weeks each at levels of (1) two five-minute periods, (2) two 10-minute periods, (3) six-five minute periods, and (4) two 10-or 15-minute periods with light snacks. Shorter workdays and elimination of Saturday work were also tried. Throughout this period daily production continued to increase, as it also did in a subsequent 12-week period when *rest periods, refreshments, and shortened workdays were eliminated*, and again when they were reinstated. Absenteeism among the six was only a third of that in the main room.

A third phase of study (the Bank Wiring Observation Room Experiment of 1931–1932) involved a group of 11 wiremen, soldermen, and inspectors who assembled terminal banks used in telephone exchanges. It became clear that the men formed a complex social group, had established their own standard of a fair day's work, and despite the piecework pay that existed, ridiculed and abused any worker who tried to work faster (or slower) than the group norm.

Mayo and others have attributed the surprising results in the first two phases to the pride of the women in being part of something important, the *esprit de corps* developed in the work group, and the satisfaction of having some control over their own destiny; the behavior of the men was attributed to the need for affection from the group (and the fear that management would lower pay rates if productivity improved). Later analysts such as Rice have criticized the studies as lacking the rigorous controls

Summary of Engineering and Management History

Ancient Civilizations	Egyptian pyramids
	China—Great Wall
	Mayan temples
	England—Stonehenge
	Alexander the Great—staffing system
	Romans—roads and aqueducts
Medieval Period	Four centuries of Dark Ages
Renaissance	Arsenal of Venice
Industrial Revolution, Eighteenth and Nineteenth Centuries	Factories
	Steam engine
Industrial Development in the United States, Nineteenth Century	Railroads, canals, steel mills
	West Point Military Academy
	Morrill Land Grant Act
	American Society for Engineering Education
Management Philosophies, Twentieth Century Scientific Management	Frederick Taylor
	Frank Gilbreth
	Lillian Gilbreth
	Henry Gantt
Administrative Management	Henry Fayol
	Max Weber
Behavioral Management	Abraham Maslow
	Hawthorne Studies
	Abilene Paradox
	Theory X and Theory Y

now demanded in scientific experiments. These studies focused the attention of an army of behavioral scientists—psychologists, sociologists, and even anthropologists (who turned their attention from the culture of remote tribes to the culture of General Motors and IBM)—on the behavior of workers individually and in groups; their work over the ensuing decades has added immeasurably to our knowledge of the art of management. The results of these studies are referred to as the Hawthorne effect, which is the tendency of persons singled out for special attention to perform as expected. While widely quoted, the legitimacy of this effect remains in question to this day.

Management Philosophies outside Western Culture

After discussion of the ancient roots of management, most of this text will focus on effective Western approaches to management. This is because we assume most of our readers are students in Western Universities seeking employment in Western companies. However, given the global nature of today's work environment (see Chapter 18), it is important that future engineering managers understand that Western management philosophies are not the only ones, and many successful management approaches have their roots in Eastern philosophies.[1]

Perhaps the two most famous examples of successful Eastern management philosophies are the dramatic rise of the Japanese economy following World War II and the current rise of the Chinese economy. Both of these changes were enabled by a fundamental difference between Eastern and Western approaches: focus on the success of the collective vs. focus on the success of the individual. In other words, the success of the Japanese rebuild was enabled by the norm of Japanese culture to focus on mutual benefit. This is further evidenced by the Toyota Production System's famous empowering any member of the factory to pull the Andon cord and stop the line when they see a problem.

At the same time that focus is transferred to the greater good, many Eastern management approaches look at the manager as nearly all powerful (notably China and India). In fact, during the author's time in India, colleagues would often explain that even middle-managers were "like a god" within the company. This view of managers as all powerful, stems from a cultural norm that seeks to save face and avoid conflicts, especially with those in authority. Bendixen and Burger note this norm is "characterized by idealism where mind and spirit are of central importance and where a state of highest perfection exists."[2] That perfection is often defined by the manager, and generally is defined in their image. The engineer working in a global setting is advised to not only be aware of these differences, but study them carefully—just because something works in one culture does not mean it will work in another!

Source: Akkaradech Maked/123RF

Sources

1. Bourgeois, L. J. and C. Black. Eastern Philosophies, Western Strategies: Strategic Intuition. *Darden Ideas to Action*. 20 June 2017. https://ideas.darden.virginia.edu/2017/06/eastern-philosophies-western-strategies-strategic-intuition/.
2. Bendixen, M. and B. Burger. Cross-Cultural Management Philosophies. *Journal of Business Research*. 1998:42:107–114.

Abilene Paradox

The Abilene paradox is the situation that results when groups take an action that contradicts what the members of the group silently agree they want or need to do. Stated another way, it is the inability of a group to agree to disagree. This is based on a story set in Abilene, Texas, by Dr. Jerry Harvey.

> Four adults are sitting on a porch in 104-degree heat in the small town of Coleman, Texas, some 53 miles from Abilene. They are engaging in as little motion as possible, drinking lemonade, watching the fan spin, and occasionally playing dominoes. The characters are a married couple and the wife's parents. At some point, the wife's father suggests they drive to Abilene to eat at a cafeteria there. The son-in-law thinks this is a crazy idea but does not see any need to upset the apple cart, so he goes along with it, as do the two women. They get in their Buick with no air-conditioning and drive through a dust storm to Abilene. They eat a mediocre lunch at the cafeteria and return to Coleman exhausted, hot, and generally unhappy with the experience. It is not until they return home that it is revealed that *none* of them really wanted to go to Abilene—they were just going along because they thought the others were eager to go.

The paradox is that not all group members are in agreement, but go along with decisions because they think the rest of the group agrees. The Abilene paradox occurs in group decision making and may happen in the workplace, with a family, or with friends.

More on behavioral management may be found in the contributions of McGregor and his Theory X and Theory Y, and Maslow's hierarchy of needs, which are discussed with leadership and human motivation in Chapter 7.

CONTEMPORARY CONTRIBUTIONS

Quality Management

Quality management (QM) is a management approach that originated in the 1950s and has steadily become more popular since the early 1980s. Quality is a description of the culture, attitude, and organization of a company that strives to provide customers with products and services that satisfy their needs. The culture requires quality in all aspects of the company's operations, with processes being done right the first time and defects and waste eradicated from operations. QM is a management philosophy that seeks to integrate all organizational functions (marketing, finance, design, engineering, production, customer service, etc.) to focus on meeting customer needs and organizational objectives. Specific topics in QM are discussed in Chapter 12. The following outline part of the quality approach:

- Meeting customer requirements
- Commitment by senior management and all employees
- Focus on processes/continuous improvement plans
- Planning quality into products and processes
- Teams
- Systems to facilitate quality control and improvement
- Employee involvement and empowerment
- Recognition and celebration
- Reducing development cycle times
- Benchmarking one's own performance and practices against others
- Six Sigma

Customer Focus

Quality and performance are judged by an organization's customers. The term **customer** refers to actual and potential users of an organization's products or services. Customers include the end users of products or services. The role of an organization's mission and vision is to align work toward meeting customer expectations. Marketing, design, manufacturing, and support must be aligned to meet customer needs. Customer-driven excellence has both current and future components: understanding the customer of today and anticipating future customer desires and needs.

Project Management

Many of the most difficult management challenges of recent decades have been to design, develop, and produce very complex systems of a type that has never been created before. Examples include the establishment of vast petroleum production systems in the waters of the North Sea or the deserts of Saudi Arabia, the collaboration of 400,000 people in the Apollo program to place people on the moon, development of complex jet aircraft, the International Space Station, the systems design for the Mission to Mars project, and software development. To create these systems with performance capabilities not previously available, there are three essential considerations for a manager to keep in balance: time (project schedule), cost (in dollars and other resources), and performance (the extent to which the objectives are achieved). Chapters 14 and 15 are devoted to discussing project management.

Globalization

People around the globe are more connected to each other than ever before. This new international system of globalization, as defined by Thomas Friedman, "has its own unique logic, rules, pressures, and incentives and it deserved its own name: 'globalization.' Globalization is not just some economic fad, and it is not just a passing trend. It is an international system." Human societies around the globe have

Engineering and Information Technology

Engineering continues to experience transformative change in the way the engineers practice due to the explosive growth of information technology (IT) over the past two generations. Not since the Industrial Revolution employed the power of steam engines has the engineering world seen such advancements. These changes began with the availability of desktop and mainframe computing power over fifty years ago. These tools brought fundamental change to the way engineers performed many aspects of their job. An example can be seen in how drafting is done. The engineer's drafting table has been replaced with CAD (computer-aided drafting / design) software. Design drawings are no longer drawn with paper and pencil with velum overlays for changes. Instead the drawings are created within the software where they can be rapidly changed and instantly sent to all stakeholders.

Today, the connected economy and Internet of Things (IoT) continues to transform engineering practice. While we're all familiar with consumer connected devices like smart speakers, connected doorbells with video monitoring, and refrigerators that can monitor what food needs to be purchased during the next shopping trip, most are less familiar with how this technology is changing the world of engineering. These changes are wide ranging and pertain to essentially every field of engineering. For example, the world of civil engineering has been changed by connected devices with GPS technology, where companies like Trimble Navigation connect earth moving equipment directly to engineering plans, dramatically reducing the need for field survey work and accelerating construction projects. For industrial engineers, these same types of connections provide real time data on the use of vehicles and inbound or outbound freight, dramatically improving their ability to design and manage supply chains and optimize equipment utilization. For mechanical engineers, modeling technology dramatically shortens the design cycle and lowers prototype costs, while electrical and power engineers can receive real-time information about grid performance and issues. While we don't know how the next technological breakthrough will impact engineering, we can be confident that it will bring some type of change.

established progressively closer contacts over many centuries, but now the pace has dramatically increased. Information and money flow more quickly than ever. Goods and services produced in one part of the world are increasingly available in all parts of the world. Workplace teams are composed of members from different parts of the world connected seamlessly by technology. International travel is more frequent. As a result, laws, economies, and social movements are forming at the international level. Globalization and opportunities are addressed more fully in Chapter 18.

Management Theory and Leadership

Management theory owes a great deal to practical executives who took the time to set down the wisdom they had accumulated in a successful management career. Henri Fayol, discussed previously, was such a man, as were Chester Barnard, who summarized his findings about people in organizations in *The*

Functions of the Executive, and Alfred P. Sloan, who documented his development of the decentralized organization with central control in *My Years with General Motors*. Just as often, these contributions are related secondhand by management writers. Peter Drucker, widely considered to be "the father of modern management," wrote many books and countless scholarly and popular articles on leadership and management. Two of the books are *The Effective Executive: The Definitive Guide to Getting the Right Things Done* and *Managing in the Next Society*. Peters and Waterman highlighted in their book *In Search of Excellence* the following: the wisdom of Walt Disney in treating theme park customers as "guests"; the emphasis of Thomas Watson, Jr., of IBM on service and customer satisfaction; the revolution in the U.S. Navy by Admiral Zumwalt "based on the simple belief that people will respond well to being treated as grownups"; and the success of MBWA ("management by walking around") by Bill Hewlett and Dave Packard at HP.

Other contributions to management theory include: *The Seven Habits of Highly Effective People* by Stephen Covey; *The One Minute Manager* by Kenneth Blanchard and Spencer Johnson; *Reengineering the Corporation* by Michael Hammer and James Champy; *Good to Great: Why Some Companies Make the Leap…and Others Don't* by Jim Collins; *Leadership 101: What Every Leader Needs to Know* by John C. Maxwell; quality management (TQM) philosophies; and the management styles of Jack Welch, formerly of General Electric. This list of contributors to management theory in the last three decades is not meant to be inclusive, rather it shows that the businessperson of today has more access to advice, some good and some not so good, than at any other time in history. Business books are better than ever. Through organized mentoring and other efforts, organizations and management are trying to preserve the wisdom that resides only in employees' heads.

DISCUSSION QUESTIONS

2-1. The practice of management is ancient, but the formal study of the body of management knowledge is relatively new. Why is this?

2-2. Stones for the pyramids were quarried far to the south (upstream on the Nile River) and were brought downstream on rafts only during the spring flood of the Nile. Discuss some of the planning and organizational implications of this immense logistic effort.

2-3. You are the production manager for a new start-up where the production floor has been up and running for six months. Things are not going smoothly. Quality is not meeting expectations and productivity issues are leading to substantial overtime work and cost overruns. Explain how the work of the Gilbreths could be applied to understand and solve these problems.

2-4. The development of cotton and woolen mills in the mill cities of England, and later New England, caused tremendous sociological change as potential workers (especially women) swarmed from rural areas to the growing industrial cities. Cite examples of similar occurrences in more recent times in developing countries.

2-5. How did Henri Fayol's approach to management compare with Weber's? How did Fayol's solutions differ? Describe Fayol's assumptions and the components of his theory of management. What remnants of Fayol's ideas exist today in management practice?

2-6. Matsushita emphasized the residual disadvantages to the United States of the teachings of Frederick Taylor. Discuss the *positive* contributions made by Taylor and his contemporaries in the scientific management movement.

2-7. What was the positive value of Max Weber's model of "bureaucracy"?

2-8. The essence of the Relay Assembly Test Room Experiments at the Hawthorne Works was that expected correlations between productivity and physical factors such as rest periods were not demonstrated. What other factors could explain the regular productivity increases observed in these experiments?

2-9. Schermerhorn defines management as the activity which "performs certain functions in order to obtain the effective acquisition, allocation, and utilization of human efforts and physical resources in order to accomplish some goal." Provide some examples of how ancient cultures met this definition.

2-10. As made clear in this chapter, engineers and engineer managers have made strong contributions to management theory and practice. List the engineers and engineer managers identified in this chapter together with their contributions, and add any others you may know of.

SOURCES

Ambrose, Stephen E., *Duty, Honor, Country: A History of West Point* (Baltimore, MD: Johns Hopkins Press, 1966).

Amrine, Harold T., Ritchey John A., and Moodie, Colin L., *Manufacturing Organization and Management*, 5th ed. (Englewood Cliffs, NJ: Prentice-Hall, Inc., 1982), pp. 15–16.

Babbage, Charles, *On the Economy of Machinery and Manufactures* (London: Charles Knight Ltd., 1832; reprinted New York: Augustus M. Kelley, Publishers, 1963), p. 132, quoted in George, *Management Thought*, pp. 76–77.

Babcock, Daniel L. and Lloyd, Brian E. "Educating Engineers to Manage Technology," *Proceedings of the 1992 International Engineering Management Conference, IEEE Engineering Management Society and American Society for Engineering Management*, p. 248.

Barnard, Chester I., *The Functions of the Executive* (Cambridge, MA: Harvard University Press, 1938).

Bedeian, Arthur G., *Management*, 2nd ed. (New York: Holt, Rinehart and Winston, 1989), p. 40.

Bedeian, Arthur G., *Management* (Hinsdale, IL: The Dryden Press, 1986), p. 29.

Blanchard, Kenneth and Johnson, Spencer, *The One Minute Manager* (New York: Berkley Books, 1981).

Breedon, R. L., *Those Inventive Americans* (Washington, DC: National Geographic Society, 1971), p. 48.

Collins, Jim, *Good to Great: Why Some Companies Make the Leap and Others Don't* (New York: Harper Business, 2001).

Covey, Stephen R., *The Seven Habits of Highly Effective People* (New York: Simon & Schuster, 1989).

Davey, Christopher J., "Engineering and Civilization," *Professional Development Handbook* (Parkville, Victoria, Australia: The Institution of Engineers, Australia, September 1992), p. 21.

Drucker, Peter F., *The Effective Executive: The Definitive Guide to Getting the Right Things Done* (New York: Harperbusiness Essentials, revised edition 2006).

Drucker, Peter F., *Managing in the Next Society* (New York: St. Martin's Press, 2002).

Durfee, W. F., "The History and Modern Development of the Art of Interchangeable Construction in Mechanisms," *Journal of the Franklin Institute*, 137:2, February 1894, quoted in George, *Management Thought*, pp. 63–64.

Fayol, Henri, *Administration Industrielle et GÈnÈrale*, Constance Storrs, trans. (London: Sir Isaac Pitman & Sons Ltd., 1949).

Friedman, Thomas L. *The Lexus and the Olive Tree* (New York: Anchor Books, 2000), p. 7.

George, Claude S. Jr., *The History of Management Thought*, 2nd ed. (Englewood Cliffs, NJ: Prentice-Hall, Inc., 1972).

Gilbreth, Frank B., quoted in Arthur G. Bedeian, "Finding The One Best Way," *Conference Board Record*, 13, June 1976, pp. 37–38. In later work, Gilbreth reduced these six motions further to 4.5 per face brick and only 2 motions per interior brick, where excess mortar need not be scraped off.

Hammer, Michael and Champy, James, *Reengineering the Corporation* (New York: HarperCollins, 1993).

Harvey, Jerry, *The Ablilene Paradox and Other Mediations on Management* (New York: John Wiley & Sons, 1988).

Hoover, Herbert C., "The Engineer's Contribution to Modern Life," reprinted in *The Profession of the Engineer* (New York: Wiley, 1929), p. 119 ff.

Johnston, Denis L., "Engineering Contributions to the Evolution of Management Practice," *IEEE Transactions on Engineering Management*, 36:2, May 1989, pp. 106–107.

Lane, Frederick C., *Venetian Ships and Shipbuilders of the Renaissance* (Baltimore, MD: Johns Hopkins Press, 1934), abstracted in George, *Management Thought*, pp. 35–41.

Maxwell John C., *Leadership 101: What Every Leader Needs to Know* (Nashville: Thomas Nelson, 2002).

Merrill, Harwood F., ed., *Classics in Management* (New York: American Management Associations, Inc., 1960), p. 13, quoted in George, *Management Thought*, p. 63.

Mihalasky, John, *The Role of Professional and Engineering Education Societies in the Development of the Undergraduate Industrial Engineering Curriculum*, Ed.D. dissertation, Columbia University, 1973 (draft copy), Chapter 2.

Peters, Thomas J. and Waterman, Jr., Robert H. *In Search of Excellence: Lessons from America's Best Run Companies* (New York: Harper & Row, Publishers, Inc., 1982).

Powell, J., *A View of Real Grievances* (publisher unknown, 1772), quoted in Asa Briggs, ed., *How They Lived*, vol. 3 (Oxford: Basil Blackwell Publisher Ltd., 1969), p. 184.

Reynolds, Terry and Seely, Bruce, "Reinventing the Wheel?" *ASEE Prism*, October 1992, p. 42.

Rice, Berkeley, "The Hawthorne Effect: Persistence of a Flawed Theory," *Psychology Today*, February 1982, pp. 70–74.

Ross, W. L., "Early Influences of the U.S. Military Academy on Engineering Technology and Engineering Graphics Education in the United States," *Proceedings of the 1991 Annual Conference, American Society for Engineering Education*, p. 1604.

Seckler-Hudson, Catheryn, *Process of Organization and Management* (Washington, DC: Public Affairs Press, 1948).

"A Secret Is Shared," *Manufacturing Engineering*, February 1988, p. 15.

Sickles, Richard V., Past President, Central Florida Society for Information Management, 2012.

Sloan, Alfred P., Jr., *My Years with General Motors* (New York: Doubleday & Company, Inc., 1964).

Taylor, Frederick Winslow *Principles of Scientific Management* (New York: Harper & Brothers, 1911), pp. 36–37.

Towne, Henry R., "The Engineer as Economist," *American Society of Mechanical Engineers Trans.*, No. 207, 1886, reprinted in Charles M. Merrick, ed., *ASME Management Division History 1886–1980* (New York: American Society of Mechanical Engineers, 1984), p. 71.

Urwick, Lyndall, "Management's Debt to Engineers," *Advanced Management*, December 1952, p. 7.

Urwick, Lyndall F. As quoted in a letter to Edmund Young, February 17, 1971, supplied by Mr. Young.

Urwick, Lyndall, "The Professors and the Professionals," after-dinner talk, Oxford Centre for Management Studies, Oxford, England, October 12, 1972.

Weber, Max, *The Theory of Social and Economic Organizations*, A. M. Henderson and Talcott Parsons, trans. and eds. (New York: The Free Press, 1947), pp. 328–334.

Wren, Daniel A., *The Evolution of Management Thought*, 3rd ed. (New York: John Wiley & Sons, Inc., copyright © 1987.

Young, Edmund, supplemental notes for "Management for Engineers" course taught at Fort Leonard Wood, MO, August 1988.

Part II

Functions of Technology Management

3

Leading Technical People

PREVIEW

There are four basic management functions defined by Fayol and discussed in Chapters 1 and 2, that are still accepted today. These basic functions of management are commonly identified as leading, planning, organizing, and controlling. Coordination ties all these functions together. President Harry S. Truman defined leadership as "the ability to get [people] to do what they don't want to do and like it." Even if you do not appreciate the sentiment of this statement, it makes clear that leadership is more than just directing others; it is getting others to engage in the work. Fortunately, leadership is truly an art form that can be learned. Just as in any other art form, there are multiple styles of leadership. Engineers need leadership skills even if not in a formal leadership role because they need to influence others to change, whether that is accepting a new product or work effectively on a project team. Engineers are trained to innovate, but many have not learned the skills to be the lead on projects, which they are often expected to do early in their careers.

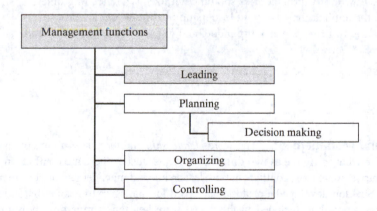

The characteristics that make leadership effective in one company for a certain situation might be ineffective in another organization or another situation. All organizations and people are different and react differently. Different leadership studies have different theoretical approaches, but the same general factors are involved:

- Characteristics and behaviors of the leader
- Attitudes, needs, behaviors, and other characteristics of the followers
- Characteristics of the organization
- Social, economic, and political climate.

This chapter begins with an examination of nature of leadership and the perceived differences in managers and leaders. Next, the traditional trait theories and their application to the engineering leader are considered. Several additional approaches are explored. These include the leadership grid, situational leadership, transactional and transformation leadership, and servant leadership.

Motivation is a key component of the leadership model. McGregor's two contrasting viewpoints (Theories X and Y) on the nature of the individual who is to be motivated are considered. Then two approaches to understanding motivation are presented: content theories and process theories. The content theories include Maslow's hierarchy of needs, Herzberg's two-factor theory, and McClelland's acquired need theory. Process theories assume that behavior is determined by expected outcomes and include Adam's equity theory, Vroom's expectancy theory, and Skinner's behavior modification.

LEARNING OBJECTIVES

When you have finished studying this chapter, you should be able to do the following:

- Describe the role of leaders and what makes leaders effective.
- Describe the nature of leadership and its significance to an organization.
- Describe how motivation theories should be utilized when leading others.
- Address the application of servant leadership in current organizations.
- Describe the full-range leadership model and when different styles of leadership within this model are effective.

LEADERSHIP

Leadership and Management.   The words *leadership* and *management* are often used interchangeably, yet they describe what many see as two different concepts. Just as there are conflicts in what the exact definition of leadership is, there exist conflicts in how leadership and management are or are not inter-related. In his seminal work, Sheldon developed a professional creed for managers to ensure that industry was run with the greatest efficiency possible. Included in this creed were key tenets regarding how management should be incorporated as a stabilizing influence on industry, one that safeguards against disruptive change. This tenet runs in conflict with the concept of leadership as the catalyst for managing and even promoting change in an organization to enable further growth and success discussed by many authors, including Collins' Level 5 leader.

This division between the meaning of management and leadership is a relatively recent split within the literature. Rost found the words used interchangeably beginning in the 1930s and continuing on in some research areas through the 1980s. The effort to split the meaning of the two words began in the late 1950s and remains unresolved. Rost notes that a key gap in these efforts to split the meaning of the two words is the tendency of researchers to denigrate management to ennoble leadership. Or as Mintzberg states:

> Ever since the distinction was made between leadership and management—leadership somehow being the important stuff and management being what surgeons call the scut work—attention focused on leadership.

To avoid this seeming capricious split, Kotter's distinction between the two can be utilized. In this definition, management, including its planning function, makes an operation run smoothly, and leadership, including direction setting (or planning), makes an organization produce or adapt to change. In this way, management and leadership are two sides to a coin and both are needed to successfully move an organization forward. Leadership could be considered the key part of what Mintzberg described in 1971 as the interpersonal work of managers. A view he echoed almost four decades later when he said:

> My view is that management without leadership is disheartening or discouraging. And leadership without management is disconnected, because if you lead without managing, you don't know what's going on. It's management that connects you to what's going on.

Nature of Leadership

Leadership is the process of getting the cooperation of others in accomplishing a desired goal. Sir William Slim, commander of the British Army that defeated the Japanese in Burma in World War II, defined leadership as that "mixture of persuasion, compulsion, and example that makes men do what you want them to do." In a subtler vein, Barney Frank said, "The great leader is the one who can show people that their self-interest is different from that which they perceived."

Paths to Leadership

People become leaders by appointment or through emergence. *Formal*, or "titular," leaders are appointed branch *managers,* committee *chairs,* or team *captains* and have the advantage of formal authority (including the power to reward and punish). A formal position only gives them the *opportunity* to prove themselves effective leaders, rather than make them leaders. Effective leaders seek influence others through persuasion whenever possible, with or without this direct authority.

Emergent, **or** *informal*, leaders evolve from their expertise or referent power as it is expressed in the process of group activity. Even as children we find certain individuals emerging as the ones whose suggestions for the games to play or the mischief to get into are followed, and throughout life we find that certain people *take the lead* and are accepted as informal leaders. When the emergent leader is then appointed or elected as a formal leader, they have a double opportunity to be effective. Recognizing this, many organizations regularly seek to evaluate potential leaders. Often this evaluation includes group situations where no leader is appointed in order to see who emerges to lead the resolution of a jointly assigned problem.

Leadership Secrets

The book *Leadership Secrets of the World's Most Successful CEOs* by Eric Yaverbaum consists of interviews with top executives discussing the proven strategies, philosophies, and tactics they use to help their organizations succeed. Each chapter features a top CEO who reveals his or her most powerful leadership technique. The proven management principles can be summarized by the following leadership strategies:

- Have a clear vision, a specific direction, and a goal for your organization.
- Communicate your vision, strategy, goals, and mission to everyone involved.
- Listen to what others tell you.
- Surround yourself with the right people, a strong team.
- Apply the Golden Rule. (Do unto others as you would have them do unto you.)
- Lead by example. Take responsibility. Make tough decisions.
- Constantly innovate to gain and to sustain competitive advantage.
- Plan everything. Leave nothing to chance.

Sources: Based on Eric Yaverbaum, *Leadership Secrets of the World's Most Successful CEOs* (Chicago: Dearborn Trade Publishing, 2004), and *Orlando Sentinel*, May 10, 2004.

Leadership Traits.   Early research into the nature of leadership tried to identify the personal characteristics, or *traits*, that made for effective leaders. This form of leadership research was commonly referred to as *the great man theory*, where leadership is a set of traits leaders have at birth. For example, Peterson and Plowman list the following 18 attributes as being desirable in a leader:

- *Physical qualities* of health, vitality, and endurance
- *Personal attributes* of personal magnetism, cooperativeness, enthusiasm, ability to inspire, persuasiveness, forcefulness, and tact
- *Character attributes* of integrity, humanism, self-discipline, stability, and industry
- *Intellectual qualities* of mental capacity, ability to teach others, and a scientific approach to problems

Harris had this list of 18 qualities and attributes evaluated by a group of 176 engineers, mostly electrical, mechanical, and aerospace engineers working for high-technology firms in the Dallas, Texas area. There were two phases to this research. In the first phase, 130 engineers, divided into three different ranges of engineering experience, were asked to rate each of the 18 characteristics individually as they perceived their necessity for effective leadership in the engineering environment. The results appear in Table 3-1.

The attribute considered *most necessary* by less experienced engineers was *ability to inspire*, whereas engineers with intermediate (6 to 15 years) experience most valued *enthusiasm* in a leader. The attribute in a leader that apparently becomes more highly valued with experience is *integrity*, rated in eighth place by young engineers, fourth place by those with intermediate experience, and first place by engineers with more than 15 years of experience.

In the second phase of the research, Harris asked an additional 46 engineers who repeated the evaluation above to rate their current engineering managers on the same scale. He then calculated the difference between the mean ranking of the perceived necessity for each quality or attribute with the mean rating of

Table 3-1   Highest- and Lowest–Ranked Qualities and Attributes in Engineering Leaders

Group I	Group II	Group III
0–5 Years Engineering Experience	6–15 Years Engineering Experience	>15 Years Engineering Experience

Highest-Ranked Qualities and Attributes

Group I	Group II	Group III
1. Ability to inspire	1. Enthusiasm	1. Integrity
2. Persuasiveness	2. Stability	2. Ability to inspire
3. Mental capacity	3. Self-discipline	3. Tact
4. Self-discipline	4. Ability to inspire	4. Stability
5. Enthusiasm	5. Integrity	5. Self-discipline
6. Tact	6. Mental capacity	6. Persuasiveness
7. Stability	7. Persuasiveness	7. Industry
8. Integrity	8. Cooperativeness	8. Enthusiasm
9. Cooperativeness	9. Ability to teach	9. Mental capacity

Lowest-Ranked Qualities and Attributes

Group I	Group II	Group III
18. Health	18. Health	18. Health
17. Forcefulness	17. Vitality	17. Forcefulness
16. Personal magnetism	16. Forcefulness	16. Ability to teach
15. Humanism	15. Personal magnetism	15. Personal magnetism
14. Vitality	14. Humanism	14. Humanism
13. Endurance	13. Endurance	13. Cooperativeness
12. Industry	12. Industry	12. Vitality
11. Scientific approach to problems	11. Scientific approach to problems	11. Scientific approach to problems
10. Ability to teach	10. Tact	10. Endurance

Source: E. Douglas Harris, "Leadership Characteristics: Engineers Want More from Their Leaders," *Proceedings of the Ninth Annual Conference*, American Society for Engineering Management, Knoxville, TN, October 2–4, ASEM, 1988, pp. 209–216. Used with Permission.

current engineering managers. He found that engineering managers exceeded the perceived need the most in the following categories (identified by the *t*-score of the difference of means):

- 6.95 Health
- 4.12 Endurance
- 3.79 Scientific approach to problems
- 3.69 Vitality
- 3.67 Forcefulness

On the other hand, these engineering managers were least successful in meeting expectations in the following categories:

- −9.16 Ability to inspire
- −5.36 Tact

- −4.82 Persuasiveness
- −4.17 Stability
- −2.88 Enthusiasm

Harris summarizes his research: "The results quite clearly show that engineers want and expect excellent leaders. The results also show that they are not getting what they want." When Harris repeated this research with European engineers, he obtained similar results, except that he found engineers in Europe were even less satisfied with their managers than were engineers in Texas.

Connolly discusses studies showing that neither appointed nor informal leaders need be much above the average intelligence of the group. He shows that the development and acceptance of emergent leaders are facilitated by social skills, by technical skills in the specific tasks facing the group, and by being at the hub of a communication net.

Myers–Briggs Preferences.   The *Myers–Briggs Type Indicator (MBTI)* measures personal preferences on four scales, each made up of two opposite preferences:

1. *Extraversion E* (focuses on the outer world of people and things, gains energy from interactions) versus *Introversion I* (focuses on the inner world of ideas and impressions)
2. *Intuition N* (focuses on the future, with a view toward patterns and possibilities) versus *Sensing S* (focuses on the present and on concrete information gained from the senses)
3. *Thinking T* (bases decisions on logic and on objective analysis of cause and effect) versus *Feeling F* (bases decisions on values and on subjective evaluation of person-centered concerns)
4. *Judging J* (prefers to have things settled—a planned and organized approach to life) versus *Perceiving P* (prefers to keep options open—a flexible and spontaneous approach to life)

Engineers and scientists frequently are evaluated as *ENTJ* or *INTJ*; successful engineering managers often are *ENTJ*; researchers in technical areas (and the engineering deans who are often chosen from them) are *INTJ*. Only about two percent of the total population test as being in these two categories. This helps explain why engineering faculty, with preferences toward organized, logical, and theoretical presentations often fail to *reach* those engineering students whose preferred modes of learning differ. It also helps explain why many political decisions on technical issues just *don't make sense* to the logical engineer. The difficulty with trait or preference theories is that for every characteristic proposed, one can find no shortage of undeniably effective leaders who seem weak in that area.

People/Task Matrix Approaches

The Leadership Grid.   Robert R. Blake and Jane S. Mouton developed the leadership grid, also called the managerial grid, which is an approach to analyzing the style of management (that is, collective leadership) in terms of two dimensions: *concern for people* and *concern for production* (now *concern for results*). The latest version of this approach is the leadership grid, shown in Figure 3-1. This approach assumes that *(9, 9) team management*, in which individual objectives are achieved in the process of achieving organizational goals, is the ultimate in effective management. The grid can be used with related analyses and interventions to achieve *organizational development* by helping the management of client organizations identify their current management style, and then work toward the recommended (9, 9) style. This approach

Dale Carnegie

Dale Carnegie in his famous book, *How to Win Friends and Influence People*, gave suggestions on how to change people without offending or arousing resentment. Techniques may change, but principles endure. He stated that a leader's job is to change the employees' attitudes and behavior. Here are some of his suggestions:

- Begin with praise and honest appreciation.
- Call attention to people's mistakes indirectly.
- Talk about your own mistakes before criticizing the other person.
- Ask questions instead of giving direct orders.
- Let the other person save face.
- Give the other person a fine reputation to live up to.
- Use encouragement.
- Make the other person happy about doing the thing you suggest.

Source: Carnegie, Dale, *How to Win Friends and Influence People*, http://www .westegg.com/unmaintained/carnegie/win-friends.html, 5 October 2005.

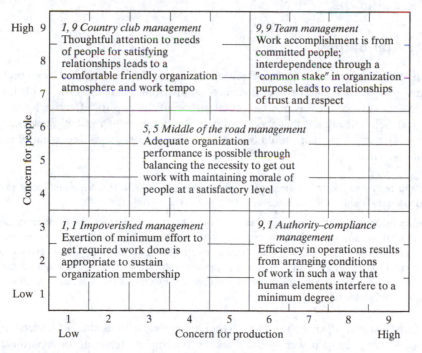

Figure 3-1 The Leadership grid figure. From *Leadership Dilemmas—Grid Solutions*, by Robert R. Blake and Anne Adams McCanse. Gulf Publishing Company, Houston, TX, copyright 1991 by Scientific Methods Inc., p. 29; reproduced by permission of the owners.

assumes that concern only for people (1, 9) leads to a workplace that is enjoyable but not productive, that concern only for results (9, 1) leads to a nonresponsive Theory X workforce, that settling for *adequate* performance and morale in (5, 5) middle-of-the-road management (once called organization man or bureaucratic management) leads only to mediocrity, and that low concern for both people and results is a sign of *impoverished management*.

Hersey and Blanchard Life Cycle Theory and Situational Leadership.   Hersey and Blanchard proposed an extension of the earlier two-dimensional models developed by the Michigan and Ohio State studies, in which the most effective leadership progresses with time through four quadrants based on the readiness of the follower. In this case, readiness is measured on two dimensions, skill (or competency) and will (or commitment). The leader then uses the correct levels of directing behavior (high directing for low skill) and supporting behavior (high support for low will). For example, in teaching a child a simple task such as tying a shoe, a parent initially concentrates on the details of the task (high directing, low supporting). Then, while continuing to correct task errors, the parent praises the child for successes (high directing, high supporting), continuing praise after the task has been learned (low directing, high supporting). Finally, when the task has been ingrained, it no longer requires the attention of the parent (low directing, low supporting), and attention can be shifted to more advanced tasks. In this approach, what level of support the employee requires will depend on the specific task and situation. This theory set the foundation for the contingency-based approaches.

Contingency Approaches

The leadership models described in the two-dimensional approaches (except for Hersey and Blanchard's) imply that only two factors (one dealing somehow with people, and the other with task or production) are important, and that there is one best combination of the two for effective leadership. In 1958, Tannenbaum and Schmidt argued that there really is a continuum of available leadership styles, and one's choice within this continuum should be contingent on the situation. Boone and Bowen assess the significance of this work:

> With the appearance of this article the perspective of *contingency theory*, the dominant theme in management and organizational theory for the next twenty years, was introduced.

> Contingency theory basically argues that there is no one right way to manage. The manager must develop a reward system, a leadership style, or an organizational structure to be appropriate for the unique combination of such factors as the nature of the subordinates, the technology of the business and the tasks that result, the rate of change in the organization, the degree of integration of functions required, the amount of time the manager has to accomplish the assignment, the quality of the manager's relationship with subordinates, and so forth.

Leadership Continuum.   Tannenbaum and Schmidt proposed a continuum of leadership style extending from complete retention of power by the manager to complete freedom for subordinates (they now prefer the term *nonmanagers* to *subordinates*). Although they identify seven *styles of leadership* along the continuum, others have emphasized these four:

1. *Autocratic (Telling)*. Manager makes decisions with little or no involvement of nonmanagers.
2. *Diplomatic (Selling)*. Manager makes decisions without consultation but tries to persuade nonmanagers to accept them (and might even modify them if they object strongly).
3. *Consultative (Consulting)*. Manager obtains nonmanagers' ideas and uses them in decision making.
4. *Participative (Joining)*. Manager involves nonmanagers heavily in the decision (and may even delegate the decisions to them completely).

Tannenbaum and Schmidt proposed that a manager should consider three types of forces before deciding what management style to employ.

1. *Forces in the manager*. The manager's value system regarding leadership and personal leadership inclinations, confidence in the nonmanagers, and feelings of security (or "tolerance for ambiguity") in an uncertain situation.
2. *Forces in the subordinate (or nonmanager)*. Greater delegation can be provided when nonmanagers have a need for independence, are ready to assume responsibility, can tolerate ambiguity, are interested in the problem, understand and relate to the goals of the organization, have the necessary knowledge and experience, and have learned to expect a share in decision making.
3. *Forces in the situation*. The type of organization and the amount of delegation common in it, the experience and success the nonmanagers have had in working together as a group, the nature and complexity of the problem, and the pressure of time.

Shackleton

Sir Ernest Shackleton set sail from the South Georgia Islands in the South Atlantic in December, 1914. His main objective of the trip was to cross the Antarctic on foot. This is a distance of 1,800 miles, and had never been accomplished. He had been to the Antarctic twice before, but had never accomplished his objective of reaching the South Pole. On this particular trip, he never even reached the continent; his ship became stuck in the ice of the Weddell Sea one day's sail from the landing site. The Antarctic winter was so cold that the men could hear the ice freeze. Ten months later, the ship was crushed by the ice and the men were left stranded. And now many call Shackleton one of the greatest leaders of the world. What did he do to earn this title?

For 19 months, Shackleton used leadership skills that are as important today as they were then. With these skills, he led the 27 crew members to safety, alive and well. He developed and unified his team despite their different backgrounds and abilities. Elements of his leadership included optimism, communication, flexibility, strong example, and encouraging enjoyment. While marooned, he encouraged the men to play sports on the ice. These activities kept the crew fit and their spirits up. What is apparent in Shackleton's leadership is that after 19 months of being stranded on the ice in one of the world's most inhospitable places, his entire crew survived and had not lost their spirit.

Source: Margot Morrell and Stephanie Capparell, *Shackleton's Way*, 2001.

The Hersey and Blanchard model discussed previously is really a situational model in which leadership styles are selected from a 2 × 2 matrix rather than a linear continuum, based on *forces in the subordinate* (specifically, maturity). In essence, the subordinates can choose the leadership style: If they want to participate in decision making, they need to demonstrate the necessary maturity and the skilled leader will encourage them to do so.

Servant Leadership

In the 1970s a new leadership style was defined. **Servant leadership** is a practical philosophy that supports people who choose to serve first, and then lead as a way of expanding service to individuals and institutions. The term servant leadership was coined by Robert K. Greenleaf, a retired AT&T executive, in his book, *Servant as Leader*, published in 1970. This approach emphasizes the leader's role as steward of the resources, including both human and financial, provided by the organization. Initially considered a fad, servant or service leadership has gained acceptance over the past 35 years. Servant leadership is characterized by the belief that leadership development is an ongoing, continuously improving process. Servant leadership encourages collaboration, trust, foresight, listening, and the ethical use of power and empowerment. Servant leaders may or may not hold formal leadership positions.

Unlike other leadership approaches with a top-down hierarchical style, servant leadership emphasizes collaboration, trust, empathy, and the ethical use of power. Some of the actions the servant leaders do are:

- Devote themselves to serving the needs of the organization's members.
- Focus on meeting the needs of those that they lead.
- Develop employees and facilitate personal growth.
- Coach others and encourage self-expression.
- Listen and build a sense of community.

Transformational Leadership and the Full Range Model

Transformational leadership has been called the new paradigm of leadership and is generally considered to have its foundation in the work completed by Burns in political leadership in the late 1970s. At that time, Burn's (1978, 20–21) explained that transformational leadership:

> ...occurs when one or more persons engage with others in such a way that leaders and followers raise one another to higher levels of motivation and morality. Their purposes, which might have started out as separate but related, as in the case of transactional leadership, become fused. Power bases are linked not as counterweights but as mutual support for common purpose. [...] But transforming leadership ultimately becomes moral in that it raises the level of human conduct and ethical aspiration of both the leader and led, and thus it has a transforming effect on both.

Bass was the first to publish a multi-factor definition of transformational leadership. In his definition, which has become the dominant definition in the research space, transformational leadership has four dimensions:

- *Idealized Influence* (initially termed Charisma)—The degree to which the leader behaves in admirable ways that cause followers to identify with and trust the leader. This trait is about the leader providing a role model for the followers.
- *Inspirational Motivation*—The degree to which a leader articulates a vision that appeals and inspires followers. These leaders challenge followers with high standards, communicate optimism about future goals, and provide meaning for the task at hand.

- *Intellectual Stimulation*—The degree to which a leader stimulates new ideas and creative solutions from their followers by challenging assumptions and encouraging risk taking.
- *Individualized Consideration* (or Individualized Attention)—The degree to which the leader understands the individual needs of each of their followers and attends to those needs.

Based on the work of Avolio and Bass, it is the combination of the four behaviors that leads to successful transformational leadership behavior that motivates others to do more than they thought possible.

Transformational leadership is only one component of the framework Bass developed, which he named the full-range leadership model. It includes not only transformational, but also transactional leadership and laissez-faire leadership (or the absence of leadership). Transactional leaders use conventional reward and punishment to gain compliance from their followers. Bass's definition has two components:

- *Contingent Reward*—In this system a bargain is struck, and a contract signed (literally or figuratively) between leader and subordinate. From that time forward, the employee's efforts (transactions) are actively monitored and when the terms of the contract are met, positive reward in the form of praise, salary increases, or promotion is provided. When the terms of the contract are not met, penalization occurs. When utilized consistently, contingent reward can be an effective form of leadership; however, it is seldom maintained at the level of consistency required for sustained performance.
- Management by Exception—This form of management is far more passive than those above. In this approach, a bargain is still struck, but as long as the contract is honored by the employee, there tends to be little feedback provided to the employee. Instead there is a mode of silence when all is well, and when something drops below standard, there is a reaction, including negative feedback. This mode of leadership can be effective in teaching new employees what not to do; however, it has minimal effect in teaching employees what to do. The behavior follows directly from the role of manager as controller.

A great deal of study has shown that transformational leadership is most effective in times of change, when organizations need to react substantially to changes in the market or face the potential of not existing. Transactional leadership is often effective in organizations in stable industries. Given the level of change facing organizations in today's economy (Chapter 18), aspiring engineering managers are urged to understand and seek to practice transformational leadership behaviors.

Other Viewpoints.   Other authors have tried to characterize leaders in almost innumerable ways. For example, Drucker presents a new paradigm—that there is no one right way to manage. Different groups in the work populations must be managed differently at different times. Thus, leadership styles are constantly changing.

For brevity in describing effective leadership, it is difficult to top the following, attributed to the Chinese philosopher Lao Tsu (about 600 B.C.), and etched in copper in the office of Jack Smith, former CEO of General Motors:

> *A leader is best*
> *when people barely know he exists.*
> *Not so good when people obey and acclaim him.*
> *Worse when they despise him.*
> *But of a good leader, who talks little,*
> *when his work is done and his aim fulfilled,*
> *they will say, "We did it ourselves."*

MOTIVATION

Introduction

To have an effective technical organization, one needs to understand the nature of motivation, which is an important part of leadership. Berelson and Steiner have defined ***motive*** as "an inner state that energizes, activates, or moves (hence 'motivation'), and that directs or channels behavior toward goals." Robbins defines **motivation** in an organizational sense as "the willingness to exert high levels of effort to reach organizational goals, conditioned by the effort's ability to satisfy some individual need."

Campbell et al. define motivation in terms of three measures of the resulting behavior:

1. The *direction* of an individual's behavior (measured by the choice made when several alternatives are available)
2. The *strength* of that behavior once a choice is made
3. The persistence of that behavior

Dale Carnegie states that "there is only one way under high heaven to get anybody to do anything. And that is by making the other person want to do it." Therefore, we need to learn why people want to do things and how they can be persuaded (or motivated) to do those things that will enhance organizational goals.

McGregor's Theory X and Theory Y.   The way leaders try to motivate someone depends on our assumptions about their basic nature. Douglas McGregor postulated two contrasting sets of assumptions about the average worker, calling them Theory X and Theory Y. In his Theory X, he painted a dismal picture of the nature of the average person and its implications for the task of management:

> The conventional conception of management's task in harnessing human energy to organizational requirements can be stated briefly in terms of three propositions. In order to avoid the complications introduced by a label, let us call this set of propositions **Theory X**:

1. Management is responsible for organizing the elements of productive enterprise—money, materials, equipment, people—in the interest of economic ends.
2. With respect to people, this is a process of directing their efforts, motivating them, controlling their actions, modifying their behavior to fit the needs of the organization.
3. Without this active intervention by management, people would be passive—even resistant—to organization needs. They must therefore be persuaded, rewarded, punished, controlled—their activities must be directed. This is management's task.

Behind this conventional theory there are several additional beliefs—less explicit, but widespread:

4. The average person is by nature indolent—they work as little as possible.
5. They lack ambition, dislikes responsibility, prefer to be led.
6. They are inherently self-centered, indifferent to organizational needs.
7. They are by nature resistant to change.
8. They are gullible, not very bright, the ready dupe of the charlatan and the demagogue.

McGregor, suggests that such behavior is not necessarily inherent in human beings, and he concludes that the carrot-and-stick approach of relying on wages for motivation "does not work at all once [people

Leadership from *Good to Great*—The Level 5 Leader

Jim Collins and his research team spent five years looking for companies that made substantial improvements in their performance over time. They finally settled on 11 companies and performed a robust study that gathered large amounts of empirical data to compare these companies to 17 others who had not made the transition to greatness. While the companies that achieved greatness were all in different industries, each engaged in versions of the strategies Collins's team identified. The core lessons from the book include finding a strategy that will be enduring and easy to defend, using technology to accelerate momentum that already exists, and confronting uncomfortable truths. Throughout all of these companies, enabling the success of these strategies was a leader of a certain type.

Collins defines these people as Level 5 leaders and noted that ten out of eleven good-to-great company leaders or CEOs came from inside the company. They were not outsiders hired in to 'save' the company. They were either people who worked many years at the company or were members of the family that owned the company. Aside from their origins, these Level 5 leaders were present in every company during pivotal transition years. Level 5 is the Executive who builds enduring greatness through a paradoxical blend of personal humility and professional will. This humility included a display of compelling modesty. The leaders were self-effacing and understated. In contrast, two thirds of the comparison companies had leaders with large personal egos that contributed to the demise or continued mediocrity of the company.

The paradox is that Level 5 leaders were fanatically driven. They seemed to almost be infected with an incurable need to produce sustained results. They were resolved to do whatever it takes to make the company great, no matter how big or hard the decisions. Perhaps most importantly, Level 5 leaders set up their successors for even greater success in the next generation.

Source: Jim Collins, *Good to Great: Why Some Companies Make the Leap…and Others Don't* (New York: Harper Business, 2001) and http://www.jimcollins.com/.

have] reached an adequate subsistence level and [are] motivated primarily by higher needs." (See section *Content Theories*.) He continues:

> For these and many other reasons, we require a different theory of the task of managing people based on more adequate assumptions about human nature and human motivation. I am going to be so bold as to suggest the broad dimensions of such a theory. Call it **Theory Y**, if you will.
>
> 1. Management is responsible for organizing the elements of productive enterprise—money, materials, equipment, people—in the interest of economic ends [identical to (1) for Theory X].
> 2. People are *not* by nature passive or resistant to organizational needs. They have become so as a result of experience in organizations.
> 3. The motivation, the potential for development, the capacity for assuming responsibility, the readiness to direct behavior toward organization goals are all present in people. Management does not have to put them there. It is the responsibility of management to make it possible for people to recognize and develop these human characteristics for themselves.

4. The essential task of management is to arrange organizational conditions and methods of operation so that people can achieve their own goals *best* by directing *their own* efforts toward organizational objectives.

McGregor summarized by saying that "Theory X places exclusive reliance upon external control of human behavior, while Theory Y relies heavily on self-control and self-direction. It is worth noting that this difference is the difference between treating people as children and treating them as mature adults."

Since McGregor published his work, Theory Y approaches have proven to be at the core of most successful companies and continue to gain wide support. A driving factor for this tidal shift in management is the transition of large swaths of the global workforce from manual to knowledge work. The change in the workforce led management guru Peter Drucker to indicate that Theory X approaches were obsolete over four decades ago. He also noted these approaches are ineffective for knowledge workers in any setting. In short, effective knowledge work only occurs when these workers are self-directed, which is not possible under Theory X. Despite this, Theory X remains prevalent and many of us have witnessed these behaviors from our own "worst manager."

Content versus Process Theories.   Theories trying to explain how people are motivated are commonly divided into two categories. *Content theories* are based on human needs and people's (often unconscious) efforts to satisfy them. *Process theories*, on the other hand, assume that behavioral choices are made more rationally, based on the expected outcomes. We examine each category in turn, with special emphasis on their application to the technical professional.

Content Theories

Maslow's Hierarchy of Needs.   One of the earliest and most influential content theories is the concept of Abraham H. Maslow that "human needs arrange themselves in hierarchies of prepotency. That is, the appearance of one need usually rests on the prior satisfaction of another." Maslow identified the following five needs, which are often portrayed in stair-step function as in Figure 3-2.

Physiological Needs.   At the lowest level of the hierarchy are physiological needs. People concentrate on these needs before continuing up the hierarchy to satisfy higher-order needs. In the workplace these include basic wages or salary, and reasonable working conditions.

Security/Safety Needs.   Next in the hierarchy, workers need job security, safe working conditions, protection against threats, and a predictable work environment. Also included at this level are the job benefits—medical, unemployment, and disability insurance—as well as retirement plans.

Affiliation Needs.   After the lower levels of the hierarchy, physiological and security, have been met, affiliation needs become a motivator for the worker. In the workplace these include compatible coworkers and a pleasant supervisor. These needs may be met outside the workplace where there is a need for interaction with others and being part of a group.

Esteem Needs.   Esteem needs are met by self-respect or self-esteem, and the esteem of others. Praise, recognition, and promotion within the company satisfy these needs. In some situations, this includes the location of a person's office.

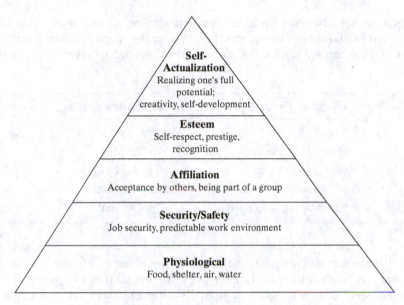

Figure 3-2 Maslow's hierarchy of needs.

Self-Actualization Needs. The highest level is the desire to become everything one is capable of becoming (to become actualized in what one is potentially). This need may be addressed through creative or challenging work or special assignments. Leaving a company and going into business for yourself is a form of this highest level.

Maslow believed that this was an approximate order of need satisfaction for most people, although there were exceptions. The *mad scientist* working alone in the corner appears to value self-esteem and self-fulfillment needs ahead of the need for love and affection. Porter et al. believe that Maslow is at least right in distinguishing between lower-order (physiological and safety) needs and the other three:

> There is strong evidence to support the view that unless the existence needs are satisfied none of the higher-order needs will come into play. There is also some evidence that unless security needs are satisfied, people will not be concerned with higher-order need.... There is, however, little evidence to support the view that a hierarchy exists once one moves above the security level.

Throughout most of human history the concern for survival (meeting the physiological needs) has been paramount. Industrial workers in the developed nations today, however, commonly find their physiological needs satisfied and most security needs met through fringe benefits, except where layoff is a threat. The higher-order needs can be fulfilled either at work or in society. Self-actualization, for example, can be achieved through hobbies, education and personal growth, and charitable or religious activities in the community, as well as through achievement at work. The approval of friends and respect of the community can substitute in part for lack of recognition at work. The challenge to effective management is to find ways in which the higher-order needs of the individual can be satisfied in the process of achieving the objectives of the organization.

Herzberg's Two-Factor Theory.   Frederick Herzberg studied the factors affecting job attitudes and found that they could be divided into two groups: those that provided motivation when they were present, and those *hygiene* factors that led to job dissatisfaction when they did not meet expectations.

Hygiene Factors	Motivator Factors
Salary	Recognition
Working conditions	Work itself
Company policies	Responsibility
Relationship with boss	Advancement
Relationship with peers	Achievement

Herzberg's hygiene factors correspond well to the lower three of Maslow's needs (physiological, safety/security, and love/relationship) and the motivators with the Maslow's upper two needs (esteem and self-fulfillment). Herzberg considered salary as primarily a hygiene factor, and certainly it leads a person to be dissatisfied when salary is less than what he or she thinks is merited, or when he or she is given a smaller raise than the employee at the next desk received. However, salary is *a way of keeping score*, and a healthy raise can be clear recognition for one's work, and in that sense, motivating. Bonuses and profit sharing can be motivating as well. For example, Worthington Industries has a profit-sharing program that can amount to half of an employee's total compensation. John H. McConnell, Worthington's founder, chairman, and CEO, reports:

> Our people care about quality. If customers don't accept our shipments, part of the cost comes out of each of our pockets. So, people take the time to do their job right the first time. Our rejection rate is less than one percent, compared to an industry average of three to five percent.

Herzberg developed the methodology of *job enrichment* to increase the content of motivators in a job. Examples of job-enrichment actions include reducing the number and frequency of controls, making the worker responsible for checking his or her own work, establishing a direct relationship between the worker and the customer or user of that work (whether internal or external), and in other ways, increasing authority and autonomy.

Job enrichment and the underlying two-factor theory have attracted many disciples who have applied it in a wide variety of environments. There have been quite a few critics as well. Myers believes that people may be categorized as either *motivation seekers*, who respond well to job enrichment, or *mainte-nance seekers*, who "are motivated primarily by the nature of their environment and tend to avoid motivational opportunities...are chronically preoccupied and dissatisfied with maintenance [hygiene] factors surrounding their job...[and] realize little satisfaction from accomplishment and express cynicism toward the positive values of work and life in general."

Leadership Failures: Ethical Leadership and Motivation

Unfortunately, it seems we can't look at the news on any given day without stories of the ethical failures of a leader in business or politics being prominently featured. Whether it is the investigation of a U.S. cabinet member for misuse of public funds, the cover-up by the leaders of Facebook regarding the hacking of customer information, the U.S. public choosing to re-elect members of congress who are under indictment (of both parties), or the arrest of Nissan Motors Chairman Carlos Ghosn in Japan for improperly reporting income; it seems we cannot trust our leaders to do what it right. It is important to recognize that these failures are not limited to people with a background in business, finance, or politics; the incentive system that led to the leadership failures at Wells Fargo was designed while an engineer was CEO and the technology cheats that led to Volkswagen's "Dieselgate" were all designed and enabled by engineers.

Why are these failings so common? A major factor is certainly a breakdown of ethics, especially a focus on ethical egoism as discussed in Chapter 16. But part of it is also a question of motivation and perhaps too great a focus on Maslow's esteem needs or Herzberg's Recognition and Achievement motivators. In short, the adoration that top management seeks from investors and their peers. If motivational factors are a major contributor, engineering managers should be prepared to correct this issue. One action they can take is to refocus the organization to consider the motivation of all members of the organization. The decisions that led to the scandals discussed here put the very existence of an organization at risk (for a case study, students are encouraged to perform a web search for Enron). This threatens the security/safety needs of all employees. Along the way, unethical actions almost certainly lead to conflict between management and employees, leading to degradation of most of Herzberg's hygiene factors, including working conditions, company policies, and relationships with management and peers. When the motivations of this broader group of stakeholders are considered, the path forward should become clear to all members of the organization.

McClelland's Trio of Needs.   David McClelland and others have proposed that there are three major motives or needs in work situations:

1. *Need for achievement* is the drive or desire to excel—to accomplish something better than has been done in the past. People with a high need for achievement tend to be entrepreneurs, setting moderately difficult goals, taking moderate risks to achieve them, and taking personal responsibility for getting things done. Although McClelland estimates that only about 10 percent of the American population has a high need for achievement, he has shown that the need for achievement can be increased through proper training.

   McClelland also believed that the higher the need for achievement was in the total society, the greater would be the prosperity of the country. Civilizations and nations that expect a lot from their youth have high-achieving societies. The academic achievement demanded of most children in Japan—and of college-bound children in many other countries—presents a serious challenge to American primary and secondary schools.

2. *Need for power* is the desire to control one's environment, including resources and people. Persons with a high need for power are more likely to be promoted to managerial positions and are likely to be successful managers if they master self-control and use their power for the good of the organization rather than solely for personal ends.
3. *Need for affiliation* is the need for human companionship and acceptance. People with a strong need for affiliation want reassurance and approval, are concerned about other people, and perform well as coordinators, integrators, and counselors, and in sales positions.

The need for affiliation might be compared with Maslow's third level (love), the need for power with his fourth level (ego or esteem), and the need for achievement with the fifth level (self-actualization). However, McClelland's point is that different people have different needs, not just the same needs in a clear hierarchy of importance. For example, an engineer with a high need for achievement may achieve success in technical assignments in the process of satisfying this need, and they might be promoted into a management position as a result. If this need for achievement is combined with a low need for power, the engineer will often peak earlier in their career and at a lower level, since the need for achievement can be satisfied by the work itself rather than (as with the need for power) requiring continuing promotions. Again, engineering jobs that put a premium on coordination and cooperation, such as today's team management organizations or the matrix organizations common in project management (Chapter 15), certainly require a blend of need for achievement and for affiliation.

Process Theories

Process theories treat human needs as just one part of the mechanism that people use in choosing their behavior, and these theories place greater emphasis on the expectation of favorable consequences or rewards. We consider four theories in this group: equity theory, expectancy theory, the Porter–Lawler extension of the first two, and behavior modification.

Equity Theory.   Developed by J. Stacey Adams, this theory is based on the simple belief that people want to be treated fairly relative to the treatment of others. Adams describes this comparison in terms of input/outcome ratios. Inputs are a person's contribution to the organization in terms of education, experience, ability, effort, and loyalty. Outcomes are the obvious rewards of pay and promotion and the subtler ones of recognition and social relationships. A person who feels under rewarded compared with someone else may (1) put forth less effort, (2) press for a higher salary (or a bigger office or a reserved parking place), (3) distort the perceived ratio by rationalizing, or (4) leave the situation (quit or transfer). Conversely, a person who feels over rewarded *may* be motivated to contribute more.

Expectancy Theory.   Formulated in 1964 by Victor Vroom, expectancy theory relates the effort a person puts forth to the expectation of achieving some desired goal. As illustrated in Figure 3-3, this involves a combination of two *expectancies*:

Effort-to-performance expectancy is a person's perception of the probability that his or her effort will lead to high performance, usually in meeting an organizationally desired goal. The ability to achieve high performance (first-order outcomes in the model) is considered a function of individual ability and the environment (tools, resources, and opportunity), in addition to the effort applied.

Performance-to-outcome expectancy, also known as *instrumentality*, is the person's perception that attaining the performance just described will lead to intrinsic and extrinsic rewards (second-order outcomes).

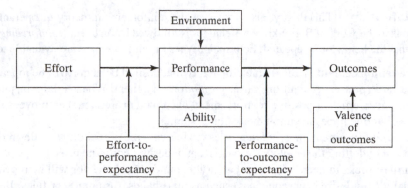

Figure 3-3 Expectancy theory of motivations.

Intrinsic rewards are intangibles such as a feeling of accomplishment or sense of achievement; extrinsic awards are tangible results such as pay or promotion. Both the effort-to-performance and the performance-to-outcome expectancy may be rated on a scale from 0.0 (no relationship) to 1.0 (certainty).

Valence measures the strength of a person's desire for these outcomes (which may be positive or negative) and is related to the individual needs we have already considered. According to this theory, motivation can be calculated as the product of the values assigned to these three factors. For example, a student's motivation to study for a final examination may be a function of (1) the expectation that study (effort) can lead to a good grade in the final, (2) the expectation that a good grade in the final can influence the grade for the course, and (3) the value the student places on earning high grades.

Although it is difficult to make use of this model quantitatively, it offers some qualitative suggestions to the manager. Effort-to-performance expectancy may be increased by assuring that the person understands the objectives that they are asked to achieve, and has the training, self-confidence, and organizational support to achieve them. Performance-to-outcome expectancy can be increased by trying to discover what the person values as an outcome, and trying to provide that reward for good performance. Choate believes that U.S. industry is at risk because it does a poor job in providing this motivation:

> Dan Yankelovich, for example, reports that only 9 out of 100 American workers think that if they work harder and smarter they'll get any benefit out of it. By contrast, 93 out of 100 Japanese workers think that if they make an extra effort, they'll get extra benefits out of it. That's an important difference.

The Porter–Lawler Extension.   Lyman W. Porter and Edward E. Lawler proposed an extension of the expectancy model involving employee satisfaction. It may be compared with Figure 3-2 as follows:

- Personal effort, abilities and traits, and role perceptions (the employee's belief that certain tasks need to be done to do their job effectively) determine performance.
- Performance, in turn, leads to intrinsic and extrinsic rewards, as in the expectancy model.
- The perceived equity (fairness) of these rewards determines the satisfaction the employee gains from the work.
- This satisfaction colors the value placed on the rewards anticipated for future cycles of work, and therefore, it influences future effort.

Behavior Modification.   This theory, also known as the reinforcement theory or operant conditioning, has its foundations in the work of B. F. Skinner. Behavior is followed by an event (*reinforcement*) that affects the probability that the behavior is repeated. Four major types of reinforcement are available to the manager:

1. **Positive reinforcement** increases the probability that desired behavior will be repeated by providing a reward (praise, recognition, raise, promotion, or other). When a dolphin jumps through a hoop it is given a fish as positive reinforcement; even with professional employees the principle is the same, but one hopes the implementation is subtler.

2. **Negative reinforcement**, or *avoidance*, seeks to increase the probability that desired behavior will be repeated by letting the employee escape from undesired consequences.

3. **Punishment** seeks to decrease the probability that undesired behavior will be repeated by imposing penalties (undesired consequences) such as reprimands, discipline, or fines. Because punishment often leads to resentment and even poorer performance, managers try to use it as a last resort.

4. **Extinction** seeks to decrease the probability that undesired behavior will be repeated by ignoring it and withholding positive reinforcement. For example, an employee's inappropriate remark at a meeting might be ignored, but the next common-sense suggestion made may be answered with the comment, "Good thinking." (The reverse is also true: Repeated failure to recognize desired behavior can lead an employee to think that it is not important and to stop doing it.)

MOTIVATING AND LEADING TECHNICAL PROFESSIONALS

Now that the general theories of human motivation and of leadership have been presented, they will be applied to the technical professional. First, the nature of the professional is recalled; then, what motivates scientists and engineers; and finally, the significance of these factors in the effective leadership of technical professionals is considered.

General Nature of the Technical Professional

A number of authors (for example, Kerr et al. and Rosenbaum) have examined the special characteristics of technical professionals (without distinguishing between scientists and engineers). They are typically described as follows:

- Having a *high need for achievement* and deriving their motivation primarily from the work itself. As such, they are most productive when they can achieve their professional goals in the process of pursuing organizational goals.
- Desiring *autonomy* (independence) over the conditions, pace, and content of their work. To achieve this, they need to participate in goal setting and decision making as it affects their work.
- Tending to *identify first with their profession* and second with their company. As professionals, they look to their peers (whether inside or outside the organization) for recognition, ethical standards, and collegial support and stimulation.
- Seeking to *maintain their expertise*, gained through long and arduous study, and stave off obsolescence (see Chapter 17) through continuing education, reading the literature, professional

All Engineers Need Leadership Skills

Engineers need to be influential. At all levels of an organization, engineers should play a significant role in driving innovations that will benefit customers and increase profits.

Engineers are trained to innovate, but unfortunately, many have not learned the skills necessary to influence others and to develop ideas that increase profits. Engineers, then, need to know how to articulate their thoughts so that others will be inspired to build on them. They need to learn how to drive projects and ideas to create innovations that customers will value.

Following are seven reasons why technical professionals need leadership skills:

- **Technical acumen alone is not influential.** Technical gurus without leadership skills have limited influence.
- **Leadership is not just for managers.** Leading and managing require different skill sets. Some leadership experts might argue this point, but most agree that leadership differs from "management."
- **Engineers lead projects.** Even engineers who aren't "project leads" provide a certain amount of direction, and they need to influence others to help get their work done.
- **Engineers can guide less-experienced peers.** Guidance is providing direction one of the three basic definitions of leadership (the other two are influence and authority).
- **Engineers need to help their managers' business succeed.** You may not be inspired to help your manager be more successful as an individual, but you must be dedicated to helping your business achieve success. If not, find another job.
- **Engineers can influence decision makers in their organizations.** Engineers understand technology better than nontechnical managers, and they understand the details better than most technical managers.
- **Everyone should be interested in building character.** Leadership is mostly character and a little bit of skill. People listen to people who have integrity and who apply it well on the job.

Source: Adapted from Gary Hinkle, *IEEE USA Today's Engineer*, April 2007. http://www.todays engineer.org/2007/Apr/leadership-skills.asp, September 2012.

society activity, and especially through work assignments that keep them working at the state of the art.

Differences among Technical Professionals

Scientists versus Engineers. Scientists and engineers often differ in significant ways. Allen and others have identified some of these differences:

- Even as undergraduates, science students place higher value on independence and learning for its own sake; engineers are more concerned with professional preparation, success, and family life.
- The *true scientist* is commonly assumed to have a doctorate; the typical engineer generally begins professional practice with a Bachelor of Science degree, and they typically earns a master's degree later.

- The scientist puts a high value on professional autonomy and publication of results; the engineer is a team worker and places little value on publication.
- Although both groups desire career development and advancement, the scientist depends heavily on reputation with peers outside the company; the engineer's advancement is tied more to activities within the company. The engineer, therefore, is motivated more by organizational goals, more comfortable with applied assignments, and more likely to seek tangible rewards within the organization.
- Science grows through evolutionary additions to the literature, to which the scientist wants to be free to add. *[T]he technologist's principal legacy to posterity is encoded in physical, not verbal, structure.* Further, the engineer is more likely to be working with developments that are considered proprietary information by the organization and thus has less opportunity to publish results.

Leading Technical Professionals

Dimensions of Technical Leadership.   Rosenbaum believed that to facilitate achievement of individual and group goals, successful technical leaders should master "five strategic dimensions":

1. *Coach for peak performance.* "Listen, ask, facilitate, integrate, provide administrative support;" act as a sounding board and supportive critic; help the professional manage change.
2. *Run organizational interference.* Obtain resources, act as advocate for the professional and his or her ideas, and minimize the demands of the bureaucracy (time and paperwork) on the professional.
3. *Orchestrate professional development.* Facilitate career development through challenging assignments; foster a business perspective in professionals; find sources where new areas of knowledge are required.
4. *Expand individual productivity through teamwork.* Make sure teams are well oriented regarding goals and roles, and that they get the resources and support they need.
5. *Facilitate self-management.* Assure that technical professionals are empowered to make their own decisions by encouraging free two-way information flow, delegating enough authority, and providing material and psychological support.

Leading as Orchestration.   McCall has evaluated a number of studies of the relationship between a formal leader and a follower group of professionals, mostly in R&D settings. He concludes that in such groups "effective supervisory leadership is more orchestration than direct application of authority. It seems a matter of creating and/or maintaining (or at least not destroying) conditions that foster scientific productivity." While the supervisor is not the only factor determining group effectiveness, McCall identifies four general areas where the leader can make a difference:

1. *Technical competence.* "The supervisor's technical competence is related both to scientific productivity and the scientists' willingness to comply with management directives…. Leaders of productive groups serve many roles that depend on technical expertise, including the following: recognizing good ideas emerging inside and outside the group; defining the significant problems; influencing work goals on the basis of expertise; and providing technical stimulation."

2. *Controlled freedom.* "In general, leaders of productive groups create controlled freedom, a condition in which decision making is shared but not given away, and autonomy is partially preserved."

3. *Leader as metronome.* McCall views this image as "perhaps the best statement of the subtlety of leadership in professional groups," and quotes Sayles and Chandler as describing the job of project manager as one that "widens or narrows limits, adds or subtracts weights where tradeoffs are to be made, speeds up or slows down actions, increases emphasis on some activities and decreases emphasis on others."

4. *Work challenge.* Since challenging work is one of the most important things to a professional, the technical manager is measured by the extent to which he or she can provide challenging assignments. The professional's view of what is challenging must be reconciled with the needs of the organization, and the challenge to the supervisor is not just making wise assignments, but structuring them as much as possible to provide the desired challenge and then persuading the individual of their importance.

Breakpoint Leadership.   McCall confined himself in the preceding statements largely to direct supervision of a group of technical professionals, especially in R&D. Then he added:

> At some point on the way up the managerial ladder, a different kind of leadership demand occurs. When influencing other parts of the organization is as important, or more important, than influencing a subordinate group, leadership is a breakpoint. Effectiveness is no longer measured simply as group productivity, but involves such things as impact on organizational direction, influence across organizational and even hierarchical boundaries, and securing and protecting organizational (and external) resources and support....For many professionals the first breakpoint leadership role is that of project manager.

Use of Motivational Theories by Engineers

Utley and Westbrook conducted a survey of 408 engineering managers in Tennessee to determine which motivational theories they were familiar with, and which theories they were actually using. Although the survey is not recent, and more management theories have been introduced in the last few years, this study remains relevant. As Table 3-2 shows, they were most familiar with three related concepts not discussed in this chapter: management by objectives (MBO, discussed in Chapter 4), quality (discussed in Chapter 12), and the findings of Peters and Waterman (*In Search of Excellence*, introduced toward the end of Chapter 2). Of the theories discussed in this chapter, they were most familiar (in descending order) with Maslow's hierarchy, McGregor's Theory X and Theory Y, Herzberg's two-factor theory, the Managerial Grid (now the Leadership Grid), and the Tannenbaum and Schmidt leadership continuum; they thought they were making use of them in about the same order. Engineering managers in government agencies favored use of MBO and quality more than did managers in government contractor organizations and private industry; other management concepts were used with about the same frequency in the three types of organizations. Top-level managers were more familiar with these motivational concepts than were lower-level managers. Managers at all levels in high-technology companies were more likely to use motivational concepts than were managers in lower-technology companies.

Table 3-2   Familiarity and Use of Motivational Theories by Engineering Managers[a]

Theory	High Tech n = 229 F	U	Medium Tech n = 179 F	U	Top Level n = 95 F	U	Mid Level n = 162 F	U	First Level n = 151 F	U	Total n = 408 F	U	Ranking n = 17 F	U
Herzberg	50	28	39	16	58	31	45	21	37	21	45	23	6	4
Maslow	69	40	59	34	76	49	67	40	55	26	64	37	4	2
McGregor	60	26	51	18	64	28	59	22	48	19	56	23	5	4
Managerial Grid	44	17	40	12	52	20	38	14	40	13	42	15	7	7
Likert System IV[b]	9	1	11	1	8	1	10	1	11	2	10	1	*	*
In Search of Excellence[c]	66	20	64	15	87	41	63	18	54	15	65	22	3	6
McClelland	21	7	15	3	16	6	19	3	19	7	18	5	*	*
Porter and Lawler	31	15	26	9	32	15	28	14	28	10	29	13	11*	8
Likert linking pin[d]	17	7	13	4	13	5	15	4	18	7	16	6	*	11*
Vroom[e]	10	0	8	2	9	0	9	1	10	1	10	1	*	*
Argyris[f]	16	4	11	2	14	3	10	2	17	4	13	3	*	*
MBO[g]	87	60	85	50	96	68	89	59	77	43	86	55	1	1
Quality circles[h]	86	38	78	34	85	46	83	40	81	26	83	36	2	3
Hersey and Blanchard	19	9	17	5	19	5	17	7	19	9	18	7	*	10
Tannenbaum and Schmidt	36	12	27	8	35	16	33	8	28	9	32	10	8	9
Ouchi[i]	31	5	30	8	45	15	27	4	25	3	30	6	10*	11*
Drucker*	33	6	28	5	41	11	28	4	27	4	31	6	9*	11*

[a]F, familiar with theory; U, use theory.

[b–i] Theories not discussed in the chapter are (b) Likert's four systems varied from System I (very autocratic) to System IV (very participative); (c) *In Search of Excellence* by Peters and Waterman was introduced in Chapter 2; (d) Likert's linking pin concept involves forming teams, starting at the top, for good decision making and communication; (e) Expectancy theory; (f) Argyris's study of personality development concluded that classical management concepts such as chain of command and span of control tended to encourage childlike, dependent behavior; (g) management by objectives, introduced in Chapter 4; (h) quality circles are introduced in Chapter 12; (i) Ouchi's Theory Z applied Japanese management systems to American companies.

*"Drucker's survival principles—modern businessperson's attitude" was the last "motivational principle" in the survey. Although Peter Drucker is perhaps the most widely read contemporary management theorist and author, his survival principles are not related to motivation. The number checking this concept may relate to the approximate error level in the survey, and lower scores on the other theories are not considered significant.

Source: Dawn R. Utley and Jerry D. Westbrook, "A Survey of Management Concepts in Technical Organizations," *Proceedings of the Ninth Annual Conference, American Society for Engineering Management*, Knoxville, TN, October 2–4, 1988, ASEM, 1988, pp. 345–351. Used with permission.

DISCUSSION QUESTIONS

3-1. Provide your definition of leadership. Is this different than your definition of management? Why or why not?

3-2. Describe a manager you have known and characterize their leadership style.

3-3. From your analysis of the findings of Harris (Table 3-1), why do you think engineers look for different qualities in their managers as they (the engineers) grow in experience?

3-4. How would McGregor classify Taylor's management style? Defend your answer.

3-5. Dennis is a new project manager for a company that develops system solutions for large projects. Dennis has not yet developed working relationships with those in his work group. In fact, some in his group are somewhat skeptical of his abilities. The organization structure at the company allows most of the power in projects to rest with the functional managers, not the project managers. Dennis feels a little uncomfortable dictating the details of tasks he assigns, preferring to allow those with much more experience in the jobs to plan them accordingly. Even though Dennis is "the boss," he is much younger than the majority of the members in his work group and feels limited in what direct effect he might have on their performance. Dennis has been thinking hard over the weekend about how to start the group off on a new project Monday morning and needs your help. Explain to Dennis how he might apply the behaviors from the Managerial Continuum to his discussion with the team on Monday. Which behaviors do you recommend? Why?

3-6. Schermerhorn defines management as the activity which "performs certain functions in order to obtain the effective acquisition, allocation, and utilization of human efforts and physical resources in order to accomplish some goal." How would a Theory X manager ensure that they meet this definition?

3-7. Six months into his new job, Bob, a new engineering graduate, is performing just well enough to avoid being placed on a corrective action plan. When hired, he was carefully selected and had the abilities required to do the job really well. At first Bob was enthusiastic about his new job, but now he isn't performing up to this high potential. As his supervisor, describe how you would apply Situational Leadership to improve Bob's performance.

3-8. Explain how Herzberg's two factor theory built on the work of Maslow. Does Herzberg argue that employers should aspire to have their employees at the highest level of Maslow's hierarchy? Why or why not?

3-9. Which of the 16 concepts of Utley and Westbrook (Table 3-6, ignoring Drucker) were you familiar with before reading this book? Which do you now feel would be useful to you as an engineering manager? Discuss why.

3-10. Would you expect the factors motivating an engineer to change as they proceed through a career? In what ways? How can the engineering manager make use of these changes?

3-11. Managers using Theory X or Y approaches run the risk of having the approach become a self-fulfilling prophecy. Explain what this means and why it might be true.

3-12. Explain how Herzberg's two factor theory built on the work of Maslow. Does Herzberg argue that employers should aspire to have their employees at the highest level of Maslow's hierarchy? Why or why not?

3-13. Herzberg specifically classed *salary* as a hygiene factor, not a motivator. How would you classify it? Why?.

3-14. *Job enrichment* seeks to make work more meaningful and give employees more control over their work. Discuss the negative responses of the blue-collar production workers toward this initiative. Why do you think workers have this attitude?

3-15. Would you expect the need for affiliation among managers to be somewhat dependent on the culture in which they grew up and/or work? If so, give an example.

3-16. Provide another example of the validity of equity theory as a motivator of human performance at work or elsewhere.

3-17. The vignette "Leadership Failures" highlighted some recent cases of poor leader behavior. Find two more recent examples and explain what happened. Explain how you think these issues might be prevented in the future.

SOURCES

Adams, J. Stacey, "Towards an Understanding of Equity," *Journal of Abnormal and Social Psychology*, November 1963, pp. 422–436.

Allen, Thomas J., *Managing the Flow of Technology: Technology Transfer and Dissemination of Technological Information within the R&D Organization* (Cambridge, MA: The MIT Press, 1977), pp. 35–41.

Avolio, B. J., and B. M. Bass, *Developing Potential across a Full Range of Leadership: Cases On Transactional and Transformational Leadership* (Marwah, NJ: Lawrence Erlbaum Associates, 2002).

Bass, B. M., Leadership *and Performance Beyond Expectations* (New York: Free Press, 1985).

Bennis, Warren, "On Becoming a Leader" (Reading, MA: Addison-Wesley Pub. Co., 1989).

Berelson, B. and Steiner, George A., *Human Behavior: An Inventory of Scientific Findings* (New York: Harcourt, Brace, & World, 1964), p. 240.

Blake, Robert R. and Mouton, Jane S., *The Managerial Grid III: The Key to Leadership Excellence* (Houston, TX: Gulf Publishing Company, 1985).

Boone, Louis E. and Bowen, Donald D., *The Great Writings in Management and Organizational Behavior*, 2nd ed. (New York: Random House, Inc., Business Division, 1987), p. 124.

Burns, J. M. *Leadership* (New York: Harper & Row, 1978), pp. 20–21.

Campbell, John P., Dunnette, Marvin D., Edward, E. Lawler, III, and Weick, Karl E., Jr., *Managerial Behavior, Performance, and Effectiveness* (New York: McGraw-Hill Book Company, 1970), p. 340.

"Can Jack Smith Fix GM?" *Business Week*, November 1, 1993, p. 126.

Carnegie, Dale, *How to Enjoy Your Life and Your Job* (New York: Simon & Schuster Inc., 1970). Excerpts from *How to Win Friends and Influence People*.

Choate, Pat, "Where Does Quality Fit in with the Competitiveness Debate?" *Quality Progress*, February 1988, p. 26.

Connolly, Terry, *Scientists, Engineers, and Organizations* (Monterey, CA: Brooks/Cole Engineering Division, a Division of Wadsworth, Inc. of Belmont, CA, 1983), p. 129.

Cribbin, James J., *Leadership: Strategies for Organizational Effectiveness* (New York: American Management Associations, Inc., 1981), pp. 36–37.

Drucker, Peter F., "Beyond Stick and Carrot: Hysteria over the Work Ethic," *Psychology Today*, November 1973, pp. 89, 91–92.

Drucker, Peter F., *Management Challenges for the 21st Century* (Woburn, MA: Butterworth-Heinemann, 1999).

Fein, Mitchell, "Job Enrichment: A Reevaluation," *Sloan Management Review*, 15:2, Winter 1974, pp. 69–88.

Greenleaf, Robert K., *Servant Leadership: A Journey into the Nature of Legitimate Power and Greatness* (Mahwah, NJ: Paulist Press, 1983) 1–5.

Harris, E. Douglas, "Leadership Characteristics: Engineers Want More from Their Leaders," *Proceedings of the Ninth Annual Conference, American Society for Engineering Management*, Rolla, MO, October 2–4, 1988, pp. 209–216.

Hersey, Paul and Blanchard, Kenneth H., *Management of Organizational Behavior: Utilizing Human Resources*, 4th ed. (Englewood Cliffs, NJ: Prentice-Hall, Inc., 1982), pp. 150–175.

Herzberg, Frederick, "One More Time: How Do You Motivate Employees?" *Harvard Business Review*, 46:1, January–February 1968, p. 57.

Kerr, S., Glinow, Von, M. A., and Schriesheim, J., "Issues in the Study of Professionals in Organizations: The Case of Scientists and Engineers," *Organizational Behavior and Human Performance*, 18, 1977, pp. 329–345.

Kotter, J. P., *A Force for Change: How Leadership Differs from Management* (New York: Free Press, 1990).

Maccoby, Michael, "Understanding the Difference between Management and Leadership." *Research Technology Management*, 43:1, January–February 2000. pp. 57–59.

Maslow, Abraham H., "A Theory of Human Motivation," *Psychological Review*, 50, 1943, pp. 370–396.

McCall, Morgan W., Jr., "Leadership and the Professional," in Connolly, *Scientists, Engineers, and Organizations*, pp. 332–335.

McClelland, David C., *The Achieving Society* (Princeton, NJ: D. Van Nostrand Company, 1961), *Power: The Inner Experience* (New York: Irvington Publishers, Inc., 1975), and *Human Motivation* (Glenview, IL: Scott, Foresman and Company, 1985).

McClelland, David C., "Achievement Motivation Can Be Learned," *Harvard Business Review*, November–December 1965, pp. 6–24.

McGregor, Douglas M., "The Human Side of Enterprise," *Management Review*, November 1957, reprinted by permission of publisher, 1957 American Management Association, New York. All rights reserved.

Mintzberg, H., Debunking management myths: A brief interview with Henry Mintzberg. Interview by Martha Mangelsdorf, MIT *Sloan Management Review*, 51, 1 (Fall 2009): p. 12.

Mintzberg, Henry, "Managerial Work: Analysis from Observation," *Management Science*, 18:2, 1971, pp. 97–110.

Myers, M. Scott, "Who Are Your Motivated Workers?" *Harvard Business Review*, 42:1, January–February 1964, pp. 73–88.

Peterson and Plowman, E. G., *Business Organization and Management* (Homewood, IL: Richard D. Irwin, Inc., 1957), pp. 50–62.

Porter, Lyman W., Lawler, Edward E., III, and Hackman, J. Richard, *Behavior in Organizations* (New York: McGraw-Hill Book Company, 1975), p. 43.

Porter, Lyman W. and Lawler, Edward E., III, *Managerial Attitudes and Performance* (Homewood, IL: Richard D. Irwin, Inc., 1968), p. 165.

Raudsepp, Eugene, "Why Engineers Work," *Machine Design*, February 4, 1960.

Robbins, Stephen P., *Management*, 4th ed. (Englewood Cliffs, NJ: Prentice-Hall, Inc., 1994), p. 465.

Rosenbaum, Bernard L., "Leading Today's Professional," *Research Technology Management*, 34:2, March–April 1991, p. 30.

Rost, J. C., "Leadership and Management," in *Leading Organizations: Perspectives for a New Era*, edited by Gill, Robinson, and Hickman (Thousand Oaks, CA: Sage, 1998), pp. 97–114.

Sayles, L. R. and Chandler, M. K., *Managing Large Systems* (New York: Harper & Row, Publishers, Inc., 1971).

Shannon, *Engineering Management* (New York: John Wiley & Sons, Inc., 1980), p. 207.

Sheldon, Oliver, *The Philosophy of Management* (New York: Pitman Publishing Ltd, 1923).

Skinner, B. F., *Science and Human Behavior* (New York: The Free Press, 1953) and *Contingencies of Reinforcement* (New York: Appleton-Century-Crofts, 1969).

Tannenbaum, Robert and Schmidt, Warren H., "How to Choose a Leadership Pattern," *Harvard Business Review*, 36:2, March–April 1958, pp. 95–101, reprinted (with "retrospective commentary") in *Harvard Business Review*, 51:3, May–June 1973, pp. 162–180.

Utley, Dawn R. and Westbrook, Jerry D., "A Survey of Management Concepts in Technical Organizations," *Proceedings of the Ninth Annual Conference, American Society for Engineering Management*, Rolla, MO, October 2–4, 1988, pp. 345–351.

Vroom, Victor H., *Work and Motivation* (New York: John Wiley & Sons, Inc., 1964).

"Worthington's Human Resources: Building Quality on the Strength of Its People," *Quality*, February 1988, p. 22.

STATISTICAL SOURCEBOOK

The following is a useful source Web site. (March, 2019)

http://www.cpp.com. Information on the Myers-Briggs Type Indicator may be found here.

4

Planning and Forecasting

PREVIEW

Planning is the next management function. Planning plays an important role in any business venture—large or small. It can make the difference between the success or failure of an organization. Strategic planning has become more important to the engineering manager because technology, competition, and ongoing changes have made the business environment less stable and less predictable.

There are as many strategic planning models as there are strategic planning experts. There are many models that have a different number of steps, but the overall purpose remains the same: set goals, conduct the work, study the effect of the work, and then realign goals to improve the implementation of the plan. Companies, organizations, educational institutions, nonprofits, committees, and individuals use a broad range of information, formats, and styles in their strategic planning models.

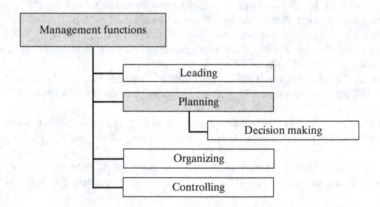

This chapter begins with the development of a model of the planning/decision-making process. Then the importance of defining an organization's mission, goals, objectives, and strategies is considered. Next, management by objectives and several planning concepts are introduced. An important part of planning is forecasting. Quantitative and qualitative methods are discussed. The end of the chapter considers strategies for managing technology.

NATURE OF PLANNING

Importance of Planning

The importance of planning in warfare was emphasized by Hsun Tzu (298–238 B.C.) in *The Art of War* (the world's oldest military treatise):

> The general who wins a battle makes many calculations in his temple ere the battle is fought. The general who loses a battle makes few calculations before hand. It is by attention to this point that I can see who is likely to win or lose.

Planning provides a method of identifying objectives and designing a sequence of programs and activities to achieve these objectives. Amos and Sarchet define planning simply as deciding in advance what to do, how to do it, when to do it, and who is to do it; from this definition, planning must obviously precede doing.

The Planning/Decision-Making Process

There is a basic, logical method for solving problems that may be called, depending on the application, the planning process, the decision-making process, or the scientific method. Although there seem to be as many diagrams of this process as there are authors, a typical representation of this process follows the general logic of Figure 4-1.

A problem cannot be solved until it is recognized. When the "roof falls in" or the workers go on strike, the existence of a problem will be obvious. On the other hand, chronic (perennial) problems or opportunities often go unrecognized, because they are part of the normal operations people have adjusted to. Video rental companies, for example, did not realize that they had a convenience problem until Netflix delivery and then various streaming services recognized the opportunity; now Blockbuster and other rental companies no longer exist.

Once a problem is recognized, the nature of the desired solution must be defined carefully in terms consistent with the overall objectives and strategy of the organization. Assumptions about the environment

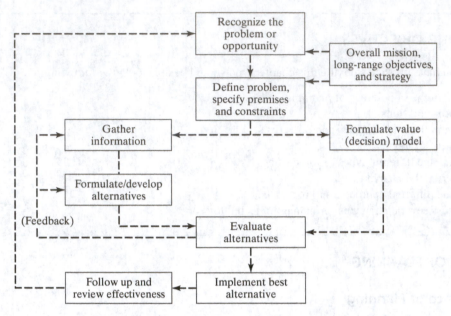

Figure 4-1   The planning/decision-making process.

(*premises*) need to be stated, and the solution found will be valid only if these assumptions prove true. Finally, the constraints or limitations bounding the solution must be defined.

Information bearing on alternative solutions is then gathered, and alternative solutions are formulated. This is the most creative step in problem solving, since alternatives that are not considered are lost. Simply stating an alternative is not enough—each concept must be fleshed out in enough detail that its benefits and disadvantages can be effectively evaluated. At the same time, some "value model" or measure of merit—quantitative or qualitative—needs to be defined, against which alternatives can be evaluated. The solution that best satisfies this value model is then recommended.

Seldom is the decision process as linear as the preceding paragraph suggests. Identification of alternatives often leads to a search for more information. Evaluation often leads to modification or combination of alternatives to find a new one that combines the advantages of several. After the solution is put into effect (implemented), it is important to check back later to determine if the problem as stated was really resolved by the solution. Often, there will be unexpected secondary effects that, once realized, need to be defined as a new problem, and the process begins anew. Problem solving/decision making is, therefore, more often an *iterative* process, involving feedback at several steps before the best resolution is found.

THE FOUNDATION FOR PLANNING

Strategic Planning

A successful enterprise needs to develop effective strategies for achieving its mission, and **strategic planning** is the organized process for selecting these strategies. Customer focus impacts and should integrate an organization's strategic plan, its value creation process, and business results. One approach that corporations use in strategic planning is identifying the businesses they are in and the ones they want to be in, and defining a strategy for getting from the first set to the second.

A clear vision of the basic purpose or mission for which it exists is essential to the long-term success of any enterprise. Drucker cites the old German tradition of the *Unternehmer* (entrepreneur or, literally, "undertaker"), who alone or with a few top managers was all that was needed to understand the underlying purpose of a firm in earlier, simpler times. In today's complex and dynamic economy, however, the basic mission of the organization must be communicated clearly and repeatedly to the many managers and professionals whose actions determine whether these purposes are achieved.

Drucker provides an example of a vision of central purpose in Theodore N. Vail's statement (made around 1910) of the American Telephone and Telegraph Company that "our business is service." Drucker believes that without this vision of the social responsibility of a natural monopoly, the "Bell System" would have inevitably been nationalized in the 1930s, as was the case in other countries. Only in the late 1960s, when technology provided alternatives to Bell's monopoly, was it time to pursue a revised purpose.

Strategic planning suggests ways (strategies) to identify and to move toward desired future states. It determines where an organization is going over the next year or more, how it is going to get there, and how it will know if it got there or not. It consists of the process of developing and implementing plans to reach goals and objectives. In short, strategic planning is a disciplined effort to produce fundamental decisions and actions that shape and guide what an organization is, what it does, and why it does it, with a focus on the future, both internally and externally. The identification of the organization's vision and mission is the first step of any strategic planning process, as shown in Figure 4-2. A **vision statement** describes in graphic terms where the organization wants to position itself in the future. A **mission statement** resembles a vision statement, but has more details and should define what the organization does and for whom. The mission statement sets forth what the company is attempting to do, and is usually what the public sees. The role of an organization's mission and vision is to align work toward meeting customer expectations. Not all companies have both a mission and vision.

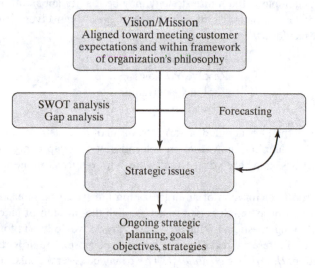

Figure 4-2   Strategic planning.

Mission—Microsoft

Our mission is to empower every person and every organization on the planet to achieve more.

Source: http://www.microsoft.com, May 19, 2019.

It is difficult to develop future strategies for the business without knowing the current status and their success at this point. At this time, an analysis of the status needs to be made. One tool that is often used is the **SWOT** (Strengths, Weaknesses, Opportunities, and Threats) **analysis**. A summary of SWOT analysis is shown in Figure 4-3. This analysis should address all factors that are key to the organization's future success. Strengths and weaknesses are basically internal to an organization and may include the following:

- Management
- Marketing
- Technology
- Research
- Finances
- Systems

The external opportunities and threats may be in some of the following areas:

- Customers
- Competition
- New technologies
- Government policies

Once the SWOT review is complete, the future strategy may be readily apparent, or as is more likely the case, a series of strategies or combinations of tactics will suggest themselves. Use the SWOTs to help identify possible strategies as follows:

> Build on **S**trengths
> Resolve **W**eaknesses
> Exploit **O**pportunities
> Avoid **T**hreats

The resulting strategies can then be modeled to form the basis of a realistic strategic plan. The SWOTs identified will assist in the planning, as well as in determining, the gap analysis. A gap analysis is a technique used to analyze/assess where you currently are with respect to where you would like to be in the future.

The basic vision, purpose, or mission of an organization must next be interpreted in terms of goals and objectives. **Goals** give purpose and direction to accomplish the mission of an organization. The goal statement answers the following questions: What do we do; why do we do it; and for whom do we do it? It is used as a continual point of reference regarding the scope or purpose. The **objectives** further clarify the goal and answer the question, *How do we go about it?* There may be several goals, and each goal may have

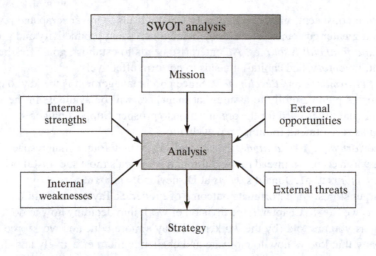

Figure 4-3 Overview of SWOT Analysis.

several objectives. **Strategies** are statements about the way objectives are to be achieved. They are relevant only to the extent that they help meet the objectives. Different organizations may interchange the words *vision* and *mission*, and *goals* and *objectives*. The importance is not in the words, but in the meaning.

One must distinguish between two types of goals that often coexist: the *official goals* that management says it is pursuing in its public statements, and the *operative goals* that it actually is pursuing. Managers of rural electric cooperatives in the United States, when asked by this author for their underlying goals, often replied, *to provide the best possible service at the lowest possible cost.* This is hardly an operative goal, since *best service* implies a high level of staffing of maintenance and repair crews, and "lowest cost" implies a lesser level. Only by examining the level actually being maintained could one deduce the operative goal.

Peter Drucker believes that objectives need to be established in all areas on which the organization's survival depends. He distinguishes eight such *key result areas*:

1. *Market share.* Market share is the ratio of dollar sales of an enterprise in a particular market to the total sales of all competitive products and services in that market. Firms with a small market share usually are less profitable because they have fewer sales over which to spread the fixed costs of operation, and managers often decline to enter or remain in a specific market unless they can either achieve a satisfactory market share or can define a smaller "market segment" in which they can be a leader.

2. *Innovation.* Most successful companies, especially in the areas of technology, are continually searching for new products and services. 3M, for example, had traditionally required its 40 divisions to achieve at least 25 percent of sales be from products introduced in the last five years. Nonetheless, some successful companies deliberately choose to be followers and to provide low-cost, high-volume products without the high expense of being first. Samsung has historically followed this path, but recently shifted to market leadership in some areas as their capabilities improved.

3. *Productivity and quality.* Productivity measures an organization's ability to produce more goods and services per unit of input (labor, materials, and investment), while *quality* is a measure of product or service performance against objectives. While higher quality is often thought of as higher cost, the

two are not inconsistent, since higher quality usually leads to lower scrap and rework losses, fewer returns, and greater customer satisfaction, thereby increasing productivity and profitability.

4. *Physical and financial resources.* An enterprise needs to establish goals for the resources (plant, equipment, inventory, and capital) it needs to perform effectively.

5. *Manager performance and development.* Since good management is the key to enterprise success, effective firms plan carefully to assure that managers will be available in the years ahead in the quality and quantity needed for the organization to prosper. Supporting goals are then developed in areas such as recruitment, training, and evaluation.

6. *Worker performance and attitude.* Peters and Waterman found that respect for the individual employee was a common thread running through America's most successful businesses. Personnel are *crew members* at McDonald's, *hosts* at Disneyland, *ambassadors* at Six Flags, and *team members* at Target stores. An unfortunate outcome of Frederick Taylor's scientific management revolution was, as we saw in Chapter 2, the division of work into deciding how to do it (by management) and doing as you are told (by the workers). Today's more educated workforce has much to offer the company that knows how to motivate and challenge them effectively and this deciding–doing split is eroding in many organizations.

7. *Profitability.* The profitability of an enterprise is essential to its continuation, and the desired level should be set explicitly as an objective against which to measure enterprise success.

8. *Social responsibility.* Every enterprise has responsibilities as a *corporate citizen* that extend beyond the legal and economic requirements. These include responsibilities to customers, employees, suppliers, community, and society as a whole. The organization that does not at least take responsibility for its effect on the environment deserves to be penalized by society.

Management by Objectives

In the same 1954 work, Drucker formulated the concept of **management by objectives** (MBO). Since then, MBO has been widely adopted to translate broad organizational goals and objectives like those discussed previously into specific individual objectives. MBO can (and usually should) be employed between manager and employee at every level in the organization. The steps in MBO are generally as follows.

First, both manager and employee should have an understanding of the goals and objectives of the overall organization and those of the manager's group.

The manager and employee then meet to establish objectives for the employee's attention over the next time period (typically quarterly or annual) that are consistent with group objectives. These objectives should require some effort to attain, yet not be beyond reach. They should be quantifiable if feasible (e.g., reduce scrap by 20 percent); if not feasible, they should be verifiable (e.g., write a new quality assurance plan) so that it is possible to determine at the end of the period whether or not the objective has been achieved. The relative amount of input from the manager and the employee in negotiating these objectives may vary, but the result should be mutual agreement. In agreeing to an objective proposed by the manager, the employee may identify specific resources or authority that needs to be supplied by the manager to make it possible, and this is to the advantage of both sides. Objectives should not be confined to tasks for the sole benefit of the manager, but should also include developmental objectives designed to strengthen the employee's capabilities (more on development focus is presented in Chapter 17).

The employee then proceeds, over the ensuing period, to carry out his or her job with an emphasis on achieving these objectives. Naturally, if problems occur or priorities change, the manager and employee can meet at any time and may modify the objectives, but they should not be changed without such agreement.

At the end of the period, the manager and employee meet again to evaluate the employee's success in meeting assigned goals. This should be a constructive process, not an excuse for placing blame. This review session should end by mutually establishing a new set of objectives for the following period, of which some may be extensions of earlier objectives and some may be new objectives.

Advantages claimed for MBO include greater commitment and satisfaction on the part of employees, enforced planning and prioritizing of future activities on the part of both managers and employees, and a more rational method of performance evaluation based on contribution to organizational objectives.

Disadvantages include the time and paperwork involved, misuse when managers simply assign (rather than negotiate) objectives, and the gamesmanship of employees who try to negotiate easy goals. There is also a tendency for employees to focus on the relatively few, verifiable, MBO objectives negotiated to the detriment of the other objectives, both qualitative and quantitative, that a professional must also keep in balance.

While annual salary reviews should not be scheduled at the same time as the periodic MBO evaluation, there needs to be some relationship between the ability to meet objectives and the reward system to make MBO effective. Finally, MBO will not be a success unless it has the initiating and continuing support of higher management.

SOME PLANNING CONCEPTS

Responsibility for Planning

Planning is a continuing responsibility of every manager. The higher managers rise in an organization, the more time they must spend in planning, and the further into the future they must try to foresee. Most large organizations have staff offices for planning. The planning staff can coordinate the overall planning effort, gather and analyze information on the economy, markets, and competition, and perform other assigned tasks. The ultimate responsibility for planning, however, must rest with top and middle management. Line management must establish the objectives for the enterprise, devise the overall strategy, provide planning guidelines, and periodically review and redirect the planning effort. To have purpose, plans must lead to action, and managers are unlikely to carry out with any enthusiasm plans they have not *bought into* by being part of the planning process.

Planning Premises

An essential for effective planning is establishment of the premises, or assumptions, on which planning is to be based. Weihrich and Koontz define planning premises as "the anticipated environment in which plans are expected to operate. They include assumptions or forecasts of the future and known conditions that will affect the operation of plans." Examples of planning premises include assumptions about future economic conditions, government decisions (regulation, tax law, and trade policy, for example), the nature of competition, and future markets.

Planning for Technology at AAON

When done well, planning can enable businesses business success. Whether the target is growth, higher profits, gaining control of a market niche, or all of the above, a well-developed and well-executed plan is key to success. One company known for their success in planning for the use of new technology and then using that new technology to obtain and maintain market leadership is AAON, Inc. of Tulsa, OK. AAON is a world leader in creating comfortable and healthy indoor environments through engineering and manufacturing semi-custom heating, ventilation, and air conditioning equipment (HVAC). While many engineering students might not immediately think of air conditioning as needing cutting edge technology, it is just such an approach that has enabled AAON to become a world leader in their market. This is an approach that begins with planning.

AAON's market niche is the design and manufacture of semi-custom units for a wide array of commercial applications. Custom manufacturing operations typically require substantially higher labor costs per unit since it is not standardized or automated (Chapter 11 discusses this further). Avoiding these higher costs through successful planning for use of automation is a key to their success. By planning for new products based on potential for market needs and then planning how those units will be manufactured, AAON has consistently been able to either acquire or design and build fabricating and testing equipment that allows them to automate large segments of production that would be handled manually in most operations. This includes large automated metal bending equipment and custom-designed testing operations which enable AAON to consistently be first to market with new technologies that better meet the needs of their customers. Through extensive planning about the potential future directions of the industry, AAON designed and has recently opened the Norman Asbjornson Innovation Center (named after the company's founder). The center provides environmental, acoustical, and HVAC load testing capabilities available nowhere else in the world. These capabilities will enable AAON to continue to be a market leader for a considerable time into the future.

AAON Inc.

Source: https://aaon.com/About
https://aaon.com/Documents/Technical/AAON_NAIC_ONLINE_180820.pdf.

In managing technology, it is essential to establish planning premises about the future of technology and competition; see the discussion "Planning for Technology at AAON." Where there are uncertainties about critical premises, prudent managers develop *contingency plans* that can be implemented if indicators show a change in the environmental conditions from those on which mainstream planning is based. Modest changes in current plans may be needed to add flexibility so that a switch to a contingency plan can be made quickly if needed.

Planning Horizon

The *planning horizon* asks how far into the future one should plan. This varies greatly, depending on the nature of the business and the plan. The vendor of pins and pennants outside a baseball stadium need not plan beyond the current season. In contrast, after Julius Rosenwald bought out Sears in 1895, he built a continuing business of mail-order service to American farmers. His planning period needed to look far enough ahead to encompass a return on this long-term investment. Weihrich and Koontz summarize this difference in the "*commitment principle*: Logical planning encompasses a future period of time necessary to fulfill, through a series of actions, the commitments involved in decisions made today." High-technology products may have short effective lives, therefore, shorter planning horizons.

The decision of a utility company to build a nuclear power plant or begin a new mining operation in the United States, on the other hand, must consider at least 10 years' time to obtain necessary approvals and build, and several decades' operation to recover the investment. Many utility companies that made such decisions based on 1970 projections of energy-use growth had no way to foresee the energy conservation following the OPEC oil crisis of 1973 or the increasing public attack on nuclear reactors, and they had to cancel partially built reactors at costs of billions of dollars. The planning horizon can vary from days to years depending on the level of the manager.

Systems of Plans

Usually, not just one plan is involved, but a system of them. In 1916 Henri Fayol divided his *Plan of Action in a Large Mining and Metallurgical Firm* into *yearly forecasts* and "ten-yearly forecasts," the latter redone every five years. Current practice is not much different, involving *strategic plans* of from 3 to 15 years into the future and *operating plans*, usually one year in duration. Complex programs will require not just a single plan, but a system of plans, each describing a related activity. For example, a complex project might require most or all of the plans listed in Table 4-1.

Table 4-1   A System of Plans for a Complex Project

Project statement of work	Production plan
Work breakdown structure	Tooling plan
Project schedule	Make-or-buy plan
Project budget	Quality assurance plan
Specifications	Facilities plan
Management plan	Training plan
Configuration management plan	Logistics support plan
Security plan	Reliability plan
System test plan	Transportation plan

Policies and Procedures

Policies are guides for decision making that permit implementation of upper-management objectives, with room for interpretation and *discretion* by subordinates. *Rules*, in contrast, do not permit discretion. Policies have a hierarchy of levels, just as plans do. For example, the president of an engineering firm might establish a policy that approval of *small* design changes should be delegated to an *appropriate* level to reduce the number of matters demanding higher-management attention. The chief engineer of a project might then implement this by establishing a supporting policy that engineering design supervisors could approve design changes costing up to $5,000 involving technical specialties in their area of responsibility (which, in turn, involves judgment on the part of the supervisors). To be effective, policies should be clear, flexible, and communicated throughout the organization; involve participation in their development; and be reviewed regularly.

A **procedure**, on the other hand, is a prescribed sequence of activities to accomplish a desired purpose. Procedures tell you *if you want to do this, do it this way*. For example, while the decision to approve (release) a design drawing requires technical judgment, the established *procedure* for doing so must be followed to assure that appropriate individuals have had a chance to approve the drawing and that its official existence is communicated to those who need to be informed of it. *Standard operating procedures* (SOPs) are examples of procedures used at the operating (working) level.

FORECASTING

An essential input to effective planning is **forecasting**—predicting what the future will be like. Planning provides the strategies, given certain forecasts, and forecasting estimates the results, given the plan. Planning is what the organization ought to do, and forecasting relates to what happens if the firm tries to implement given strategies in a possible environment. There must be alternatives to the plan based on the forecasts. For example, if one of your strategies is to encourage people to drive to your destination, there should be an alternative strategy if the price of fuel skyrockets.

The engineering manager must be concerned with both future markets and future technology, and must therefore understand both sales and technological forecasting. The most important premise or assumption in planning and decision making is the level of future sales (or, for nonprofit activities, of future operations). Almost everything for which we plan is based on this assumption—the production level (which determines how many people we must hire and train, or if production declines, lay off); the need for new facilities and equipment; the size of the sales force and advertising budget; new funding for purchases; and for investment in inventory and accounts receivable.

Qualitative Methods

Jury of Executive Opinion.   This is the simplest method, in that the executives of the organization (typically, the heads of various divisions) each provide an estimate (educated guess) of future volume, and the president provides a considered average of these estimates. This method is inexpensive and quick and may be entirely acceptable if the future conforms to the assumptions the executives have used in estimating.

Delphi Method.   Another use of expert opinion is the Delphi method, which begins with the present state of technology and extrapolates into the future, assuming some expected rate of technical progress.

A common forecasting method is the use of a panel of experts in the technical field involved. Panels are sometimes consulted using the Delphi method (named after the Oracle at Delphi in ancient Greek mythology). Each expert is asked independently when some future technical breakthrough (such as the launching of a space shuttle or the feasibility of a new product, for example) will occur, if ever. Averages of the estimates are then reported back to panel members, who are then given a chance to modify their estimate or explain why they hold their belief. A final value is adopted after several such iterations. One of the advantages of this technique is that it eliminates the effects of interaction among experts.

Sales Force Composite.   In this commonly used method, members of the sales force estimate sales in their own territory. Regional sales managers adjust these estimates for their opinion of the optimism or pessimism of individual salespeople, and the general sales manager *massages* the figures to account for new products or factors, of which individual sales people are unaware. Since the field sales force is closest to the customer, this method has much to recommend it. However, if there is any suggestion that the estimate a salesperson provides will drive the goal they must achieve, the sales force may find it in their own best interest to *play games* with the figures.

Users' Expectation.   When a company sells most of its product to a few customers, the simplest method for forecasting budgets is to ask the customers to project their needs for the future period. The customers depend for their own success on reliable sources of supply, and so communication is in the best interest of both parties. This might be done by market testing or market surveys. For consumer goods, though, not only is such information expensive to obtain, but consumers often do not know what they will purchase in the future.

Choice of Method.   Companies with effective planning will combine a variety of methods to arrive at the best sales forecast. Qualitative estimates from the sales force and customer surveys may be compared against quantitative estimates obtained from moving average or regression models (discussed in the following sections). Finally, the chief executive, with the assistance of other top officers, will establish a sales forecast to be used in future planning.

Quantitative Methods

Simple Moving Average.   Where the values of a parameter show no clear trend with time, a forecast F_{n+1} for the next period can be taken as the simple average of some number n of the most recent actual values A_t:

$$F_{n+1} = \frac{1}{n}\sum_{t=1}^{n} A_t \qquad\qquad (4\text{-}1)$$

Example

If sales for years 2018, 2017, 2016, 2015 ($n = 4$) were 1,600; 1,200; 1,300; and 1,100 respectively, sales for 2019 would be forecast as

$$F_{2019} = \frac{1{,}600 + 1{,}200 + 1{,}300 + 1{,}100}{4} = 1{,}300$$

Weighted Moving Average.   The preceding method has the disadvantage that an earlier value (2014, for example) has no influence at all, but a value n years in the past (2015) is weighted as heavily as the most recent value (2018). We can improve on our model by assigning a set of weights w_t that total unity (1.0) to the previous n values:

$$F_{n+1} = \sum_{t=1}^{n} w_t A_t, \text{ where } \sum_{t=1}^{n} w_t = 1.0 \qquad\qquad \textbf{(4-2)}$$

Example

Using weights of 0.4, 0.3, 0.2, and 0.1 for the most recent ($n = 4$) past years in our preceding example yields

$$
\begin{aligned}
F_{2019} &= 0.4A_{2018} + 0.3A_{2017} + 0.2A_{2016} + 0.1A_{2015} \\
&= 0.4\,(1{,}600) + 0.3\,(1{,}200) + 0.2\,(1{,}300) + 0.1\,(1{,}100) \\
&= 1{,}370
\end{aligned}
$$

Exponential Smoothing.   The weighted moving average techniques have the disadvantage that you (or your computer) must record and remember n previous values and n weights for each parameter being forecast, which can be burdensome if n is large. The simple exponential smoothing method continuously reduces the weight of a value as it becomes older, yet minimizes the data that must be retained in memory. In this technique, the forecast value for the next period F_{n+1} is taken as the sum of (1) the forecasted value F_n for the current period, plus (2) some fraction α of the difference between the actual (A_n) and forecasted (F_n) values for the current period:

$$F_{n+1} = F_n + \alpha(A_n - F_n) = \alpha A_n + (1 - \alpha)\,F_n \qquad\qquad \textbf{(4-3)}$$

Expanding the right-hand equation by using similar expressions for F_n, F_{n-1}, and so on, gives the following:

$$
\begin{aligned}
F_{n+1} &= \alpha A_n + (1 - \alpha)\left[\alpha A_{n-1} + (1 - \alpha)F_{n-1}\right] \\
&= \alpha A_n + \alpha(1 - \alpha)A_{n-1} + \alpha(1 - \alpha)^2 A_{n-2} + \alpha(1 - \alpha)^3 A_{n-3} + \cdots
\end{aligned}
$$

This last equation shows that the weight put on past values continues to decrease, but never becomes zero. To start the forecast sequence, the first forecast must be set equal to the actual value of the preceding year.

For the data used in the two preceding examples, the forecasts are as shown in Table 4-2 for two selected values of α:

For example, if $\alpha = 0.3$,

$$
\begin{aligned}
F_{2017} &= 0.3A_{2016} + 0.7F_{2016} \\
&= 0.3\,(1{,}300) + 0.7\,(1{,}100) = 1{,}160 \\
F_{2018} &= 0.3A_{2017} + 0.7F_{2017} \\
&= 0.3\,(1{,}200) + 0.7\,(1{,}160) = 1{,}172
\end{aligned}
$$

Table 4-2 Exponential Smoothing Calculation

Year(t)	Forecast F_t		Actual Value A_t
	$\alpha = 0.3$	$\alpha = 0.6$	
2015			1,100
2016	1,100	1,100	1,300
2017	1,160	1,220	1,200
2018	1,172	1,208	1,600
2019	1,300	1,443	

Regression Models

Regression models are a major class of *explanatory forecasting models*, which attempt to develop logical relationships that not only provide useful forecasts, but also identify the causes and factors leading to the forecast value. Regression models assume that a *linear relationship* exists between a variable designated the *dependent (unknown) variable* and one or more other *independent (known) variables*.

The simple regression model assumes that the independent variable I depends on a single dependent variable D. Figure 4-4 gives an example in which the parameter we have been forecasting in the moving average calculations is taken as D, and time, for convenience, expressed as (year–2014), is taken as I.

The regression problem is to identify a line

$$\boxed{D = a + bI} \tag{4-4}$$

such that the sum of the squares of the deviations between actual and estimated values (the vertical line segments in the figure) is minimized. The two constants in this least squares equation are found from

$$\boxed{\begin{aligned} b &= \frac{n \sum (D_i I_i) - \sum I_i \sum D_i}{n \sum (I_i^2) - (\sum I_i)^2} \\ a &= \sum \frac{D_i}{n} - b \sum \frac{I_i}{n} = \overline{D} - b\overline{I} \end{aligned}} \tag{4-5}$$

where $\overline{D}$ and $\overline{I}$ are the mean values of D and I, respectively, and indicate a summation from $i = 1$ to n.

Regressions can be used to calculate the best fit to a straight line on a normal graph (as in Figure 4-3).

The resulting forecasts can be seriously in error if the assumptions on which they are based prove to be in error. Electric utility companies came to believe over a period of several decades that the demand for electricity would increase about 7 percent per year. Although this may not sound like a large growth rate, it means that demand should increase by a factor of about $(1.07)^{10} = 2.0$ (i.e., double) every 10 years. Since large power plants take at least that long to plan and construct, when the oil crisis struck in 1973 and consumers drastically reduced their use of now-expensive power through conservation, utility

Example

The trend line in Figure 4-3 is calculated using the data in Table 4-3:

$$b = \frac{4(8{,}500) - 6(5{,}200)}{4(14) - (6)^2} = 140$$

$$a = \frac{5{,}200}{4} - 140\left(\frac{6}{4}\right) = 1{,}300 - 140(1.5) = 1{,}090$$

And a value for 2019 is forecast:

$$D_{2019} = 1{,}090 + (2009 - 2005)(140) = 1{,}090 + 560 = 1{,}650$$

companies were left with billions of dollars of capacity under construction that would not be useful for a long time. At the beginning of this century, telecommunications companies made a similar error. This over-expansion caused many of these companies to fail, but led to lower internet service costs today.

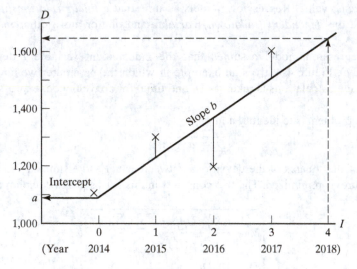

Figure 4-4 Simple regression model.

Table 4-3 Data for Regression Calculation

	I	D	DI	I^2
	0	1,100	0	0
	1	1,300	1,300	1
	2	1,200	2,400	4
	3	1,600	4,800	9
Σ	6	5,200	8,500	14
Mean	1.5	1,300		

Norm Augustine gives a convincing but facetious example of the problems with such extrapolation in his delightful *Augustine's Laws*. For example, he plots the logarithm of the cost of military aircraft against the year in which they were first operational, from the Wright Flyer (about \$3,000 in 1912) to the F-15 (about \$20 million in 1975), and concludes the following:

> In the year 2054, the entire [U.S.] defense budget will purchase just one tactical aircraft. The aircraft will have to be shared by the Air Force and the Navy $3\frac{1}{2}$ days each per week except for leap year, when it will be made available to the Marines for the extra day.

Multiple Regression.   In multiple regression, the dependent variable D is assumed to be a function of more than one independent variable I_j, such as

$$D = c_0 + c_1 I_j + \frac{c_2}{I_2} + c_3 I_3^2 + \cdots \tag{4-6}$$

The dependent variable can be assumed to be proportional directly or inversely, proportional to a power or a root, or proportional in some other way to the independent variables, as is suggested in the preceding equation. Past values of dependent and independent variables are then used in *regression analysis* to reduce the independent variables to the most important ones and to find the values for the constants c_i that give the best fit. For example, a manufacturer of replacement automobile tires might find that the demand for tires varied with the cost of gasoline, the current unemployment rate, sales of automobiles two years before, and the weight of those automobiles. Dannenbring and Starr provide one (of many) convenient sources of further information on multiple regression and other explanatory forecasting models.

Technological Forecasting

Engineers usually are involved in planning environments where technology is changing, and it is essential that planning be done according to the best estimate of the technology that will be available in the future. Shannon bases his belief in the feasibility of technological forecasting on three premises: (1) technological events and capabilities grow in a very organized manner; (2) technology responds to needs, opportunities, and the provision of resources; and (3) new technology can be anticipated by understanding the process of innovation. According to Marvin Cetron, a technological forecast is a prediction, based on confidence that certain technical developments can occur within a specified time with a given level of resource allocation. Two types of technological forecasting should be considered: **normative** and **exploratory**.

In **normative technological forecasting**, one works backward from the future to the present. A desired future goal is selected, and a process designed to achieve this goal is developed. For example, the U.S. government might decide that it is essential to have power available from nuclear fusion in commercial quantities in the year 2040, and will work backward to establish a schedule for a full-scale demonstration plant, a smaller pilot plant, the research tasks that must precede them, and finally develop the overall budget and schedule required to reach the normative goal. President Kennedy's 1961 decision to land a man on the surface of the moon before the end of that decade certainly required normative technological forecasting.

An **exploratory technological forecast** is the Delphi method described earlier in the chapter. For example, Japan's Science and Technology Agency polled some 2,800 of the nation's leading specialists in research institutes and universities to identify R&D goals for the next 30 years. The five most important goals (and the year they were expected to be achieved) were as follows:

1. Commercialization of technologies to eliminate air pollutants such as nitrogen oxides (2003)
2. Invention of a computer to operate faster than 10 teraflops (10 trillion calculations per second) (2004)
3. Discovery of the major development mechanism of cancer (2010)
4. Commercialization of effective methods to prevent the spread of cancer (2007)
5. Diffusion of global-scale environmental preservation technologies (2011)

The reader can see that few (if any) of these objectives were achieved by the expected timeline. However, these goals supported resource allocation that moved research forward in the desired direction.

One useful model for technological forecasting is the technology S-curve, shown in Figure 4-5. The performance gained from a new technology tends to start slowly, then rises almost exponentially as many scientists and engineers begin applying themselves to product improvement. Ultimately, as the technology becomes mature, performance gains become more and more difficult to attain, and performance approaches some natural limit. Gains then require the use of an entirely new technology.

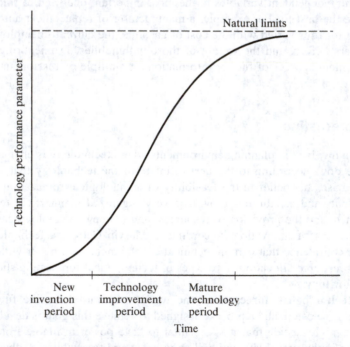

Figure 4-5 Technology S-curve (From Frederick Betz, *Managing Technology: Competing through New Ventures, Innovation, and Corporate Research*, Prentice-Hall, Inc., Englewood Cliffs, NJ, 1987, p. 62; reprinted by permission of Prentice-Hall, Inc., copyright 1987).

For example, the efficiency of the incandescent lamp (measured in lumens per watt) was improved from Edison's original carbon filament by the development of today's tungsten filament in 1910 by General Electric Research Laboratories, but the continuing improvement in lighting efficiency has come from the invention and improvement of the fluorescent lamp, mercury and sodium vapor lamps, and most recently the Light Emitting Diode (LED). Similarly, improvement of the cost per bit of memory storage (the inverse of performance) over time has shifted to a new and more efficient curve each time a denser integrated circuit with more memory capacity was developed. The technology forecaster might have predicted in advance that this process would continue, making new applications requiring extensive computer memory storage more and more practical.

Another tool for forecasting is the internet, which has become a powerful tool at every company's disposal for evaluating the competition, predicting the market, and establishing trends. One method is to have company web pages equipped with counters to keep track of visitors and set "cookies" to gather additional information. This information allows a company to evaluate their customers' habits and determine the most appropriate and beneficial way to deal with each one. This activity also provides a database of information for trending and predicting future responses. There are also online services to help a company to determine demographics for any potential product or service. One service allows the user to analyze the demographics of the internet population according to selectable characteristics. The vast amount of information available at the fingertips has certainly helped simplify the task of evaluating the competition. In addition to their product specifications, many companies post their financial information for perusal. There are also bulletin boards, chat rooms, and polling sites full of public opinions. Government, and even some private organizations, provide statistical data and evaluations. The best part is that most of this information is free, and even if the sites do charge a fee, at least they are easy to access and readily available.

The methods listed here are not the only methods used for forecasting, and what is used depends on what is being forecast and what data is available. Combining methods is effective when different forecasting methods are available. Ideally, one should use as many as five different methods and combine their forecasts using a predetermined mechanical rule. Lacking strong evidence that some methods are more accurate than others, one should use a simple average of forecasts.

STRATEGIES FOR MANAGING TECHNOLOGY

Invention and Innovation

American folklore idealizes its inventors—people who have come up with an idea for a novel product or process. But that is not enough. Betz gives as an example the development of the Xerox copier. Chester Carlson, who had experience as a carbon chemist, a printer, and a patent lawyer, sought a better way to copy legal documents. He conceived the idea of projecting an image of the work to be copied onto paper coated with ink, to hold the ink electrostatically where dark spots (letters) were projected, and then to bake these letters into the paper. He succeeded in obtaining a crude image in 1935, but the invention was neither efficient nor economical. Nonetheless, it was enough to apply for a patent, which was issued in 1942.

Carlson went from company to company looking for support for the process; he was turned down repeatedly. Finally, a group at Battelle Memorial Institute, a nonprofit R&D organization, agreed in 1945 to try to develop the process into a commercially practical one in return for a share of the royalties. Joseph Wilson, president of the small Haloid Corporation, took the risk of producing the first copiers based on Carlson's patented designs and Battelle's developments. That company grew and grew as a result and changed its name to Xerox after its principal product, the office copier.

The contributions of Battelle and Haloid constitute **technological innovation**, *the introduction into the marketplace of new products, processes, and services based on new technology*. Without innovation, inventions create little benefit. Reduction of ideas to successful products and processes is a difficult task. Chapter 9 demonstrates that many different ideas must be considered and developed enough for careful evaluation to produce one successful, profitable new product.

Producing the first successful product is often not enough; innovation must continue to keep the product line competitive. In July 1981, Adam Osborne became the first to market the personal computer as a complete package (computer, disk drive, monitor, printer, and software), and it was an instant success. However, his monitor was only five inches (diagonally) in size and would hold only 52 characters across (versus 80 on a typed page). Competitors (especially Kaypro) moved quickly to correct this deficiency, and by September 1983, the Osborne Company was in bankruptcy. The lesson is clear—product innovation and improvement must be a continuing part of technology strategy.

Myron Tribus believes the competitiveness of an enterprise depends on two distinct thrusts, each having two characteristics: (1) doing new and better things (invention and innovation, just discussed), and (2) doing things in new and better ways (quality and productivity). He elaborates:

> Of the four characteristics, invention, innovation, quality, and productivity, the last three require the collaboration of many people. Invention, on the other hand, is usually the product of one or at the most two or three minds. Management to enhance invention, therefore, is somewhat different from management to enhance innovation, quality, and productivity. Invention is not subject to scheduling.

The earlier discussion of AAON shows examples of the first characteristic being used to drive the second.

Managing Technological Change

Top management in technological enterprises must constantly be aware of the technologies underlying their business and the potential for change. Business history is replete with stories of companies that failed to recognize new technology that would replace their key products in time to take corrective action. Most of the companies that led in production of vacuum tube-based electronics are no longer important today in electronics; other companies led the transistor revolution and replaced them; in turn, a new group of firms have come to the fore with products based on large-scale integrated circuits. A few companies (Microsoft, for example) have maintained leadership or have at least been vigorous early followers as technology changed, and they have prospered because they have succeeded in integrating technology strategy into business strategy.

One of the most compelling examples of how the advent of new technology can quickly alter the way we do business is the birth of the internet. With the rise of this new technology came opportunity and convenience that no company or person could afford to ignore. According to the World Internet Usage and Population Statistics, there were 345.7 million internet users in North America in 2018, a 219 percent

Technological Change

Switzerland had long been the world leader in watch production and sales. With a large array of well-known brands extending from economical to luxury products, Swiss watchmakers had enjoyed a reputation of fine craftsmanship. Between 1968 and 1969, the world's first quartz watch prototypes were developed by the Swiss. The quartz watch was commercially available in 1970. Despite this beginning, the Swiss watch industry failed to see the market potential in the quartz watch, and they continued their focus on the mechanical watch. Between the mid-1970s and 1983, the Swiss watch industry saw its portion of the world watch market drop from 30 percent to 10 percent in number of units sold. The Swiss watch industry eventually recovered from the effects of the quartz revolution, and today Switzerland is once again the world's leading watch exporter in terms of total value.

Source: http://invention.smithsonian.org/centerpieces/quartz/global/switzerland.html, September 2005.

increase since 2000. Worldwide there are 4.2 billion internet users, a number that has grown over 1000 percent since 2000. More details of internet usage may be found in Table 4-4.

To remain competitive, every company experienced a need to establish an entire new group within their structure to develop and maintain a website, and to provide an interface between their organization and the rest of the world. With the development of powerful search engines, people began leaning heavily on the internet for information when investigating a need. As a result, companies lagging behind in this area quickly became inferior in their visibility level. This avenue of visibility became particularly important for engineering-related products or services because of their specialized nature. Clients typically research the specifications of available products after their need has been determined. Companies such as Omega and Hart Scientific began providing detailed specifications and pricing information online for the engineer's perusal. Having the current information readily accessible on the engineer's desktop helps ensure that a company's product is considered.

Many companies also recognized the internet as an inexpensive and timely method for keeping their customers informed. They began using the internet to advertise, announce public safety issues and recalls,

Table 4-4 World Internet Usage and Population Statistics

World Regions	Growth 2000–2018 (percentage)	Internet Users (millions) June 2018	Penetration (percentage of population)
Africa	10,199	464.9	36.1
Asia	1,704	2,062.2	49.0
Europe	570	705.1	85.2
Latin America/Caribbean	2,235	438.2	67.2
Middle East	4,894	164.0	64.5
North America	219	345.7	95.0
Oceania/Australia	273	28.4	67.5
WORLD TOTAL	**305.5**	**2,267,233,742**	**32.7**

Source: Data from Internet World Stats, https://www.internetworldstats.com/stats.htm, December 2018.

and provide technical support, software upgrades, online manuals, interactive troubleshooting knowledge bases, product information, and contacts 24 hours a day. Customers came to expect this service, and any company not providing it began to be overlooked. Web pages and smart phone applications now allow companies to market to and service customers anywhere, anytime. This is particularly valuable in the international market because of the different time zones and corresponding business hours.

DISCUSSION QUESTIONS

4-1. Some say planning has primacy among the managerial functions. Do you agree? Why or why not?

4-2. Develop your own model of the steps in the planning process.

4-3. Find the mission and vision statements for two large companies using their web page. Then use the internet archive at archive.org to find these statements from 10 and 20 years ago. Have they changed? Why do you think this is the case?

4-4. Select a company whose apparent mission or purpose has changed over the course of its history, and describe the change.

4-5. Select a company or industry for which the strategic management of technology is important. Describe some of the *base*, *key*, and *pacing* technologies that are important for their strategic management of technology.

4-6. Pick a company with which you are familiar, and estimate from its actions what the objectives of its management appear to be in each of Drucker's eight "key result areas."

4-7. Briefly outline the concept of management by objectives and the steps involved in implementing this technique in organizations.

4-8. For what types of employees or positions do you think management by objectives should prove particularly effective? Ineffective?

4-9. For a given product and company (such as automobiles from Ford), list a set of premises (assumptions) regarding such matters as the economy, competition, materials, labor, customer demand, and others, that should govern their planning over the next five years.

4-10. What length of planning horizon would you recommend for planning (a) the forest resources of a large paper company, (b) the construction of a new automobile plant, and (c) the creation of a new housing development of 15 homes?

4-11. Create an extended list of the plans and decisions for a large manufacturing company that will depend on the sales forecast for the next year.

4-12. Use the discussion of AAON to outline how another company you are familiar with might manage technology to obtain a competitive advantage.

PROBLEMS

4-1. Sales of a particular product (in thousands of dollars) for the years 2015 through 2018 have been $48,000, $64,000, $67,000, and $83,000, respectively.
  (a) What sales would you predict for 2019, using a simple four-year moving average?
  (b) What sales would you predict for 2019, using a weighted moving average with weights of 0.50 for the immediate preceding year and 0.3, 0.15, and 0.05 for the three years before that?

4-2. Using exponential smoothing with a weight of 0.6 on actual values:
   (a) If sales are $45,000 and $50,000 for 2017 and 2018, what would you forecast for 2019? (The first forecast is equal to the actual value of the preceding year.)
   (b) Given this forecast and actual 2019 sales of $53,000, what would you then forecast for 2020?

4-3. In Problem 4-1, taking actual 2015 sales of $48,000 as the forecast for 2016, what sales would you forecast for 2017, 2018, and 2019, using exponential smoothing and a weight on actual values of **(a)** 0.4 and **(b)** 0.8?

4-4. In Problem 4-1, what sales would you forecast for 2019, using the simple regression (least squares) method?

SOURCES

Amos, John M. and Sarchet, Bernard R., *Management for Engineers* (Englewood Cliffs, NJ: Prentice-Hall, Inc., 1981), p. 51.

Augustine, Norman R., *Augustine's Laws and Major System Development Programs*, revised and enlarged (Washington, DC: American Institute of Aeronautics and Astronautics, 1983), p. 55.

Betz, Frederick, *Managing Technology: Competing through New Ventures, Innovation, and Corporate Research* (Englewood Cliffs, NJ: Prentice-Hall, Inc., 1987).

Dannenbring, David G. and Starr, Martin K., *Management Science: An Introduction* (New York: McGraw-Hill Book Company, 1981), Chapter 19.

Drucker, Peter F., *Management: Tasks, Responsibilities, Practices* (New York: Harper & Row, Publishers, Inc., 1974), pp. 75–77.

Drucker, Peter F., *The Practice of Management*. © 1954 (© renewed 1982) by Peter F. Drucker; reprinted by permission of HarperCollins Publishers, Inc.

Fraker, Susan, "High-Speed Management for the High-Tech Age," *Fortune,* March 5, 1984, pp. 62–68.

Nagahama, Hajime, "Technopolicy," *Look Japan,* May 1993, reported in *The Futurist,* September–October 1993, p. 8.

Peters, Thomas J. and Waterman, Robert H., Jr., *In Search of Excellence: Lessons from America's Best Run Companies* (New York: Harper & Row, Publishers, Inc., 1982).

Roman, Daniel D., "Technological Forecasting in the Decision Process," *Academy of Management Journal*, 13(2), June 1970, pp. 127–138.

Shannon, Robert E., *Engineering Management* (New York: John Wiley & Sons, Inc., 1980), p. 43.

Tribus, Myron, "Applying Quality Management Principles in R&D," *Engineering Management Journal*, 2(3), September 1990, p. 29.

Weihrich, Heinz and Koontz, Harold, *Management: A Global Perspective,* 10th ed. (New York: McGraw-Hill Book Company, 1993).

Young, Edmund, "Management Thoughts for Today from the Ancient Chinese," *Management Bulletin*, October 1980, p. 29.

STATISTICAL SOURCEBOOK

The following is a useful source website. (March, 2019)

   http://www.internetworldstats.com/Internet World Stats is an International website that features up to date world Internet Usage, Population Statistics, Travel Stats and Internet Market Research Data, for over 233 individual countries and world regions.

5

Decision Making

PREVIEW

Decision making is an essential part of planning. Decision making and problem solving are used in all management functions, although usually they are considered a part of the planning phase. This chapter presents information on decision making and how it relates to the first management function of planning. A discussion of the origins of management science leads into one on modeling, the five-step process of management science, and the process of engineering problem solving.

Different types of decisions are examined in this chapter. They are classified under conditions of certainty, using linear programming; risk, using expected value and decision trees; or uncertainty, depending on the degree with which the future environment determining the outcomes of these decisions is known. The chapter continues with brief discussions of integrated databases, management information and decision support systems, and expert systems, and closes with a comment on the need for effective implementation of decisions.

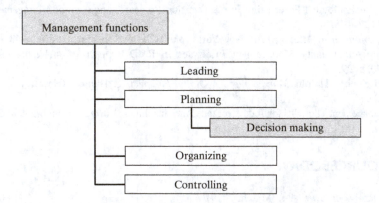

NATURE OF DECISION MAKING

Relation to Planning

Managerial decision making is the process of making a conscious choice between two or more rational alternatives in order to select the one that will produce the most desirable consequences (benefits) relative to unwanted consequences (costs). If there is only one alternative, there is nothing to decide. The overall planning/decision-making process has already been described at the beginning of Chapter 4, and there we discussed the key first steps of setting objectives and establishing assumptions. In this chapter, we consider the process of developing and evaluating alternatives and selecting from among them the best alternative, and we review briefly some of the tools of management science available to help us in this evaluation and selection.

If planning is truly "deciding in advance what to do, how to do it, when to do it, and who is to do it" (as proposed by Amos and Sarchet), then decision making is an essential part of planning. Decision making is also required in designing and staffing an organization, developing methods of motivating subordinates, and identifying corrective actions in the control process. However, it is conventionally studied as part of the planning function, and it is discussed here.

Occasions for Decision

Chester Barnard wrote his classic book *The Functions of the Executive* from his experience as president of the New Jersey Bell Telephone Company and of the Rockefeller Foundation, and in it he pursued the nature of managerial decision making at some length. He concluded that

> the occasions for decision originate in three distinct fields: (a) from authoritative communications from superiors; (b) from cases referred for decision by subordinates; and (c) from cases originating in the initiative of the executive concerned.

Barnard points out that occasions for decisions stemming from the "requirements of superior authority . . . cannot be avoided," although portions of it may be delegated further to subordinates. Appellate cases (referred to the executive by subordinates) should not always be decided by the executive. Barnard explains that "the test of executive action is to make these decisions when they are important, or when they cannot be delegated reasonably, and to decline the others."

Barnard concludes that "occasions of decision arising from the initiative of the executive are the most important test of the executive." These are occasions where no one has asked for a decision, and the executive usually cannot be criticized for not making one. The effective executive takes the initiative to think through the problems and opportunities facing the organization, conceives programs to make the necessary changes, and implements them. Only in this way does the executive fulfill the obligation to *make a difference* because they are in that chair rather than someone else.

Types of Decisions

Routine and Nonroutine Decisions.   Pringle et al. classify decisions on a continuum ranging from routine to nonroutine, depending on the extent to which they are *structured*. They describe *routine* decisions as focusing on well-structured situations that

> recur frequently, involve standard decision procedures, and entail a minimum of uncertainty. Common examples include payroll processing, reordering standard inventory items, paying suppliers, and so on. The decision maker can usually rely on policies, rules, past precedents, standardized methods of processing, or computational techniques. Probably 90 percent of management decisions are largely routine.

Indeed, routine decisions usually can be delegated to lower levels to be made within established policy limits, and increasingly they can be programmed for computer decision if they can be structured simply enough—for example, whether a bagged product meets weight requirements. *Nonroutine* decisions, on the other hand, deal with unstructured situations of a novel, nonrecurring nature, often involving incomplete knowledge, high uncertainty, and the use of subjective judgment or even intuition, where no alternative can be proved to be the best possible solution to the particular problem. Such decisions become more and more common the higher one goes in management and the longer the future period influenced by the decision is. Unfortunately, almost the entire educational process of the engineer is based on the solution of highly structured problems for which there is a single textbook solution. Engineers often find themselves unable to rise in management unless they can develop the tolerance for ambiguity that is needed to tackle unstructured problems. Students often find it helpful to think about how the bulk of an engineering education is focused on the black and white, while an engineering management education includes a key focus on working through the "gray ball of goo" to distill the black and white.

Objective Versus Bounded Rationality.   Simon defines a decision as being objectively rational if *in fact* it is the correct behavior for maximizing given values in a given situation. Such rational decisions are made "(a) by viewing the behavior alternatives prior to decision in panoramic [exhaustive] fashion, (b) by considering the whole complex of consequences that would follow on each choice, and (c) with the system of values as criterion singling out one from the whole set of alternatives." Rational decision making, therefore, consists of *optimizing*, or *maximizing*, the outcome by choosing the single best alternative from among all possible ones, which is the approach suggested in the planning/decision-making model at the beginning of Chapter 4. However, Simon believes that actual behavior falls short of objective rationality in at least three ways.

1. Rationality requires a complete knowledge and anticipation of the consequences that will follow on each choice. In fact, knowledge of consequences is always fragmentary.
2. Since these consequences lie in the future, imagination must supply the lack of experienced feeling in attaching value to them. But values can be only imperfectly anticipated.
3. Rationality requires a choice among all possible alternative behaviors. In actual behavior, only a few of these possible alternatives ever come to mind.

Managers, under pressure to reach a decision, have neither the time nor other resources to consider all alternatives or all the facts about any alternative. A manager "must operate under conditions of *bounded rationality*, taking into account only those few factors of which he or she is aware, understands, and regards as relevant." Administrators must *satisfice* by accepting a course of action that is satisfactory or "good enough," and get on with the job rather than searching forever for the "one best way." Managers of engineers and scientists, in particular, must learn to insist that their employees go on to other problems when they reach a solution that *satisfices*, rather than pursuing their research or design beyond the point at which incremental benefits no longer match the costs to achieve them.

Level of Certainty.   Decisions may also be classified as being made under conditions of **certainty**, **risk**, or **uncertainty**, depending on the degree with which the future environment determining the outcome of these decisions is known. These three categories are compared later in this chapter.

MANAGEMENT SCIENCE

Origins

Quantitative techniques have been used in business for many years in applications such as return on investment, inventory turnover, and statistical sampling theory. However, today's emphasis on the quantitative solution of complex problems in operations and management, known initially as *operations research* and more commonly today as *management science*, began at the Bawdsey Research Station in England at the beginning of World War II with an interdisciplinary team led by P. M. S. Blackett of the University of Manchester. The formation of this group seems to be commonly accepted as the beginning of operations research.

Some of the problems this group (and several that grew from it) studied were the optimum depth at which antisubmarine bombs should be exploded for greatest effectiveness (20 to 25 feet) and the relative merits of large versus small convoys (large convoys led to fewer total ship losses). Soon after the United States entered the war, similar activities were initiated by the U.S. Navy and the Army Air Force. With the immediacy of the military threat, these studies involved *research* on the *operations* of existing systems. After the war, these techniques were applied to longer-range-military problems and to problems of industrial organizations. With the development of more and more powerful electronic computers, it became possible to model large systems as a part of the design process, and the terms *systems engineering* and *management science* came into use. Management science has been defined as having the following "primary distinguishing characteristics":

1. A systems view of the problem—a viewpoint is taken that includes all of the significant interrelated variables contained in the problem.
2. The team approach—personnel with heterogeneous backgrounds and training work together on specific problems.
3. An emphasis on the use of formal mathematical models and statistical and quantitative techniques.

What Is Systems Engineering?

The International Council on Systems Engineering (INCOSE) defines systems engineering is an interdisciplinary approach and means to enable the realization of successful systems. It focuses on defining customer needs and required functionality early in the development cycle, documenting requirements, then proceeding with design synthesis and system validation while considering the complete problem: Operations, Performance, Test, Manufacturing, Cost & Schedule, Training & Support, Disposal.

Systems Engineering integrates all the disciplines and specialty groups into a team effort forming a structured development process that proceeds from concept to production to operation. Systems Engineering considers both the business and the technical needs of all customers with the goal of providing a quality product that meets the user needs.

Models and Their Analysis

A **model** is an abstraction or simplification of reality, designed to include only the essential features that determine the behavior of a real system. For example, a three-dimensional *physical* model of a chemical processing plant might include scale models of major equipment and large-diameter pipes, but it would not normally include small pipings or electrical wirings. The *conceptual* model of the planning/decision-making process in Chapter 4 certainly does not illustrate all the steps and feedback loops present in a real situation; it is only indicative of the major ones.

Most of the models of management science are *mathematical* models. These can be as simple as the common equation representing the financial operations of a company:

$$net\, income \,=\, revenue \,-\, expenes \,-\, taxes$$

They may also involve a very complex set of equations. As an example, the Urban Dynamics model was created by Jay Forrester to simulate the growth and decay of cities. This model consisted of 154 equations representing relationships between the factors that he believed were essential: three economic classes of workers (managerial/professional, skilled, and "underemployed"), three corresponding classes of housing, three types of industry (new, mature, and declining), taxation, and land use. The values of these factors evolved through 250 simulated years to model the changing characteristics of a city. Even these 154 relationships still proved too simplistic to provide any reliable guide to urban development policies.

Management science uses a five-step process that begins in the real world, moves into the model world to solve the problem, and then returns to the real world for implementation. The following explanation is, in itself, a conceptional model of a more complex process:

Real World	Simulated (Model) World
1. Formulate the problem (defining objectives, variables, and constraints).	
	2. Construct a mathematical model (a simplified yet realistic representation of the system).
	3. Test the model's ability to predict the present from the past, and revise until you are satisfied.
	4. Derive a solution from the model.
5. Apply the model's solution to the real system, document its effectiveness, and revise further as required.	

The **scientific method** or scientific process is fundamental to scientific investigation and to the acquisition of new knowledge based upon physical evidence by the scientific community. Scientists use observations and reasoning to propose tentative explanations for natural phenomena, termed *hypotheses*. **Engineering problem solving** is more applied and is different to some extent from the scientific method.

Scientific Method	Engineering Problem Solving Approach
• Define the problem.	• Define the problem.
• Collect data.	• Collect and analyze the data.
• Develop hypotheses.	• Search for solutions.
• Test hypotheses.	• Evaluate alternatives.
• Analyze results.	• Select solution and evaluate the impact.
• Draw conclusion.	

The Analyst and the Manager

To be effective, the management science analyst cannot just create models in an *ivory tower*. The problem-solving team must include managers and others from the department or system being studied—to establish objectives, explain system operation, review the model as it develops from an operating perspective, and help test the model. The user who has been part of model development, has developed some understanding of it and confidence in it, and feels a sense of *ownership* of it is most likely to use it effectively.

The manager is not likely to have a detailed knowledge of management science techniques or the time for model development. Today's manager should, however, understand the nature of management science tools and the types of management situations in which they might be useful. Increasingly, management positions are being filled with graduates of management (or engineering management) programs that have included an introduction to the fundamentals of management science and statistics. Regrettably, all too few operations research or management science programs require the introduction to organization and behavioral theory that would help close the manager–analyst gap from the opposite direction.

TOOLS FOR DECISION MAKING

Categories of Decision Making

Decision making can be discussed conveniently in three categories: decision making under certainty, under risk, and under uncertainty. The *payoff table*, or *decision matrix*, shown in Table 5-1 will help in this discussion. Our decision will be made among some number m of alternatives, identified as A_1, A_2, $\cdots$, A_m. There may be more than one future "state of nature" N. (The model allows for n different futures.) These future states of nature may not be equally likely, but each state N_j will have some (known or unknown) probability of occurrence p_j. Since the future must take on one of the n values of N_j, the sum of the n values of p_j must be 1.0.

The *outcome* (or payoff, or benefit gained) will depend on both the alternative chosen and the future state of nature that occurs. For example, if you choose alternative A_i and state of nature N_j takes place (as it will with probability p_j), the payoff will be outcome O_{ij}. A full payoff table will contain m times n possible outcomes.

Let us consider what this model implies and the analytical tools we might choose to use under each of our three classes of decision making.

Decision Making Under Certainty

Decision making under **certainty** implies that we are certain of the future state of nature (or we assume that we are). (In our model, this means that the probability p_1 of future N_1 is 1.0, and all other futures have zero probability.) The solution, naturally, is to choose the alternative A_i, which gives us the most favorable outcome O_{ij}. Although this may seem like a trivial exercise, there are many problems that are so complex that sophisticated mathematical techniques are needed to find the best solution.

Table 5-1   Payoff Table

Alternative	State of Nature/Probability					
	N_1	N_2	$\cdots$	N_j	$\cdots$	N_n
	p_1	p_2	$\cdots$	p_j	$\cdots$	p_n
A_1	O_{11}	O_{12}	$\cdots$	O_{1j}	$\cdots$	O_{1n}
A_2	O_{21}	O_{22}	$\cdots$	O_{2j}	$\cdots$	O_{2n}
$\cdots$	$\cdots$	$\cdots$	$\cdots$	$\cdots$	$\cdots$	$\cdots$
A_i	O_{i1}	O_{i2}	$\cdots$	O_{ij}	$\cdots$	O_{in}
$\cdots$	$\cdots$	$\cdots$	$\cdots$	$\cdots$	$\cdots$	$\cdots$
A_m	O_{m1}	O_{m2}	$\cdots$	O_{mj}	$\cdots$	O_{mn}

Linear Programming. One common technique for decision making under certainty is called **linear programming**. In this method, a desired benefit (such as profit) can be expressed as a mathematical function (the value model or **objective function**) of several variables. The solution is the set of values for the independent variables (**decision variables**) that serves to maximize the benefit (or, in many problems, to minimize the cost), subject to certain limits (**constraints**). Steps include:

- State the problem.
- What are the decision variables?
- Objective function
- Constraints

Example

Consider a factory producing two products, product X and product Y. The problem is this: If you can realize $10.00 profit per unit of product X and $14.00 per unit of product Y, what is the production level of x units of product X and y units of product Y that maximizes the profit P each day? Your production, and therefore your profit, is subject to resource limitations, or *constraints*. Assume in this example that you employ five workers—three machinists and two assemblers—and that each works only 40 hours a week.

- Product X requires three hours of machining and one hour of assembly per unit.
- Product Y requires two hours of machining and two hours of assembly per unit.

State the problem: How many of product X and product Y to produce to maximize profit?

Decision variables: Let x = number of product X to produce per day
Let y = number of product Y to produce per day
Objective function: maximize $P = 10x + 14y$
Constraints: $3x + 2y \leq 120$ (hours of machining time)
$x + 2y \leq 80$ (hours of assembly time)

As illustrated in (Figure 5-1), you can get a profit of:

- $350 by selling 35 units of X or 25 units of Y
- $700 by selling 70 units of X or 50 units of Y
- $620 by selling 62 units of X or 44.3 units of Y; or (as in the first two cases as well) any combination of X and Y on the *isoprofit line* connecting these two points.

Since there are only two products, these limitations can be shown on a two-dimensional graph (Figure 5-2). Since all relationships are linear, the solution to our problem will fall at one of the corners. To find the solution, begin at some feasible solution (satisfying the given constraints) such as $(x,y) = (0,0)$, and proceed in the direction of "steepest ascent" of the profit function (in this case, by increasing production of Y at $14.00 profit per unit) until some constraint is reached. Since assembly hours are limited to 80, no more than 80/2, or 40, units of Y can be made, earning $40 \times \$14.00$, or $560 profit. Then proceed along the steepest allowable ascent from there (along the assembly constraint line) until another constraint (machining hours) is reached. At that point, $(x,y) = (20,30)$ and profit $P = (20 \times \$10.00) \times (30 \times \$14.00)$, or $620. Since there is no remaining edge along which profit increases, this is the optimum solution.

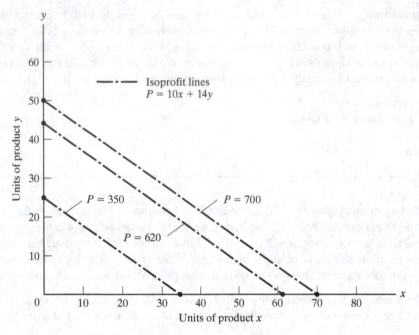

Figure 5-1 Linear program example: isoprofit lines.

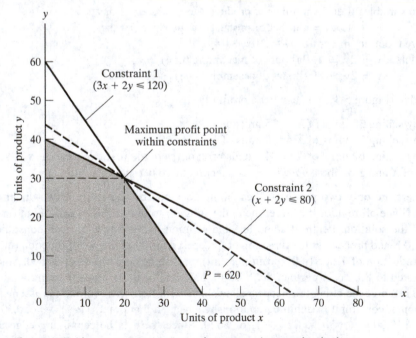

Figure 5-2 Linear program example: constraints and solution.

Computer Solution.   About 50 years ago, George Danzig of Stanford University developed the *simplex method*, which expresses the foregoing technique in a mathematical algorithm that permits computer solution of linear programming problems with many variables (dimensions), not just the two (assembly and machining), as in the previous example. Now, linear programs in a few thousand variables and constraints are viewed as *small*. Problems having tens or hundreds of thousands of continuous variables are regularly solved; tractable integer programs are necessarily smaller, but are still commonly in the hundreds or thousands of variables and constraints. Today there are many linear programming software packages available.

Another classic linear programming application is the oil refinery problem, where profit is maximized over a set of available crude oils, process equipment limitations, products with different unit profits, and other constraints. Other applications include assignment of employees with differing aptitudes to the jobs that need to be done to maximize the overall use of skills; selecting the quantities of items to be shipped from a number of warehouses to a variety of customers while minimizing transportation cost; and many more. In each case there is one best answer, and the challenge is to express the problem properly so that it fits a known method of solution.

Decision Making Under Risk

Nature of Risk.   In decision making under **risk** one assumes that there exist a number of possible future states of nature N_j, as we saw in Table 5-1. Each N_j has a known (or assumed) probability p_j of occurring, and there may not be one future state that results in the best outcome for all alternatives A_i. Examples of future states and their probabilities are as follows:

- Alternative weather (N_1 = rain; N_2 = good weather) will affect the profitability of alternative construction schedules; here, the probabilities p_1 of rain and p_2 of good weather can be estimated from historical data.
- Alternative economic futures (boom or bust) determine the relative profitability of conservative versus high-risk investment strategy; here, the assumed probabilities of different economic futures might be based on the judgment of a panel of economists.

Expected Value.   Given the future states of nature and their probabilities, the solution in decision making under risk is the alternative A_i that provides the highest **expected value** E_i, which is defined as the sum of the products of each outcome O_{ij} times the probability p_j that the associated state of nature N_j occurs:

$$E_i = \sum_{j=1}^{n} (p_j O_{ij}) \qquad\qquad \text{(5-1)}$$

Example

For example, consider the simple payoff information of Table 5-2, with only two alternative decisions and two possible states of nature. Alternative A_1 has a constant cost of $200, and A_2 a cost of $100,000 if future N_2 takes place (and none otherwise). At first glance, alternative A_1 looks like the clear winner, but consider the situation when the probability (p_1) of the first state of nature is 0.999 and the probability (p_2) of the second state is only 0.001. The expected value of choosing alternative A_2 is only

$$E(A_2) = 0.999(\$0) - 0.001(\$100,000) = -\$100$$

Table 5-2   Decision Making Under Risk

	N_1 $p_1 = 0.999$	N_2 $p_2 = 0.001$
A_1	$\$-200$	$\$-200$
A_2	$\$0$	$\$-100,000$

Note that this outcome of $\$-100$ is not possible: the outcome if alternative A_2 is chosen will be a loss of either $0 or $100,000, not $100. However, if you have many decisions of this type over time and you choose alternatives that maximize expected value each time, you should achieve the best overall result. Since we should prefer expected value E_2 of $\$-100$ to E_1 of $\$-200$, we should choose A_2, other things being equal.

But first, let us use these figures in a specific application. Assume that you own a $100,000 house and are offered fire insurance on it for $200 a year. This is twice the "expected value" of your fire loss (as it has to be to pay insurance company overhead and agent costs). However, if you are like most people, you will probably buy the insurance because, quite reasonably, your attitude toward risk is such that you are not willing to accept loss of your house! The insurance company has a different perspective, since they have many houses to insure and can profit from maximizing expected value in the long run, as long as they do not insure too many properties in the path of the same hurricane or earthquake.

Example

Consider that you own rights to a plot of land under which there may or may not be oil. You are considering three alternatives: doing nothing ("don't drill"), drilling at your own expense of $500,000, or "farming out" the opportunity to someone who will drill the well and give you part of the profit if the well is successful. You see three possible states of nature: a dry hole, a mildly interesting small well, and a very profitable gusher. You estimate the probabilities of the three states of nature p_j and the nine outcomes O_{ij}, as shown in Table 5-3.

The first thing you can do is eliminate alternative A_1, since alternative A_3 is at least as attractive for all states of nature and is more attractive for at least one state of nature. A_3 is therefore said to *dominate* A_1.

Next, you can calculate the expected values for the surviving alternatives A_2 and A_3:

$$E_2 = 0.6(-500,000) + 0.3(300,000) + 0.1(9,300,00) = \$720,000$$
$$E_3 = 0.6(0) + 0.3(125,000) + 0.1(1,250,000) = \$162,500$$

and you choose alternative A_2 if (and only if) you are willing and able to risk losing $500,000.

Table 5-3   Well Drilling Example—Decision Making Under Risk

Alternative	State of Nature/Probability			
	N_1: Dry Hole $p_1 = 0.6$	N_2: Small Well $p_2 = 0.3$	N_3: Big Well $p_3 = 0.1$	Expected Value
A_1: Don't drill	$ 0	$ 0	$ 0	$ 0
A_2: Drill alone	$-500,000$	300,000	9,300,000	720,000
A_3: Farm out	0	125,000	1,250,000	162,500

Decision trees provide another technique for finding expected value. They begin with a single *decision node* (normally represented by a square or rectangle), from which a number of decision alternatives radiate. Each alternative ends in a *chance node,* normally represented by a circle. From each chance node radiate several possible futures, each with a probability of occurring and an outcome value. The expected value for each alternative is the sum of the products of outcomes and related probabilities, just as calculated previously.

Example

Figure 5-3 illustrates the use of a decision tree in the simple insurance example.

   The conclusion reached is identical mathematically to that obtained from Table 5-2.

   Decision trees provide a very visible solution procedure, especially when a sequence of decisions, chance nodes, new decisions, and new chance nodes exist. For example, if you are deciding whether to expand production capacity in December 2013, a decision a year later, in December 2014, as to what to do then will depend both on the first decision and on the sales enjoyed as an outcome during 2014. The possible December 2014 decisions lead to (a larger number of) chance nodes for 2015. The technique used starts with the later year, 2008 (the farthest branches). Examining the outcomes of all the possible 2015 chance nodes, you find the optimum second decision and its expected value, in 2014, for each 2014 chance node—that is, for each possible combination of first decision in December 2013 and resulting outcome for 2014. Then you use those values as part of the calculation of expected values for each first-level decision alternative in December 2013.

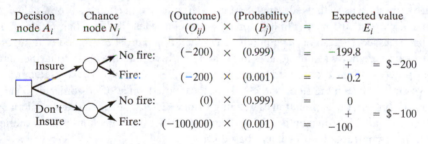

Figure 5-3 Example of a decision tree.

Queuing (Waiting-Line) Theory.   Most organizations have situations where a class of people or objects arrive at a facility of some type for service. The times between arrivals (and often the time required for serving each arrival) are not constant, but they can usually be approximated by a probability distribution. The first work in this field was by the Danish engineer A. K. Erlang, who studied the effect of fluctuating demand for telephone calls on the need for automatic dialing equipment. Table 5-4 lists some other common examples of waiting lines.

Table 5-4 Typical Waiting-Line Situations

Organization	Activity	Arrivals	Servers
Airport	Landing	Airplanes	Runway
Court system	Trials	Cases	Judges
Hospital	Medical service	Patients	Rooms/doctors
Personnel office	Job interviews	Applicants	Interviewers
Supermarket	Checkout	Customers	Checkout clerks
Toll bridge	Taking tolls	Vehicles	Toll takers
Tool room	Tool issue	Machinists	Tool room clerks

Queuing Theory in Our Lives: Smart Retail and the Internet of Things (IoT)

Although it may sound abstract, queuing theory impacts our lives nearly every day. Whether it is improving rush hour traffic flow with ramp meters, speeding your time out of the grocery store through efficient checkout, or letting you ride more rides at an amusement park through guest management, appropriate applications of queueing theory allows us to do more in less time. Historically, successfully using queuing theory to improve customer experience has suffered from a lack of accurate data about customer behavior within a store. This leads to the common experience of arriving at checkout, only to find a long line until someone from the store calls for more help "to the front."

Today, we're witness to a transformation of this experience through new data collected through our devices (e.g., smart phones) and IoT. Through smart retail, stores large (e.g., Target and Walmart) and small are providing us with useful apps for our smart phones, in exchange for the data the app collects. For many, when a customer is in the store, that data includes tracking their movements and what items they have passed and might have purchased. When this data is combined with that of previous visits, the store can predict when customers in the store will come to the check-out lines and how long the checkout process might take (based on predicted number of items). These predictions enable the store to move staff between the floor and checkout areas in advance of customer demand, leading to shorter lines, all through the application of queuing theory. Of course, even these modern applications may soon be eliminated as Amazon and others experiment with no-checkout retail experiences using the next edition of smart retail with camera-based systems!

Simulation

There are many situations where the real-world system being studied is too complex to express in simple equations that can be solved by hand or approximated in a reasonable time. In other situations, safety or the cost of prototyping requires other approaches to be considered. A common approach in such cases is to construct a computer program that simulates certain aspects of the operation of the real system by mathematically describing the behavior of individual parts and the interactions between the parts. The computer model is an approximation of the real system. The computer model can be executed repeatedly under various conditions to study the behavior of the real system. In many cases, stochastic activities can be inserted in the model in the form of probability distributions. In other cases, random variability is limited, such as when using simulation to test a new system.

There are three categories of computer simulations—live, virtual, and constructive. Live simulations have real people and real equipment operating in a simulated environment. An example is live training exercises conducted by the military. Virtual simulations have real people using simulated equipment. An example would be a driving simulator or computer games, such as a flight simulator. Constructive simulations have simulated people and equipment, such as what might be found in a model of a factory production layout or airport screening operation. Live and virtual simulations are typically used where safety is an important consideration. Constructive simulations are typically used where cost, decision making, and prototyping limit implementing the real system.

Live and virtual simulations can be complex, requiring specially developed software and often expensive equipment, as well as special facilities such as virtual reality rooms. Stochastic and deterministic variables are selectively used in these simulations. Stochastic variables are often used in live and virtual training applications where it might be advantageous to have an opponent's behavior unpredictable. In testing, deterministic variables are preferred so that the simulated system can be evaluated under tightly controlled conditions. The languages most commonly used in live and virtual simulations are C and C++. Constructive simulations are also complex and are especially useful when many runs need to be made (as they can run faster than real time) or when it is not practical to use actual humans as participants in the simulation. Although programs for constructive simulations can be written in common languages such as Python, JAVA, or C++, special-purpose simulation languages such as Arena, or FlexSim are powerful and more efficient for this purpose

Because computer simulations are approximations of real-world activities, there is inherent uncertainty in their results. For this reason, computer simulations must be carefully verified and validated to ensure that they accurately reflect the characteristics of the real-world system in the range of interest. Additionally, stochastic variables are often used to introduce the variability of real-world parameters. The outcome of a single run of a simulation program with many probabilistic values is generally not significant, but can be economically rerun 100 or 1,000 times to develop a probability distribution of the final outcome. Conditions simulated in the model can then be changed and the modified model exercised again until a satisfactory result is obtained. The policy expressed in the most successful version of the model can then be tested in the real world; its success there will depend largely on how well the critical factors in the real world have been captured in the model.

Currently, computer simulations are widely used by the military for training, health care, entertainment, design, logistics, etc. A general trend sees increased use of connecting individual simulations to represent very complex systems and increased human interaction during the execution of the simulation. For example, constructive and virtual simulations are being combined by oil companies and NASA to facilitate better understanding of complex design and logistics issues, while allowing human interaction with the model as it runs.

Source: Brian Goldiez, Deputy Director, Institute for Simulation & Training, University of Central Florida.

The essence of the typical queuing problem is identifying the optimum number of servers needed to reduce overall cost. In the tool room problem, machinists appear at random times at the window of an enclosed tool room to sign out expensive tools as they are needed for a job, and attendants find the tools, sign them out, and later receive them back. The production facility is paying for the time of both tool room attendants and the (normally more expensive) machinists, and therefore it wishes to provide the number of servers that will minimize overall cost. In most of the other cases in the table, the serving facility is not paying directly for the time lost in queues, but it wishes to avoid disgruntled customers or clients who might choose to go elsewhere for service. Mathematical expressions for mean queue length and delay as a function of mean arrival and service rates have been developed for a number of probability distributions (in -particular, exponential and Poisson) of arrival and of service times.

Risk as Variance.   Another common definition of risk is variability of outcome, measured by the variance or (more often) its square root, the standard deviation.

Decision Making Under Uncertainty

At times, a decision maker cannot assess the probability of occurrence for the various states of nature. Uncertainty occurs when there exist several (i.e., more than one) future states of nature N_j, but the probabilities p_j of each of these states occurring are not known. In such situations the decision maker can choose among several possible approaches for making the decision. A different kind of logic is used here, based on attitudes toward risk.

Different approaches to decision making under uncertainty include the following:

- The first, the optimistic decision maker, may choose the alternative that offers the highest possible outcome (the "**maximax**" solution);
- The second, the pessimistic decision maker, may choose the alternative whose worst outcome is "least bad" (the "**maximin**" solution);
- The third decision maker may choose a position somewhere between optimism and pessimism ("**Hurwicz**" approach);
- The fourth decision maker may simply assume that all states of nature are **equally likely** (the so-called "principle of insufficient reason"), set all p_j values equal to $1.0/n$, and maximize expected value based on that assumption.
- The fifth decision maker may choose the alternative that has the smallest difference between the best and worst outcomes (the "**minimax regret**" solution). Regret here is understood as proportional to the difference between what we actually get, and the better position that we could have received if a different course of action had been chosen. Regret is sometimes also called "opportunity loss." The minimax regret rule captures the behavior of individuals who spend their post-decision time regretting their choices.

Example

Consider two investment projects, X and Y, having the discrete probability distribution of expected cash flows in each of the next several years as shown in Table 5-5.

Expected cash flows are calculated in the same way as expected value:

a. $E(X) = 0.10(3,000) + 0.20(3,500) + 0.40(4,000) + 0.20(4,500) + 0.10(5,000)$
    $= \$4,000$
b. $E(Y) = 0.10(2,000) + 0.25(3,000) + 0.30(4,000) + 0.25(5,000) + 0.10(6,000)$
    $= \$4,000$

Although both projects have the same mean (expected) cash flows, the expected values of the variances (squares of the deviations from the mean) differ as follows (see also Figure 5-4):

$V_X = 0.10(3,000 - 4,000)^2 + 0.20(3,500 - 4,000)^2 + \cdots + 0.10(5,000 - 4,000)^2$
   $= 300,000$
$V_Y = 0.10(2,000 - 40,000)^2 + 0.25(3,000 - 4,000)^2 + \cdots + 0.10(6,000 - 4,000)^2$
   $= 1,300,000$

The standard deviations are the square roots of these values:

$$\sigma_X = \$548, \sigma_Y = \$1,140$$

Table 5-5 Data for Risk as Variance Example

Project X		Project Y	
Probability	Cash Flow	Probability	Cash Flow
0.10	$3,000	0.10	$2,000
0.20	3,500	0.25	3,000
0.40	4,000	0.30	4,000
0.20	4,500	0.25	5,000
0.10	5,000	0.10	6,000

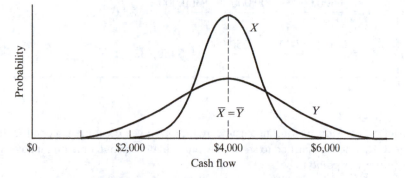

Figure 5-4 Projects with the same expected value but different variances.

Since project Y has the greater variability (whether measured in variance or in standard deviation), it must be considered to offer greater *risk* than does project X.

Example

Using the well-drilling problem as shown in Table 5-3, consider if the probabilities p_j for the three future states of nature N_j cannot be estimated. In Table 5-6, the "Maximum" column lists the best possible outcome for alternatives A_2 and A_3; the optimist will seek to "**maximax**" by choosing A_2 as the best outcome in that column. The pessimist will look at the "Minimum" column, which lists the worst possible outcome for each alternative, and he or she will pick the maximum of the minimums (**Maximin**) by choosing A_3 as having the best (algebraic) worst case. (In this example, both maxima came from future state N_3 and both minima from future state N_1, but this sort of coincidence does not usually occur.)

Table 5-6 Decision Making Under Uncertainty Example

Alternative	Maximum	Minimum	Hurwicz (α = 0.2)	Equally Likely
A_2	$9,300,000*	$-500,000	$1,460,000*	$3,033,333*
A_3	1,250,000	0*	250,000	458,333

*Preferred solution.

A decision maker who is neither a total optimist nor a total pessimist may be asked to express a "coefficient of optimism" as a fractional value α between 0 and 1 and then to use this formula:

$$\text{Maximize } \left[\alpha \text{ (best outcome)} + (1 - \alpha) \text{ (worst outcome)} \right]$$

The outcome using this "**Hurwicz**" approach and a coefficient of optimism of 0.2 is shown in the third column of Table 5-6; A_2 is again the winner.

If decision makers believe that the future states are "**equally likely**," they will seek the higher expected value and choose A_2 on that basis:

$$E_2 = \frac{-500,000 + 300,000 + 9,300.000}{3} = \$3,033,333$$

$$E_3 = \frac{0 + 125,000 + 1,250,000}{3} = \$458,333$$

The final approach to decision making under uncertainty involves creating a second matrix, not of outcomes, but of **regret**. Regret is quantified to show how much better the outcome might have been if you had known what the future was going to be.

Different decision makers will have different approaches to decision making under uncertainty. None of the approaches can be described as the "best" approach, for there is no one best approach. Obtaining a

Example

If there is a "small well" under your land and you did not drill for it, you would regret the $300,000 you might have earned. On the other hand, if you farmed out the drilling, your regret would be only $175,000 ($300,000 less the $125,000 profit sharing you received). Table 5-7 provides this regret matrix and lists in the right-hand column the maximum regret possible for each alternative. The decision maker who wishes to minimize the maximum regret (minimax regret) will therefore choose A_2.

Table 5-7   Well Drilling Example—Decision Making Under Uncertainty—Regret Analysis

Alternative	N_1: Dry Hole	N_2: Small Well	N_3: Big Well	Maximum Regret
	State of Nature			
A_1: Don't drill	$ 0	$300,000	$9,300,000	$9,300,000
A_2: Drill alone	500,000	0	0	500,000
A_3: Farm out	0	175,000	8,050,000	8,050,000

solution is not always the end of the decision-making process. The decision maker might still look for other arrangements to achieve even better results. Different people have different ways of looking at a problem.

Game Theory.   A related approach is *game theory*, where the future states of nature and their probabilities are replaced by the decisions of a competitor. Begley and Grant explain:

> In essence, game theory provides the model of a contest. The contest can be a war or an election, an auction or a children's game, as long as it requires strategy, bargaining, threat, and reward.

In other situations, game theory leads to selecting a mixture of two or more strategies, alternated randomly with some specified probability. Again, Begley and Grant provide a simple example:

> In the children's game called **Odds and Evens**, for instance, two players flash one or two fingers. If the total is 2 or 4, Even wins; if [it is] 3, Odd wins. A little analysis shows that the winning ploy is to randomly mix up the number of fingers flashed. For no matter what Odd does, Even can expect to come out the winner about half the time, and vice versa. If Even attempts anything trickier, such as alternating 1s and 2s, he can be beaten if Odd catches on to the strategy and alternates 2s and 1s.

There are many other techniques or methods for decision making. One is **Six Thinking Hats**. It is used to look at decisions from a number of important perspectives. This forces you to move outside your habitual thinking style, and helps you to get a more rounded view of a situation. Each person has a different hat, which has a different meaning. This enables teams to think together more effectively, and a means to plan thinking processes in a detailed and cohesive way.

Another technique is a children's game—**Rock, Paper, Scissors**. The game is often used as a choosing method in a way similar to **coin flipping, drawing straws**, or throwing **dice**. Unlike truly **random** selection methods, however, Rock, Paper, Scissors can be played with a degree of skill by recognizing and exploiting nonrandom behavior in opponents.

COMPUTER-BASED INFORMATION SYSTEMS

Integrated Databases and the Cloud

Just over a generation ago, most organizations held critical information (about customers, production, materials, etc.) in a number of different manners. This method of storage caused many to have incomplete information which created issues ranging from stock outs and production problems to poor customer experiences. With the advent of systems such as Enterprise Resource Planning (see Chapter 11), Customer Management Systems, and other integrated databases, these issues began to be resolved. Today, information like this is typically stored in cloud-based storage systems accessible to almost all members of an organization and even many outside. It is integrations such as these that allow us to watch a video on a streaming service seamlessly switching from our phone, to our computer, to our TV.

Management Information/Decision Support Systems

Traditionally, top managers have relied primarily on oral and visual sources of information: scheduled committee meetings, telephone calls, business luncheons, and strolls through the workplace, supplemented by the often condensed and delayed information in written reports and periodicals. The explosion of computer networks, centralized databases, synchronized cloud storage, and user-friendly software has provided a new source of prompt, accurate data to the manager. A recent survey showed that 93 percent of senior executives used a personal computer, 60 percent of them for planning and decision support.

Contemporary authors distinguish two classes of application of computer-based management systems:

Management Information Systems (MIS) focus on generating better solutions for structured problems, as well as improving efficiency in dealing with structured tasks.

On the other hand, a *Decision Support System* (DSS) is interactive and provides the user with easy access to decision models and data in order to support semistructured and unstructured decision-making tasks. It improves effectiveness in making decisions where a manager's judgment is still essential.

As one rises from front-line supervisor through middle management to top management, the nature of decisions and the information needed to make them changes (see Table 5-8). The higher the management level is, the fewer decisions may be in number, but the greater is the cost of error. A carefully constructed master database should be capable of providing the detailed current data needed for operational decisions as well as the longer-range strategic data for top management decisions.

Table 5-8 Effect of Management Level on Decisions

Management Level	Number of Decisions	Cost of Making Poor Decisions	Information Needs
Top	Least	Highest	Strategic
Middle	Intermediate	Intermediate	Implementation
First-line	Most	Lowest	Operational

Expert Systems

As part of the field of **artificial intelligence** (AI), a type of computer model has been developed with the purpose of making available to average or neophyte practitioners in many fields the skill and know-how of experts in the field. These *expert systems* are created by reviewing step by step with the experts the reasoning methods they use in a particular application and reducing these to an *inference engine* that, combined with a *knowledge base* of facts and rules and a *user interface*, may be consulted by someone newer to the field who wants guidance. These knowledge-based applications of artificial intelligence have enhanced productivity in business, science, engineering, and the military. With advances in the last decade, today's expert systems clients can choose from dozens of commercial software packages with easy-to-use interfaces.

IMPLEMENTATION

Decisions, no matter how well conceived, are of little value until they are put to use—that is, until they are implemented. Koestenbaum puts it well:

> Leadership is to know that decisions are merely the start, not the end. Next comes the higher-level decision to sustain and to implement the original decision, and that requires courage.
>
> Courage is the willingness to submerge oneself in the loneliness, the anxiety, and the guilt of a decision maker. Courage is the decision, and a decision is, to have faith in the crisis of the soul that comes with every significant decision. The faith is that on the other end one finds in oneself character and the exhilaration of having become a strong, centered, and grounded human being.

DISCUSSION QUESTIONS

5-1. Give some examples of each of the three "occasions for decision" cited by Chester Barnard. Explain in your own words why Barnard thought the third category was most important.

5-2. (**a**) Explain the difference between "optimizing" and "sufficing" in making decisions, and (**b**) distinguish between routine and nonroutine decisions.

5-3. Use a concrete example showing the five-step process by which management science uses a simulation model to solve real-world problems.

5-4. Research how large and small stores are using Internet of Things (IoT) to reduce queues. Are these experiments being successful? Why or why not?

5-5. Describe an example from an organization you know or have read about where a common database is used for a number of different purposes. Also, can you describe an example where a common database is not used for a number of different purposes?

PROBLEMS

5-1. You operate a small wooden toy company making two products: alphabet blocks and wooden trucks. Your profit is $30.00 per box of blocks and $40.00 per box of trucks. Producing a box of blocks requires one hour of woodworking and two hours of painting; producing a box of trucks takes three hours of woodworking, but only one hour of painting. You employ three woodworkers and two painters, each working 40 hours a week. How many boxes of blocks (B) and trucks (T) should you make each week to maximize profit? Solve graphically as a linear program and confirm analytically.

5-2. A commercial orchard grows, picks, and packs apples and pears. A peck (quarter bushel) of apples takes four minutes to pick and five minutes to pack; a peck of pears takes five minutes to pick and four minutes to pack. Only one picker and one packer are available. How many pecks each of apples and pears should be picked and packed every hour (60 minutes) if the profit is $3.00 per peck for apples and $2.00 per peck for pears? Solve graphically as a linear program and confirm analytically.

5-3. Solve the drilling problem (Table 5-3) by using a decision tree.

5-4. You must decide whether to buy new machinery to produce product X or to modify existing machinery. You believe the probability of a prosperous economy next year is 0.6 and of a recession is 0.4. Prepare a decision tree, and use it to recommend the best course of action. The applicable payoff table of profits ($+$) and losses ($-$) is:

	N_1: Prosperity ($)	N_2: Recession ($)
A_1 (Buy new)	+950,000	−200,000
A_2 (Modify)	+700,000	+300,000

5-5. If you have no idea of the economic probabilities p_j in Question 5-4, what would be your decision based on uncertainty using (**a**) maximax, (**b**) maximin, (**c**) equally likely, and (**d**) minimax regret assumptions?

5-6. You are considering three investment alternatives for some spare cash: Old Reliable Corporation stock (A_1), Fly-By-Nite Air Cargo Company stock (A_2), and a federally insured savings certificate (A_3). You expect the economy will either "boom" (N_1) or "bust" (N_2), and you estimate that a boom is more likely ($p_1 = 0.6$) than a bust ($p_2 = 0.4$). Outcomes for the three alternatives are expected to be (1) $2,000 in boom or $500 in bust for Old Reliable Corporation; (2) $6,000 in boom, but $5,000 (loss) in bust for Fly-By-Nite; and (3) $1,200 for the certificate in either case. Set up a payoff table (decision matrix) for this problem, and show which alternative maximizes expected value.

5-7. If you have no idea of the economic probabilities p_j in Question 5-6, what would be your decision based on uncertainty using (**a**) maximax, (**b**) maximin, (**c**) equally likely, and (**d**) minimax regret assumptions?

5-8. Your company has proposed to produce a component for an automobile plant, but it will not have a decision from that plant for six months. You estimate the possible future states and their probabilities as follows: Receive full contract (N_1, with probability $p_1 = 0.3$); receive partial contract (N_2, $p_2 = 0.2$); and lose award (no contract) (N_3, $p_3 = 0.5$); Any tooling you use on the contract must be ordered now. If your alternatives and their outcomes (in thousands of dollars) are as shown in the following table, what should be your decision?

	N_1	N_2	N_3
A_1 (Full tooling)	+800	+400	−400
A_2 (Minimum tooling)	+500	+150	−100
A_3 (No tooling)	−400	−100	0

SOURCES

Amos, John M. and Bernard R. Sarchet, *Management for Engineers* (Englewood Cliffs, NJ: Prentice-Hall, Inc., 1981), p. 51.

Barnard, Chester I., *The Functions of the Executive* (Cambridge, MA: Harvard University Press, 1938), p. 190–191.

Begley, Sharon with Grant, David, "Games Scholars Play," *Newsweek*, September 6, 1982, p. 72.

de Bono, Edward, *Six Thinking Hats* (New York: Little, Brown, & Company, 1985).

Forrester, Jay W., *Urban Dynamics* (Cambridge, MA: The MIT Press, 1969).

Grammas, Gus W., Lewin, Greg, and DuMont Bays, Suzanne P., "Decision Support, Feedback, and Control," in John E. Ullmann, ed., *Handbook of Engineering Management* (New York: John Wiley & Sons, Inc., 1986), Chapter 11.

Hicks, Philip E., *Introduction to Industrial Engineering and Management Science* (New York: McGraw-Hill Book Company, 1977), p. 42.

Koestenbaum, Peter, *The Heart of Business: Ethics, Power, and Philosophy* (Dallas, TX: Saybrook, 1987), p. 352.

Pringle, Charles D., Jennings, Daniel F., and Longnecker, Justin G., *Managing Organizations: Functions and Behaviors* (Columbus, OH: Merrill Publishing Company, 1988), pp. 131, 154.

Simon, Herbert A., *Administrative Behavior: A Study of Decision-Making Processes in Administrative Organization*, 3rd ed. (New York: Macmillan Publishing Company, 1976), p. 80.

Young, Edmund, example taken from supplemental class notes used in teaching from the manuscript of this text, September 1988.

6

Organizing

PREVIEW

After the management functions of leading and planning, the next to be presented is organizing. This chapter begins by distinguishing between the legal forms of organization: proprietorship, partnership, and corporation. The discussion then moves to the organizing process and the various logics of subdivision, or *departmentalization*. Effective spans of control are discussed as well as the nature of line, staff, and service relationships. The effect of technology on organization structure is described, and finally, current trends in organizational forms and teams are introduced.

Teams are an important part of the workforce today, and they are created either within the planning function or the organizing function, or in project management. Often, there are impromptu teams that are formed by employees spontaneously. Today many teams are virtual and they work across space, time, and organizational boundaries with links strengthened by various communication technologies.

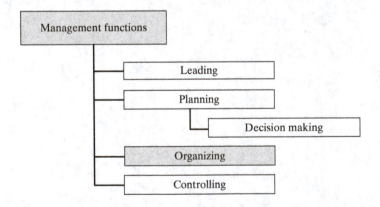

LEARNING OBJECTIVES

When you have finished studying this chapter, you should be able to do the following:

- Analyze the different forms of an organization.
- Explain different organizational structures.
- Describe the differences in line and staff relationships.
- Describe the use and value of teams.

NATURE OF ORGANIZING

Legal Forms of Organization

An organization is a group of individuals who work together toward a common goal. Here we will compare the types of legal entities into which businesses can be organized. These include the sole proprietorship, the partnership, the corporation, and the cooperative. We will then examine other aspects of organizations.

The **sole proprietorship** is a business owned and operated by one person. It is simple to organize and to shut down, has few legal restrictions, and the owner is free to make all decisions. Profit from it is taxed only once—on the Schedule C (Profit or Loss from Business or Profession) attachment to the owner's individual income tax form. However, the owner faces unlimited liability for the debts of the business, they may find it difficult to raise capital to fund growth of the business, and the duration of the business is limited to the life of the proprietor.

The **partnership** is an "association of two or more partners to carry on as co-owners of a business for profit" (Uniform Partnership Act). The partnership is almost as easy to organize as a sole proprietorship and has relatively few legal restrictions. Partnerships permit pooling the managerial skills and judgments and the financial strengths of several people who have a direct financial interest in the enterprise, but suffer the disadvantages of divided decision-making authority and potential damage to the business when partners disagree. Although a partnership files a tax return to allocate partnership profit (or loss), it does not pay taxes—the partners do so on their individual tax forms, whether they actually receive the profit or leave it in the enterprise to grow further. Normally, partners have unlimited liability for partnership debts. In a *limited partnership*, there must be at least one *general partner* with unlimited liability, but the rest may be *limited partners*, who are financially liable only to the extent of their investment in the venture.

A **limited liability company (LLC)** is a newer business structure allowed by state statute. Owners, called members, have limited personal liability for the debts and actions of the LLC. There is no maximum number of members and most states permit single member LLCs. LLCs are similar to a partnership, providing management flexibility and the benefit of pass-through taxation and are generally easy to form and dissolve. Income is taxed only once.

Corporations are legal entities owned by shareholders, who in general have no liability beyond loss of the value of their stock. Corporations have perpetual life (as long as they submit an annual report to the state in which they are chartered), and can sell stock to raise money or transfer ownership. It is more difficult and expensive to organize a corporation, but the main disadvantage is that corporate income is taxed twice: once as corporate income tax the year the profit is made, and again as personal income tax when the after-tax profit is distributed as dividends. Also, corporations are subject to many state and federal controls not affecting other forms of business. Under certain conditions, corporations with no more than 35 shareholders, all U.S. residents, may elect to be treated as "Subchapter S" corporations and avoid double taxation.

Cooperatives (co-ops) are a special type of organization owned by users or customers, to whom earnings are usually distributed tax-free in proportion to patronage. Recreational Equipment Inc. (REI) is one of the United States' most well-known co-ops. More commonly, co-ops are used in rural areas to handle distribution of farm products and electricity. In these organizations, each member of this service buys a share initially for a few dollars, and they can cast one vote to elect the board members who manage the cooperative.

While sole proprietorships are the most common form of business organization in sheer numbers, most large organizations are corporations.

Organizing Defined

Weihrich and Koontz believe that people "will work together most effectively if they know the parts they are to play in any team operation and how their roles relate to one another....Designing and maintaining these systems of roles is basically the managerial function of organizing." They continue:

> For an *organizational role* to exist and be meaningful to people, it must incorporate (1) verifiable objectives, which...are a major part of planning; (2) a clear idea of the major duties or activities involved; and (3) an understood area of discretion or authority, so that the person filling the role knows what he or she can do to accomplish goals. In addition, to make a role work out effectively, provision should be made for supplying needed information and other tools necessary for performance in that role.

> It is in this sense that we think of *organizing* as (1) the identification and classification of required activities, (2) the grouping of activities necessary to attain objectives, (3) the assignment of each grouping to a manager with the authority (delegation) necessary to supervise it, and (4) the provision for coordination horizontally (on the same or similar organizational level) and vertically (for example, corporate headquarters, division, and department) in the organization structure.

Organizing by Key Activities

Effective organizing must first consider the basic mission and long-range objectives established for the organization and the strategy conceived to accomplish them. Peter Drucker recommends first identifying the *key activities*, which he terms the "load-bearing parts of the structure." He poses three questions to help identify the key activities:

1. In what area is excellence required to achieve the company's objectives?
2. In what areas would lack of performance endanger the results, if not the survival, of the enterprise?
3. What are the *values* that are truly important to us in this company?

Once the key activities have been established, Drucker suggests "two additional pieces of work: an analysis of decisions and an analysis of relations." In *decision analysis* one must first identify what decisions are needed to attain effectiveness in key activities. Then the nature of these decisions is established in terms of their *futurity* (the period in the future to which they commit the company), the *impact* they have on other functions, their *frequency* (recurrent decisions can be made at lower levels once policies for them have been established), and the extent to which they involve ethical, social, and political considerations. *Relations analysis*, on the other hand, asks with whom the person in charge of an activity will have to work, and it seeks to assure "that the crucial relations, that is, the relationship on which depend its success and the effectiveness of its contribution, should be easy, accessible, and central to the unit."

In the 1990s, more and more organizations were restructured into teams that include the specialists needed to carry through a project or solve a problem, and that are delegated the authority (*empowered*) to make the necessary decisions; and that continues in the twenty-first century. In the modern concept of *concurrent engineering* (discussed in Chapter 10), teams of design engineers, marketing people, and production specialists work together to launch new products earlier that better meet customer needs.

Over the past decade, many U.S. companies have shifted their organizations to a greater blend of full-time and part-time employees and temporary or contract employees. A change evident even in higher education, where an ever greater percentage of faculty are in non-tenured or adjunct roles. This approach that has given birth to the free-agent or gig economy, where many workers are not directly employed by any one company but cobble together multiple sources of contract work. In 2018, Gallup estimated that 36 percent of U.S. workers were engaged in this sector of the economy, a shocking 57 million people. The implications of these shifts for a future engineering career are discussed in Chapter 17.

TRADITIONAL ORGANIZATION THEORY

Patterns of Departmentation

Organizations are divided into smaller units by using a number of different approaches. A **hierarchical organization** is an organizational structure where every entity in the organization, except one, is subordinate to a single other entity. This arrangement is a form of a hierarchy. This is the dominant mode of organization among large organizations; most corporations, governments, and organized religions are **hierarchical** organizations. This arrangement of individuals within a corporation is generally according to power, status, and job function. Figure 6-1 illustrates two of the more common methods. These approaches are used in later discussions to help us "grow" a company you founded. You enjoy woodworking and have become very talented at making unique coffee tables. They are admired by your friends and neighbors, who buy some, and then you find several local stores who want to carry them. You now are an *entrepreneur* and have a business. As demand increases, you need some help in the shop, and you hire several local people (Tom, Dick, and Mary) who will, naturally, take direction from you (Figure 6-1a).

As you grow, you find yourself away from the plant (now moved to a local industrial park) for extended periods, selling your product and arranging financing. You appoint the most experienced worker as supervisor, and later as production manager. You hire salespeople to help sell your product and, as they increase in number, appoint one as sales manager. A local certified public accountant agrees to work half-time as your finance manager, and an engineering student moonlights as your designer. You have now established a pattern of **functional departmentation**, which is the first logic of subdivision for most new organizations, and which is present at some level in almost any organization. Functional subdivision does

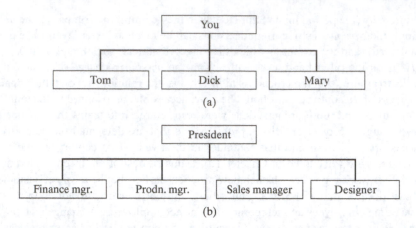

Figure 6-1   Methods of departmentalization: (a) Basic organization. (b) Functional departmentation.

not need to be to a single level (as in Figure 6-1b). Marketing is often divided into sales, advertising, and market research. Production may be broken into component production, assembly, and finishing.

As your business grows, you may also become interested in producing plastic furniture for outdoor settings. You know that production methods for plastic furniture are very different from those for wooden items, and you organize separate production shops under separate supervisors to produce the two products. Next, you discover that your patio furniture appeals to a different market, and you need a different group of salespeople. Then you find that the sales force dealing with plastic furniture needs much closer contact with your plastic production foreman than they do with salespeople selling wood furniture to the consumer, but that the chain of command through the general sales manager, then you, and finally the overall production manager makes decision making slow and difficult. You may now be ready to reorganize by *product* as in Figure 6-2a.

The wood and plastic furniture divisions will begin with their own manufacturing and marketing functions, and later you may add accounting and personnel functions to each division. Because obtaining bank loans, selling stock, and other financial activities are best handled centrally, you will need consistent personnel policies in both divisions, and you need some top-level advice on new markets and new technical advances; thus, you will need to organize a *staff* at the corporate level in addition to your product divisions.

Next, you may find that accommodating regional differences is the key to effective management, and you may then create regional (geographic) divisions. A company that builds housing developments may find, for example, that regional differences in housing styles, construction codes, marketing media, and methods of mortgage financing are very important, and you may set up separate geographic divisions, each responsible for construction and marketing in its own region. Geographic subdivision is more common at lower levels; sales forces are commonly divided by region, for example, for more efficient and more personal customer contact.

Sometimes the **type of customer** is a more important consideration than location, and departmentation by customer is advised. For example, creating weapons systems for the U.S. Department of Defense often requires state-of-the-art technology and a special understanding of military procurement and product

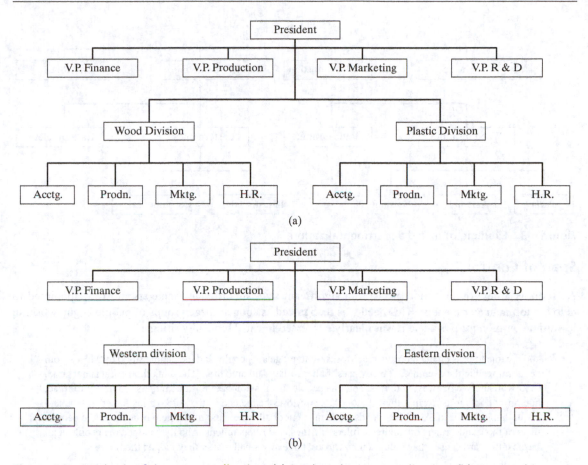

Figure 6-2 Methods of departmentalization: (a) Product departmentalization. (b) Geographic departmentalization.

use. On the other hand, products that will be used by other industries in producing their own goods need to be cost-effective and rugged; consumer goods need to be attractive in appearance and price. Many firms successful in producing military goods have tried marketing to the consumer with poor success, because it requires a different mental outlook and frequently a separate organization devoted to that market. Even when the product is the same or similar (washing machines, for example), separate sales forces may be desirable for regional sales to consumers through local distributors and for large-volume national sales to major store chains or to the federal government for military housing use.

Where manufacturing or service is carried on around the clock, operating personnel may be grouped by *shift* or *time*. Subdivision by sheer *numbers*, as in the example that follows, is indicated only when a large number of people must perform very similar and routine tasks, and this is becoming increasingly less common. As one might expect, enterprises may combine several or all of these methods in designing their organization. In Figure 6-3, functional subdivision is at the top level with product and process subdivision in manufacturing, and geographic and customer departmentation in marketing.

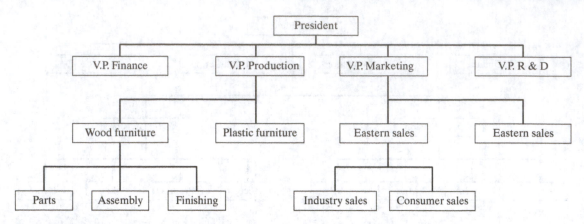

Figure 6-3   Example of mixed departmentalization.

Span of Control

As soon as a new organization grows to a significant size, subordinate managers must be appointed to help the top manager manage. This need was recognized as soon as large groups of people began working toward a common purpose, and it was clearly expressed in early biblical writings:

> And it came to pass on the morrow, that Moses sat to judge the people: and the people stood by Moses from the morning unto the evening.... And Moses' father-in-law said unto him "The thing thou doest is not good. Thou wilt surely wilt away, both thou and this people that is with thee: for this thing is too heavy for thee; thou art not able to perform it thyself alone."... So Moses hearkened to the voice of his father-in-law, and did all that he had said. And Moses chose able men out of all Israel, and made them heads over the people, rulers of thousands, rulers of hundreds, rulers of fifties, and rulers of tens. And they judged the people at all seasons: the hard causes they brought unto Moses, but every small matter they judged themselves

The question is not *whether* intermediate managers are needed, but how many. This depends on the number of people reporting directly to each manager, referred to as the *span of management* or **span of control**. For example, if a simple hierarchical organization with only 64 workers (nonmanagers) and a chief executive officer has a span of only two subordinates per manager at every level, it will need 62 managers at five intermediate (middle) levels of management between worker and CEO; with a span of four (Figure 6-4a), it will have 20 managers at two levels; with a span of eight (Figure 6-4b), it will have only a single level of eight managers.

Many armies are organized on a span of control of about four: four squads per platoon, four platoons per company, four companies per battalion, and so on. Wren reports that the span of 10 (rulers of thousands, hundreds, and tens) was adopted independently by the Egyptians, by the Roman legions (with their *centurions* commanding 100 soldiers), by the Tatars (Tartars) of Mongolia, and by the Incas of what is now Peru and Chile—peoples who had little in common other than 10 fingers to "count off."

Narrow spans of control (tall organizations) are not only expensive because of the cost of having so many managers, but the multiple levels can increase communication and decision time and stifle initiative because of the temptation of a manager with few subordinates to micromanage (interfering in decisions

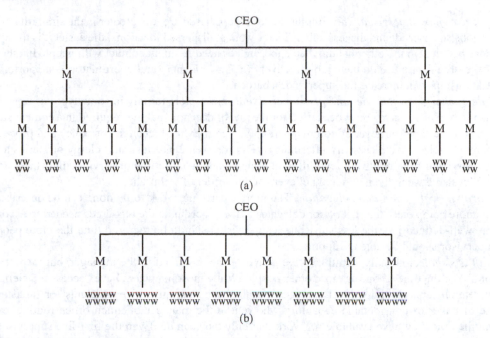

Figure 6-4 Control spans of (a) four and (b) eight compared. M, manager; w, worker.

that should be made at lower levels). Excessively wide spans, on the other hand, can leave managers with inadequate time to supervise the activities for which they are responsible and leave subordinates with inadequate access to their busy supervisor.

Factors Determining Effective Spans. What, then, determines a desirable span? Graicunas, a Lithuanian engineer and a Paris-based management consultant, stated that this depended on the number of *relationships* that existed between manager and subordinates, individually and in various combinations, and among the subordinates themselves. He calculated the number of relationships for a manager with n subordinates as

$$n\left[2^{(n-1)} + n - 1\right] \tag{6-1}$$

so that every subordinate added *more than doubled* the number of relationships the manager had to be concerned with and, Graicunas assumed, the difficulty of the job. However, many of Graicunas's relationships are not significant in a particular application, and effective span of control depends on many factors other than the simple number of subordinates. Studies of effective spans have identified the following conditions as affecting the number of people a manager can effectively supervise:

- *Subordinate training.* The more completely subordinates are trained for their jobs, the fewer demands they place on supervisors.

- *Nature of jobs supervised.* The simpler the tasks supervised are, the greater is the similarity between the jobs supervised; and the less subordinates work at dispersed locations, the easier it is to supervise more people. On the other hand, when subordinates need frequent contact with people in other parts of the organization to do their job effectively (as may planners and coordinators), supporting these relationships can increase the supervisor's burden.
- *Rate of change of activities and personnel.* Events move more rapidly in some types of organization than in others. An army must be staffed for the rate of decision making required in a combat situation (and for the rapid turnover of commanders). On the other hand, changes of policy and procedure in many churches take place only after many years (or even centuries), and clergy or clerics are well educated and have relatively few different assignments in a lifetime of service; thus, there are very few levels between the individual believer and the leader of a church.
- *Clarity of instruction and delegation.* The more clearly the work to be done can be described, and the more completely the supervisor delegates to the subordinate the resources needed to accomplish this well-defined job, the less subsequent supervision should be required (and the more people the supervisor should be able to support).
- *Staff assistance.* Usually, administrative activity is not confined to the manager, but involves some (or all) of the time of one or more other people. While most managers have "access to" clerical and secretarial support, higher-level managers usually have "administrative assistants" or "assistants to" the manager of considerable capability who relieve the manager of much office routine, expanding the time they have available for work that only they can do. Even the first-line supervisor often delegates some of the short-term leadership: the engineering design supervisor may have 20 engineers, but some will be more experienced "lead engineers" who are responsible for the day-to-day activities of younger engineers and technicians in completing a common task. In the military, senior managers commonly have deputies who are fully qualified to act in their absence, almost doubling the effective leadership potential of the office.

Effective management spans also vary by level within organizations. First-line supervisors, who are concerned with their direct subordinates, but not with lower levels, usually have larger spans than do middle managers. Spans of CEOs may vary substantially, depending on the managerial style of the incumbent. Finally, the skill and experience of the manager does, of course, have an effect on the number of people that he or she can supervise effectively.

Recent Trends in Spans.   Generally, recent trends have been to increase the *spans of control*, which ultimately decreases the number of organizational levels within a given company or organization. This shift to large spans of controls is due in part to the **information revolution**. With more automated systems, databases, and ever-increasing methods of communication, decisions can be made efficiently. Line workers and technicians no longer have a small role in a particular process, but have the ability to manage, in large part, the particular process that they are partly responsible for with the latest in technology. This trend to large spans will generally be around 20 to 30 subordinates per span, and the organization should consist of no more than five organizational levels. Key points that will result from larger spans of control are as follows:

1. Significant reduction of administrative costs
2. More effective and efficient organization communication

3. Faster decisions and closer interaction between organizational levels
4. Requirement that all levels of personnel become better trained, informed, and educated
5. Better leadership at all levels

The information technology available today touches every aspect of our lives, especially how business is conducted. The ability to gather information rapidly, process it, make precise and accurate decisions, and disseminate information, as well as increased organizational communications, has paved the way for large span of control. This can only benefit the organization as a whole, requiring better educated, involved, and trained workers, and forcing better leadership, decision making, and involvement from managers.

Line and Staff Functions and Relationships

Traditionally, the "line functions" in an organization were those that accomplished the main mission or objectives of the organization, and these were thought to include production, sales, and finance in the typical manufacturing organization. "Staff functions," on the other hand, were those that helped the line accomplish these objectives by providing some sort of advice or service. A useful distinction may be made between *personal staff*, such as the "assistant to" who does troubleshooting or special assignments for a single manager, and *specialized staff*, who serve the entire organization in an area of special competence. Examples of specialized staff organizations include personnel, procurement, legal counsel, and market research. In today's more complex knowledge-based organizations, the activities of "staff specialists" may be as essential to the ultimate success of the organization as "line workers," and these distinctions have become blurred.

Much more fruitful is examining the type of *relationship* involved in a particular transaction. **Line** relationships are superior–subordinate relationships and can be traced in a "chain of command" from the organization president through a succession of levels of managers to the lowest worker.

Staff relationships are *advisory* in nature. Four types of staff relationships, arranged in order of increasing levels of influence, are (1) providing advice only on request, (2) recommending where the staff office deems appropriate, (3) "consulting authority," in which line managers must consult (but need not obey) staff in their area, and (4) "concurring authority," in which the staff specialist has a veto authority over the line manager.

Functional (specialized) authority is a special type of staff authority over others who are not their line subordinates. It is as binding as line authority, but does not carry the right to discipline for violation. Usually, it controls "how to" accomplish some action falling in the area of responsibility of the staff office, and it is delegated to staff because of the need for uniformity or special expertise. Examples include specification of budget formats by the financial officer and of criteria for documenting research findings or for reducing product liability by the legal counsel.

Service relationships are "facilitative activities" that are centralized for economy of scale, uniformity, or special capability, but are only supportive of the main mission. Examples include custodial, security, and medical services.

A manager may, at different times, exhibit all of these relationships. For example, a human resource manager will exert line authority over direct subordinates in their office, provide staff advice to the chief executive on the need for instituting employee support programs, exercise functional authority by defining how job descriptions must be filled out, and provide a service to the entire organization by maintaining employee records.

Friction between line and staff personnel occurs for many reasons. Staff specialists may have little understanding of the problems and realities of the line organization. Line managers, on the other hand, have little understanding of the expertise of the staff specialist and the need the organization has for it. Each side needs to listen to the other with courtesy and mutual respect for the good of the whole organization. Military officers tend to have assignments alternating between command (line) and staff responsibilities, and they are often sent to "Command and Staff School" or some equivalent in midcareer; as a result, they have a better chance at understanding both sides of this relationship.

Corporate restructuring in recent decades has reduced the size of specialist staff organizations at the corporate and divisional levels. Instead, individual specialists become members of working teams (discussed later in this chapter) that, as a group, are empowered to get the work of the organization accomplished with much less need for approvals *up the chain of command*. As a result, specialists can integrate their knowledge into work as it is being done, avoiding much of the friction, misunderstanding, and wasted or repeated effort of the past.

TECHNOLOGY AND MODERN ORGANIZATION STRUCTURES

The Woodward and Aston Studies

The nature of manufacturing processes and the size of the organization have in the past exerted considerable influence on organizational design. In the 1950s, Joan Woodward and her associates studied the operations of about 100 manufacturing firms in the South Essex region of England, gathering data on manufacturing methods, organization, communication, and performance. She reported some significant differences when she organized these firms into categories of increasing complexity of manufacturing process technology. She classified 80 of these firms into three broad classes (the rest employing combinations of these or remaining unclassified):

1. *Unit*: production of units to customer's orders, prototypes, large equipment in stages, or small batches to customer's orders (sometimes known as job-shop operation)—24 firms. Production runs in this group are too small to justify specialized manufacturing equipment, procedures, or tooling, and production is normally carried out by skilled craftsmen using general-purpose equipment and their past experience.
2. *Mass*: production of large batches, often on an assembly line, and mass production—31 firms. Long production runs justify special production methods, specially designed equipment, and elaborate methods of scheduling and programming. Jobs, on the other hand, tend to be standardized and repetitive, and to use less-skilled workers, whose efforts are regulated by the speed of the assembly line. Automobiles and household appliances often are made using these methods.
3. *Process*: continuous process production systems such as those used in the petroleum and chemical industries—25 firms. These normally involve high capital investment per worker and are highly automated. Skilled workers are needed to monitor and maintain these complex production systems.

Woodward found two characteristics that increased continually as manufacturing complexity increased from unit to mass to process technologies: (1) the number of levels of management, and (2) the span of control *for chief executives*. As shown in Table 6-1, however, in many ways the unit and process technologies at the low and the high ends of the technology scale were similar to each other and different

Table 6-1   Organization of Characteristics Versus Production Technology

Production Technology	Unit	Mass	Process
Number of firms observed	24	31	25
Levels of management (mode)	3	4	6
Span of control—chief executives (median)	4	7	10
Span of control—first-line supervisors (median)	23	49	13
Typical management system	Organic	Mechanistic	Organic
Development of staff activities	Limited	High	Limited
Predicting, scheduling, and control systems	Limited	Extensive	Integral
Communications	Verbal	Written	Verbal
Pleasantness and openness of organizational climate	Greater	Less	Greater

Source: Based on Joan Woodward, *Industrial Organization: Theory and Practice* (Oxford: Oxford University Press, 1965).

from the large batch and mass production technologies in the middle of the scale. For example, the quantity production technologies showed a much higher median number of employees reporting to each first-line supervisor, had the most highly developed line–staff organization with the largest number of staff specialists, employed highly developed production control systems, used the greatest amount of formal written documentation, and had the least pleasant organizational climate. They also favored *mechanistic* management systems, which are like classic bureaucracies: centralized, formalized, standardized jobs.

Firms using unit or process production technologies, on the other hand, were more likely to favor *organic* management systems, which are more decentralized with less rigidly defined jobs, less attention to rank, and a great deal of lateral (as opposed to vertical) communication. They also enjoyed a more pleasant and relaxed organizational climate, more reliance on verbal communication, and fewer people reporting to first-line supervisors. Control systems were less necessary in unit manufacturing and integral with (built into) the continuous processing equipment.

Woodward's observations were on small to medium-sized companies (only 13 of the 100 had more than 1,000 employees, the largest being under 9,000) located in southeastern England, and these were studied in the 1950s. Whereas some later studies have supported Woodward's findings, others have produced conflicting results.

Perhaps the most prominent of the studies dissenting from Woodward's findings was by a group of scholars at the University of Aston, Birmingham, England. They investigated a group of 46 firms in the vicinity of Birmingham with from 240 to more than 25,000 employees. In studying the operations technology of these firms, they used a 10-step *production technology* scale quite similar to Woodward's. They did agree with Woodward that mass production firms had larger spans of control for first-line supervisors, with more staff specialists for control and greater distinctions between line and staff than did unit or process firms. However, they found that the size of the firm (in number of employees) correlated better with other parameters than did the type of technology. Thus the "larger" firms (25,000 employees) were more likely to have high levels of specialization, standardization, formalization, and centralization, regardless of the type of technology. Further work indicates that *both* the technological complexity of Woodward and the organizational size dimension of the Aston group must be considered in effective organization design.

TEAMS

Teams continue to be an integral and growing part of the workforce, and teamwork is essential in modern organizations. One of the principles for management of the modern enterprise is teaming. A **team** is defined as follows: a small number of people who are committed to a common goal, objectives, and approach to this goal that they are mutually accountable to reaching. Teams quite often have complementary skills that are used in the problem solving. Today, all workers must be able to work together in interdisciplinary teams to carry out and coordinate the operations of the enterprise. Since engineers tend to spend most of their time working with interdisciplinary teams, being effective at team work is a critical skill.

According to Katzenbach and Smith, there are several effective approaches to building team performance:

- Establish urgency and direction. All team members need to believe the team has clear objectives.
- Select members based on skill and skill potential, not personalities.
- Pay particular attention to first meetings and actions.
- Set some clear rules of behavior.
- Set and seize upon a few immediate performance-oriented tasks and goals.
- Challenge the group regularly with fresh facts and information.
- Spend time together.
- Use positive feedback, recognition, and reward.

The functioning of organizations in a global environment has led to the formation of **virtual teams**. These virtual teams, unlike traditional teams, must accomplish their objectives by working across distance and time by using technology to facilitate collaboration.

There are two primary categories of variables that make virtual teams more complex. These are (1) the crossing of boundaries related to time, distance, and organization, and (2) communication and collaboration, using technology. For the reasons stated, virtual teams are far more dependent on having a clear purpose than face-to-face teams. Purpose defines why a particular group works together. As important as positive relationships and high trust are in all teams they are even more important in virtual teams. The lack of daily face-to-face time, which normally offers opportunities to quickly clear things up, can heighten misunderstandings.

Virtual teams in industry work across space, time, and organizational boundaries with links strengthened by webs of communication technologies. What is new is the array of interactive technologies at their disposal. The basic elements of the virtual team process are communication, planning, and managing or implementing. See "Virtual Teams: Experiences Connecting with Technology" for more detailed discussion.

The proposal teams gathered together by aerospace companies to respond to a major military request for proposal (RFP) provide an excellent example of the "disposable organization"; as many as 1,000 people, often from several cooperating companies, may come together for one to three months for this specific purpose and then disband back to their original organizations or other teams. Project management organizations and their operations are of special importance to engineers, and two chapters in this book (14 and 15) are dedicated to project management. Particular attention is placed in Chapter 15 on *matrix management* organizations, which are frequently used in project management.

There are other modern examples of temporary or "team" organizational structures; Cleland and Kerzner provide descriptions of production teams, worker-management teams, product-design teams, quality teams, project management teams, crisis-management teams, and task forces. Product-design teams for design review and configuration management are discussed in Chapter 10, the function of quality teams in Chapter 12, and project management teams in Chapters 14 and 15.

Virtual Teams: Experiences Connecting with Technology

Since the 1970s, studies have shown that a majority of the message in inter-personal communication comes through something other than the words that were said, including tone of voice and body language (see details in Chapter 17). For this reason, there are certain discussions that will always be most effectively held in a face-to-face manner. However, between the costs (a business class flight from the United States to Europe is generally over $5,000 with a flight to Australia double that) and the time lost to travel, it is not always practical or even advisable to have meetings face-to-face. Since the turn of the century, the growth of global companies (discussed in Chapter 18) and virtual offices have steadily increased the need for virtual teams. Fortunately, the technology to support these meetings has come a long way from the days of echoey speaker phones and dedicated video conference rooms.

At this time, nearly every student has enjoyed the functionality of FaceTime or similar technologies to talk to far away friends and family. While this technology is amazing for one-on-one communication, it is not particularly robust for multi-person meetings where documents need to be shared. This is where technologies like WebEx, GoToMeeting, appear.in, and many others shine. In these systems, not only can multiple people join simultaneously with video and audio from their computer or mobile device, all members of the meeting can see the documents being discussed and edited live.

Why does this matter? First, video calls encourage members of the meeting to be engaged in the discussion—we can all see you if you start working on something else! Second, video allows a large portion of body language communications to be received by all members. In the author's experience, this is especially important with participants from other cultures, whose voice intonation is different than our own, but whose facial expressions generally still show the same frustrations or satisfaction. Finally, by everyone being able to view documents in real time, it keeps the group all on the same page and makes the meeting more effective and efficient.

Source: © yupiramos/123RF

Impact of the Information Revolution

The main feature of the information revolution is the growing economic, social, and technological role of information. But it is not "information" that fuels this impact. As Drucker stated at the turn of the century:

> It is not the effect of computers and data processing on decision-making, policymaking, or strategy. It is something that practically no one foresaw or, indeed, even talked about ten or fifteen years ago: *e-commerce*— that is, the explosive emergence of the Internet as a major, perhaps eventually *the* major, worldwide distribution channel for goods, for services, and, surprisingly, for managerial and professional jobs. This is profoundly changing economies, markets, and industry structures; products and services and their flow; consumer segmentation, consumer values, and consumer behavior; jobs and labor markets. But the impact may be even greater on societies and politics and, above all, on the way we see the world and ourselves in it.

A generation later, these words could not be more true. Modern computer and communications technologies continue to rapidly change our organizations in ways that we do not yet fully understand. Not since the introduction of the electric motor into industry has there been an innovation so universal in its scope. Lund and Hansen believe that the time horizons between design and production are collapsing because the design database, once created, is available for design analysis and evaluation, creating prototypes, control of ultimate production, and even planning and control of quality inspection. Quicker start-ups and product changes reduce the optimal size of production runs, reducing the resources tied up in in-process and finished-goods inventory. Product life cycles are being shortened in many industries. The successful firms will be those evidencing flexibility, adaptability, and a quick response to changes in the market, all characteristics that information technologies can enable.

Information technology is making many changes in the organization structure of companies as well. It is making long distance communication as simple as pressing a button. Information can be passed quickly from management to the workers or from team to team. Supervisors do not need to be near the workers under them in order to pass on orders or to check up on production. Supervisors do not even need to speak the same language as those they work with because the information can be translated quickly by using technology. Because of this rapid change in technology, workers and managers need to be more skilled in its use. Organizations may find themselves in a state of limbo as they try to adjust to the rapidly changing world created by this revolution, but if they do not adjust, they will certainly find themselves outpaced by a competitor who could make the adjustment.

Lund and Hansen also "see a diminishing of the size and importance of centralized corporate headquarters" as operating decisions are pushed to lower levels (and simpler ones are automated). Executives will be able to draw figures from central databases as needed to analyze a given situation, reducing the need for the intermediaries that gather and analyze data today. Tom Peters predicts an even greater impact—the "complete destruction of hierarchy as we have known it...the biggest change in organization in thousands of years" because of the access all employees will now have to all the company's information.

As computer-based automation replaces conventional processes, it will sharply reduce the number of workers (and their supervisors) needed per unit of output. Factory workers will be *monitoring* the production process rather than forming part of it, and they will need at least the following skills:

1. Visualization (ability to manipulate mental patterns)
2. Conceptual thinking (or abstract reasoning)

3. Understanding of process phenomena (machine fundamentals and machine/material interactions)
4. Statistical inference (appreciation of trends, limits, and the meaning of data)
5. Oral and visual communication
6. Attentiveness
7. Individual responsibility

If this view is correct, there will be little future in industry for the uneducated employee or mechanistic leadership styles.

Peter Drucker provided, in 1988, an excellent forecast of today's emerging organizational styles:

> ...[In the new information-based organization] it becomes clear that both the number of management levels and the number of managers can be sharply cut. The reason is straightforward: it turns out that whole layers of management neither make decisions nor lead. Instead, their main, if not their only, function is to serve as "relays"—human boosters for the faint, unfocused signals that pass for communication in the traditional pre-information organization.

Drucker saw four special problems for management as particularly critical in the new information-based organization:

1. Developing rewards, recognition, and career opportunities for specialists [since opportunities for promotion into the management hierarchy will drastically decrease]
2. Creating unified vision in an organization of specialists
3. Devising the management structure for an organization of task forces
4. Ensuring the supply, preparation, and testing of top management people [since the progression of middle management levels that provided this training in the past have diminished]

By late 1994, this revolution in American organizations was well under way. A special *Business Week* report, "Rethinking Work," discusses some of the salient aspects:

1. Virtual disappearance of job security, replaced by shared responsibility: employers have an obligation to provide opportunity for self-improvement; employees have to take charge of their own careers
2. Increasing demand for well-paid professional and technical workers; decreasing demand for operators, laborers, craftsmen, clerical staff, and farm workers
3. Reduced real wages (purchasing power down from 1973 to 1993 by 23 percent for high school dropouts, 15 percent for high school graduates, 8 percent for college graduates, and 5 percent for those with two years graduate work), increasing the need for the two-income family
4. Continuing "downsizing" of staff, with the surviving personnel working longer hours under higher stress
5. Increases in part-time, contract, and self-employed workers who are paid only when needed without the fringe benefits that often add 40 percent to payroll cost

These trends have only continued under a new name with the transition to the gig economy. Thomas Friedman, in the book *The World Is Flat*, describes how an organization structure will change as a result of the information revolution. First, organizations will start to become flatter as the ability to move information from the top to the front line and back becomes quicker and cheaper. Layers of management that were in an organization to accelerate the flow of information in the past will actually slow information and decision making today and will be removed. Next, there will be less and less support staff as information technology (IT) systems allow users to manage the process on their own. For example, electronic calendars now make it easy to set up a meeting with many participants across numerous time zones. Next, business structures will migrate more and more toward partnership/outsource models for non-core-related activities. A few examples of areas that are being outsourced are payroll, accounts receivable, and IT. This outsourcing allows companies to become more focused on what they do best. There will still be management in charge of all the functions to facilitate the partnership and contract, but the former structure underneath those managers will be gone. Finally, decision making will be pushed out of central offices down to front-line employees as the rate of change around new products, partnering, and customer focus accelerates. A more complete discussion of the implications of our flattening world is included in Chapter 18.

Even with these changes, American industry in the twenty-first century will remain internationally competitive, offering continuing opportunity for those who have the skills and training needed. Every company has to become transnational in the way it is run (see Chapter 18 for discussion of global organizations). However, individuals must take personal responsibility for their own careers, to assure they continue to acquire the new knowledge and skills they will need. Some of the ways the engineering professional can do this are discussed in Chapter 17.

In the 1990s, more and more organizations were restructured into teams that include the specialists needed to carry through a project or solve a problem, and that are delegated the authority (*empowered*) to make the necessary decisions. That trend has only accelerated in the twenty-first century. In the modern concept of *concurrent engineering* (discussed in Chapter 10), teams of design engineers, marketing people, and production specialists work together to launch new products earlier.

DISCUSSION QUESTIONS

6-1. You have begun a small, but growing business. What advantages and disadvantages should you consider before changing it from a sole proprietorship to a corporation? To an LLC?

6-2. Select four different businesses, and identify some of their "key activities" by posing Drucker's three questions.

6-3. Under what conditions might each of the following logics of departmentation be desirable: functional, geographic, customer, product, and process?

6-4. Using company websites, identify at least one large company that uses each of the three different departmentation approaches from question 6-3. Does this approach appear to make sense for that company? Why or why not?

6-5. The chapter discusses how teams have become more virtual in recent years. Do you see this trend continuing? Why or why not?

6-6. Distinguish between functional (specialized) staff authority and traditional line authority.

6-7. According to the Woodward and Aston studies, what conditions lead to a formalized, standardized organizational environment?

6-8. Describe from your experience (or reading) a temporary organization or task force formed to accomplish some specific purpose. How was it formed, organized, and ultimately disbanded?

6-9. If the development of the information-based organization continues to have the effect on management predicted by Drucker, what will be the impact on career expectations of engineers and other specialist professionals?

6-10. Choose an enterprise with which you are familiar that has undergone a significant recent reorganization. Compare the new and old organizations with regard to (**a**) size and influence of specialized staff, (**b**) management levels, (**c**) typical spans of control, and (**d**) responsibility delegated to nonmanagerial professionals. What other changes occurred in the reorganization?

6-11. Explain in your own words how you think virtual teams and communication technology will impact your engineering career.

SOURCES

Cleland, David I. and Kerzner, Harold, *Engineering Team Management* (New York: Van Nostrand Reinhold Company, Inc., 1986), pp. 6–17.

Drucker, Peter F., "The Coming of the New Organization," *Harvard Business Review*, January–February 1988, pp. 45–46.

Drucker, Peter F., *Management: Tasks, Responsibilities, Practices* (New York: Harper & Row, Publishers, Inc., 1974), pp. 530–549.

Duarte, Deborah L. and Snyder, Nancy, *Mastering Virtual Teams: Strategies, Tools, and Techniques That Succeed* (San Francisco: Jossey-Bass, 1999).

Friedman, Thomas L., *The World Is Flat: A Brief History of the Twenty-first Century* (New York: Farrar, Straus and Giroux, 2005).

Gallup Organization, What Workplace Leaders Can Learn From the Real Gig Economy. 2018.

Hagen, Mark R., "Teams Expand into Cyberspace," *Quality Progress*, June 1999, 32(6), pp. 90–93.

Katzenbach, Jon R. and Smith, Douglas K., "The Discipline of Teams," *Harvard Business Review*, March–April 1993.

King James Bible, Exodus, Chapter 18.

Lipnack, Jessica and Stamps, Jeffrey, *Virtual Teams* (New York: John Wiley & Sons, 1997).

Lund, Robert T. and Hansen, John A., *Keeping America at Work: Strategies for Employing the New Technologies* (New York: John Wiley & Sons, Inc., copyright © 1986), pp. 64–93.

Peters, Thomas J., commentary on the *Nightly Business Report* (public television program), June 8, 1988.

"Rethinking Work: The New World of Work," *Business Week*, October 17, 1994, pp. 76–87.

Toffler, Alvin, *Future Shock* (New York: Random House, Inc., 1970; paperback, New York: Bantam Books, 1971), pp. 125, 132–133 (Bantam).

Weihrich, Heinz and Koontz, Harold, *Management: A Global Perspective*, 10th ed. (New York: McGraw-Hill Book Company, 1993), p. 244.

Whyte, William H., Jr., *The Organization Man* (Garden City, NY: Doubleday & Company, Inc., 1956).

Woodward, Joan, *Industrial Organization: Theory and Practice* (Oxford: Oxford University Press, 1965).

Wren, Daniel A., *The Evolution of Management Thought*, 3rd ed. (New York: John Wiley & Sons, Inc., 1987), pp. 15–21.

7

Some Human Aspects of Organizing

PREVIEW

This second chapter devoted to the management function of organizing begins by considering the steps in staffing technical organizations. The first step is human resource planning, in which the type and number of people needed in the next six months to a year is established. Next is the process of personnel selection. It begins with the job application process from the employee viewpoint including discussion of effective résumés and cover letters, the employment application, campus interviews for engineering graduates, reference checks, plant visits, and the job offer. This section ends with the employer viewpoint with the process of orienting and training the new employee and appraising his or her performance.

In the second major section of the chapter, the nature of authority, the sources of authority, and power are considered. Next the system of assignment, delegation, and accountability are studied. The chapter closes with a discussion of committees and meetings: reasons for using them, problems they present, and methods of making them effective.

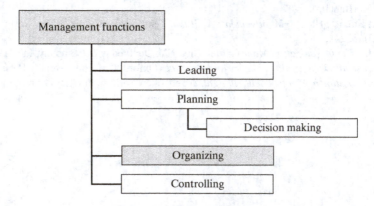

LEARNING OBJECTIVES

When you have finished studying this chapter, you should be able to do the following:

- Describe the steps in staffing technical organizations.
- Discuss the process of an employment application.
- Discuss the nature of authority and power.
- Explain the concepts of delegation.
- Describe the structure of committees.

STAFFING TECHNICAL ORGANIZATIONS

The management function of staffing involves finding, attracting, and keeping personnel of the quality and quantity needed to meet the organization's goals. Staffing is included in some management textbooks as part of the organization function and in others as a separate function, but the same steps are required. Effective staffing begins with identifying the nature and number of people needed, continues through planning how to get them with recruitment, and then selecting the best applicants. Once employees are hired, staffing continues with orienting and training them in the ways of the organization, evaluating their performance, and providing adequate opportunities for growth.

Human Resource Planning

Hiring Technical Professionals. Hiring a laborer when jobs are scarce may involve just a call to the nearest union hall, but hiring quantities of engineers and other professionals, whether new college graduates or experienced professionals with specific skills, requires advanced planning because these hiring processes typically last six months to more than a year. Planning for the overall personnel (or human resource) needs of a large high technology firm can therefore be quite complex. Following is the process used in one division of a large aerospace firm to come up with the required quantity and quality of technical personnel.

1. Document the number of technical personnel of each classification presently on hand.
2. Estimate the number of professionals of each type needed in the near future (six months to a year) to meet firm contracts and likely potential business.
3. Estimate the expected attrition in the current staff, including (a) resignations as a function of the labor market; (b) transfers out to other divisions and promotion to higher positions; and (c) retirements, deaths, and leaves of absence.
4. Establish the need for increased personnel as

$$increase\,(4) = need\,(2) - personnel\ on\ hand\,(1) + attrition\,(3).$$

Subdivide this increase (4) into (5) new college hires, (6) experienced professionals, (7) technician support, and (8) other sources.

5. Each 100 new college hires may require making 200 offers, as a result of 400 candidates visiting your plant or division, stemming from 600 campus interviews. The campus interviews, in turn, might require scheduling trips to 20 campuses to interview 10 students in each of three interview days. (The factors quoted here will vary with the economy, industry, and employer.)

6. Develop a hiring plan to acquire experienced personnel by using national and local hiring, employment agencies and "headhunters," career centers, and employee referrals.

7. Develop a plan to acquire needed technicians and technologists from two- and four-year technical institutes, B.A. and B.S. graduates in physics and math, discharged military technicians, advertisements, state and commercial employment services, and employee referrals.

8. Needs that cannot be met by sources (5), (6), and (7), especially those of too short a duration to justify permanent hiring, can be met by scheduling overtime, hiring contract (temporary) engineers, borrowing engineers from other company divisions, and contracting work to other company divisions or to other companies.

Hiring Managers.   A similar plan must be developed for staffing management positions. Figure 7-1 illustrates what the managerial staffing needs might be from one year to the next for an organization employing about 300 first-line managers, 200 middle managers, and 100 upper-level managers. Most middle- and upper-management positions are shown being filled by promotion, although a few hires at these levels will always be needed where the organization does not already have someone with the right skills. A healthy organization will have a large annual requirement for new first-line supervisors, many of whom will be promoted within the company from employees experienced in a specialty, but often with little experience in management.

Job Requisition/Description.   A manager wishing to fill a professional position normally must fill out a form known variously as a job description or job requisition, which then is approved by higher management and given to the personnel (human resources) department as guidance in its search for candidates who might be considered for the position. Table 7-1 illustrates a typical job requisition.

Table 7-1   Example of Job Description/Requisition

Job Requisition

Title of Position: Research Engineer

Educational Requirements: B.S. in chemical engineering or equivalent

Experience: At least two years in chemical processing, with pilot-plant operation and process development experience preferred.

Description of Duties:

1. Supervise pilot-plant operations for producing new organic intermediates.
2. Identify and recommend process improvements, including conversion of existing batch methods to a continuous process.
3. Work with production engineering to design manufacturing plant.

Will Report To: Manager of Chemical Process Research

Salary Range: $80,000 to $110,000 per year

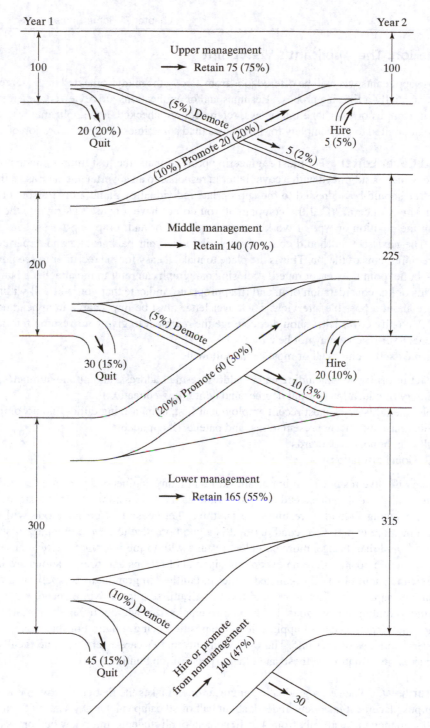

Figure 7-1 Illustration of typical annual management flow.

Finding the Job: The Applicant's Viewpoint

Selecting those applicants who will be offered jobs from among the many contacted in the search previously described is essentially a filtering process. Résumés and/or applications are reviewed, potential candidates are screened in campus or telephone interviews, references are checked, and applicants who pass through these screens are invited to the company for interviews (and sometimes testing) before job offers are made.

Résumé and Cover Letter.   For most engineering professionals, the first impression is normally made by the résumé, which is submitted with a cover letter in response to an advertisement or as an initial inquiry. The **cover letter** should be addressed to the appropriate individual by name, not "Personnel Director" or "To Whom it May Concern." (Call the company if you do not have a name.) Normally, the letter begins by identifying the position or type of work you are applying for and, if appropriate, where you heard of the opening. The next section should concisely describe how your background and experience meets or exceeds the requirements of the job. This is the place to make it easy for the recruiter to see how well qualified you are—bullet points are encouraged! A closing paragraph can refer to the attached résumé, thank the recipient for his or her consideration of it, and (if appropriate) indicate that you will call within a specified time to inquire about a possible interview. The cover letter must be impeccable in appearance, grammar, and spelling. A quality cover letter should encourage the recipient to give your résumé fair consideration; with a poor one, your résumé may not be read.

The **résumé** itself includes all or most of the following:

1. Contact information—including name, address, e-mail address, and phone number(s)
2. Summary of education (formal degrees and continuing education)
3. Employment history—most recent employment first, emphasizing duties and accomplishments
4. Publications, significant presentations, and patents (if applicable)
5. Significant honors and awards
6. Professional affiliations

Writing an effective résumé is an important skill that many engineers do not master easily. An effective résumé normally should not exceed one page. Studies have shown that professional recruiters spend only seconds reviewing individual resumes, so anything that doesn't fit on page one will likely not be seen anyway. For more experienced candidates, it is a good practice to have a one page résumé to utilize with applications and then bring a more complete résumé with to job interviews. (Résumés of candidates applying for academic positions are an exception, since publications and presentations are listed there in detail.) The résumé should be well organized, concise, faultless in grammar and spelling, and attractively printed on quality paper. An effective résumé layout will utilize a lot of white space, with clear bolding and underlining to highlight key points. It should also emphasize (without being dishonest) those parts of your education and experience most applicable to the position being sought. For this reason, an individual may need several versions of a résumé. The web, your university's career services, and faculty are all good sources for current information on these and other aspects of job hunting.

Campus Interview.   The newly graduating engineer often makes the first contact with potential employers in the campus placement interview. Indeed, about half of all campus interviews are with engineering students, even though they make up only from 4 to 10 percent of all students. Interview outcomes are a complex dynamic of the attributes of the applicant, of the interviewer, and of the situation (the physical setting and the economic demand for engineers, for example). The interviewer needs to learn enough about the applicant to recommend for or against an invitation for a site visit, and the applicant needs to learn about the employment opportunities and other advantages (and limitations) of working for the employer. The applicant is advised not only to read the potential employer's placement brochure in advance, but also to learn more about the company via their web pages and public reports, and discussions with classmates and faculty who may know something about the organization. Some students are uncomfortable in early interviews and do not sell themselves well; many colleges provide the chance for mock interviews, often using video, to help develop this skill. Engineers need to learn to conduct interviews as well, since they may find themselves interviewing candidates at their site or back on campus after a few years' experience.

Reference Checks. Before inviting an applicant for a site visit, a prospective employer commonly checks the references given in an application, or requests them if they have not already been provided. References for the new graduate include professors and supervisors from part-time jobs; for the experienced engineer they will be primarily past and (if your employer knows of your search) current supervisors and coworkers. Prior to giving a potential employer your references, each individual on the list should be asked if they will provide a positive reference for you. You should also keep your references informed of any interviews you receive and what you might be most interested in about any particular opportunity.

An increasing problem with references is the fear of liability if a bad reference is given. Some employers have a policy of confirming only that a former employee worked under a given job title during a given calendar period, although these same employers may try to get the maximum information in reference checks on people they are interviewing.

Site Visits. When a company has a strong interest in an engineer or other professional, it may extend an invitation for a visit to a chosen company location at company expense. For these visits take extra copies of your résumé with you. Some of the people interviewing you might not have seen your résumé previously. Over the course of a trip, it's common for applicants to be interviewed by as many as 20 people on the staff, including one or more supervisors with open position(s) for whom the candidate is being considered, and at least part of the visit should involve a tour of the area in which the candidate might work. The candidate's reaction to the work observed and the type of questions asked give insight into their interest and suitability for the position; at the same time the candidate can gain insight into the work being done and judge from work observed and answers to their questions whether that might be the right assignment. At some point, human resources will provide information on company benefit programs and answer questions on general company policies.

At the end of an interview or plant visit it is perfectly proper for the applicant to inquire, "When do you expect to make a hiring decision?" or "If I haven't heard from you by [date], may I call you?" A prompt letter thanking the interviewer for courtesies extended and expressing continuing interest in the company is appropriate.

Starting Salary. If an employer is interested in an applicant, sooner or later they will ask, "What salary do you expect?" Often this will occur toward the end of a site visit, and the applicant should be prepared. It is important to study surveys, talk to colleagues, and do web searches of wage structures for the type of work and location being considered. Remember, everything is not set in stone. A candidate who replies,

Interviewing Well: Tell Stories and Prepare

As a soon-to-be graduating engineer, you might assume that the best way to have a great interview is to have the best answers to the interviewer's questions. While the quality of your answers will certainly impact the interviewer's impression of you and their decision on what your next steps will be, it is only part of the equation. Perhaps of equal importance is your ability to engage the interviewer(s) in a conversation and tell them stories that illustrate your ability to successfully do whatever it is they are asking about. Highly skilled interviewers will commonly say that "a good interview is like a dance" with very active give and take between both sides.

So how do you do this, even when you're terribly nervous? Prepare: know the job you are applying for, what the company cares about (both for this position and overall), and how your skills fit into those areas. It is a good idea to review the types of questions that might be asked by the company through sites like glassdoor.com so that you are better prepared, but don't memorize answers! If they have invited you for an interview, they think you are qualified for the job, but they want you to confirm their thoughts. Tell a story of how you applied the skills they are looking for in every question. Make sure you talk about results and provide specifics. It's a good sign if the interviewer has follow-up questions to your stories—you have piqued their interest. If a question comes up that you don't feel you can answer, ask to come back to it later. Finally, as you are studying the company before your interview, make sure to develop some questions about the position and the company. It is impressive when candidates show up with a list of questions already written. It is especially effective to ask how the position might fit into specific things you have read about in terms of the company strategy. As the interviewer(s) answer your questions, make sure to take some notes. Finally, don't be surprised if the person interviewing you doesn't seem as familiar with your resume as you'd expect. Even if they have already reviewed your resume in detail, it's common for the details of one candidate to blend into others. For this reason, make sure you always bring extra copies of your resume to every interview (at this time it's OK if it is longer than one page).

"Whatever is your going rate," will probably be offered the bottom of the range. Since future salary adjustments in most companies are typically percentage adjustments to current salary, inequities in starting salary can be very costly over the course of a career.

Experienced engineers will measure their expectations based on the years since their bachelor's degree, graduate degrees if any; the quality of their experience; local cost of living; and other factors. As a group, engineers earn some of the highest starting salaries among college graduates. A bachelor's degree in engineering is required for most entry-level jobs. In Table 7-2 the average salaries are given. Salary varies by the region of the country, industry, and by metropolitan area. More summaries of engineering salaries collected by the Bureau of Labor Statistics may be found on the Internet.

Job Offer.   The employment offer is a standard format letter delineating a specific position and salary offer, reporting date, position and title, the person the candidate will report to, and often provisions for moving expenses or signing bonuses. An offer for employment is not official until this letter is received.

Table 7-2 Median Salary by Engineering Specialty, 2017

Type of Engineer	Bachelor's Degree
Aerospace/aeronautical	$113,030
Agricultural	$74,780
Biomedical	$88,040
Chemical	$102,160
Civil	$84,770
Electrical/electronics	$97,970
Environmental/environmental health	$86,800
Industrial/manufacturing	$85,880
Mechanical	$85,880
Petroleum	$132,280

Source: U. S. Bureau of Labor Statistics, http://www.bls.gov/ooh/Architecture-and-Engineering/home.htm, May 2019.

The candidate should acknowledge the offer immediately and ask about the expected time to respond. A candidate with other potential offers in process may ask for a reasonable delay. A candidate who already has a better offer from someone else can reply, "I've been offered $Y by Z company—I'd rather work for you, but this is a factor I'll have to consider in my answer." Striking a balance between demanding too much and selling oneself at too low a rate requires the candidate to have a clear understanding of his or her true worth in the current job market.

Job Application Process—Employer Viewpoint

Orientation and Training. When a new employee reports to work, the employing organization needs to help the newcomer become part of the organization by introducing him or her to the policies and values of the organization as a whole and the specific requirements of the person's new department and job. The human resources department normally has the responsibility to tell the newcomer about benefits and general company policies. This can be accomplished with a short one-on-one discussion on the first day as the new employee processes through human resources or a more formal presentation periodically for all new employees; in either case, most organizations of any size will provide every employee a current edition of an employees' handbook describing benefit programs.

Inculcating the values of the organization, such as attitudes toward ethics, quality, safety, and customers, is a more difficult task involving establishing norms. While these values can be emphasized in presentations made to new employees by management, to be given credence they must be evident in their practice by members of the organization. Some large organizations will spend from three months to a year rotating the new employee through a variety of departments and jobs to orient the individual to the organization before placing them in the first *permanent* assignment. Occasionally, a fast-growing organization will have a formal orientation program set up in which functional managers will briefly describe the nature and function of their departments. More often, the new employee will be assigned directly to a department and supervisor.

Engineering Management Applications in a Nonengineering Environment

After benchmarking successful restaurants nationwide, Pal's Sudden Service developed a new drive-through store concept designed for ultra-efficient operation and fast service and it has a process for everything organizational and operational. The company's Business Excellence Process is the key integrating element, a management approach to ensuring that customer requirements are met in every transaction. Pal's training processes support accomplishment of its objectives and improved business

results. The majority of the general staff and hourly managers range in age from 16 to 32. Typically, they view their job at Pal's as an entry into the job market, providing a first step toward a long-term career in another industry. These factors create a young, inexperienced, transient workforce that must be trained to produce quality results and make positive direct customer contacts on a regular basis. Their approach to finding good employees is as follows: Hire for Attitude—Teach the Skills.

Pal's uses a four-step model to train its employees—show, do it, evaluate, and perform again—and requires employees to demonstrate 100 percent competence before being allowed to work at a specific workstation. This may require repeating specific training modules before demonstrating that level of competence. In-store training on processes, health and safety, and organizational culture is required for new staff at all facilities via computer-based training, flash cards, and one-on-one coaching. Since 1995, the turnover rate at Pal's has decreased from nearly 200 percent to 127 percent in 2000, and it continues to fall. In comparison, the best competitor's turnover rate in 2000 topped 300 percent.

Source: Adapted from http://www.nist.gov/baldrige/pals.cfm and www.palsweb.com, September 2012.

In any event, the immediate supervisor of the new employee bears the major responsibility for introducing them to the new group and the specific job assignment. Supervisors tend to be busy with current problems the new employee cannot help with until *brought up to speed*, and so they will often hand the new hire a stack of information (either physical or digital) to read for familiarization, and then get back to the immediate problem. After several days of such isolation the new hire begins to wonder why they are there! The more astute supervisor realizes that there will always be current problems and spends some time getting the new employee started and thinking through some initial assignments that will assist in the orientation process. Often, other employees in the group will be asked to assist by taking the new employee along on visits to other departments, introducing the new hire, and in the process providing insight into current activities of the immediate group and its relationships with the larger organization. Frequently, a specific senior member of the group will be assigned primary responsibility for mentoring the new employee.

In a more comprehensive sense, orientation and training can be considered to include the total *socialization* of the new employee to the environment and culture of their new organization. Pringle et al. describe this well:

> The socialization process, culminating with the employee's transformation from an "outsider" to an organizational "insider," may require anywhere from a month to a year, depending on the particular organization and the individual. Socialization encompasses such formal and informal activities as learning the job and developing appropriate skills, forming new interpersonal relationships, and accepting the organization's culture and norms. From the organization's perspective, effective socialization results in order and consistency in behavior.

Appraising Performance. There are several reasons for requiring formal appraisal of an employee's performance. A written record of performance in some consistent form is especially important in large organizations where personnel are frequently transferred, such as the military service, and in bureaucratic organizations such as civil service, to justify terminating (firing) poor performers and rewarding exceptional ones.

Perhaps the oldest and most common technique for performance appraisal is the conventional *rating scale*, in which an employee is given a rating by checking one of five or more level-of-performance boxes for each of a series of attributes. For example, clerical and other hourly workers in the University of Missouri system are rated in five steps from "Outstanding" to "Inadequate" in each of (1) knowledge of the work, (2) quality of the work, (3) quantity of the work, (4) attendance and punctuality, (5) carrying out instructions, and (6) an overall appraisal. Sometimes, each box in the matrix of attributes and ratings has a word description to help the rater, as shown in the apocryphal example in Table 7-3.

Table 7-3 Rating Scale for Cartoon Heroes

Performance Factors	Far Exceeds Requirements	Exceeds Requirements	Meets Requirements	Needs Improvement	Does Not Meet Requirements
Quality	Leaps tall buildings with a single bound	Must take a running start to leap over tall buildings	Can leap only over short buildings	Crashes into buildings when attempting to leap over them	Cannot recognize buildings at all
Timeliness	Is faster than a speeding bullet	Is as fast as a speeding bullet	Not quite as fast as a speeding bullet	Would you believe a slow bullet	Wounds self with bullet when attempting to shoot
Initiative	Is stronger than a locomotive	Is stronger than a bull elephant	Is stronger than a bull	Takes bull by the horns	Watches bull walk by
Ability	Walks on water consistently	Walks on water in emergencies	Washes with water	Drinks water	Has water on the knee
Communications	Talks with a higher power	Talks with angels	Talks to themself	Argues with themself	Loses those arguments

The conventional rating system is easy to develop and easy to grade, but it presents a number of problems. Some raters suffer from a "halo effect," in which they assign the same rating to every category; some from a "recency effect," in which they base their rating only on the most recent part of the rating period. Raters differ in their interpretation of "outstanding" and the other categories, and some are more lenient than others in their ratings. Moreover, raters find that they are competing with their peers in trying to justify promotion or other benefits for their employees, and thus they soon recognize the competitive need to inflate ratings. In the U.S. Army at one time, over 90 percent of all officers were rated in the top *outstanding* category, and the designation of being merely *above average* threatened a military career. The *pure rating* again places no limit on the fraction of employees who can be rated *superior*. In the *forced ranking*, or *ladder*, approach only one person can be placed on each step of the ladder. The rater is forced to discriminate between employees, but has no way to identify employees considered equal or to indicate significant gaps in ability between two people in the sequence. The *modified ranking* satisfies these last two objections. In the percentile, or *forced distribution* approach, 40 percent of employees must be placed in the third category (average), 20 percent each in the second and fourth (respectively, above and below average), and only 10 percent in the first and fifth categories (superior and poor).

Most forms used for appraisal of professionals in large organizations involve a combination of methods. Ranking or forced distribution methods have some logic to them, but in a culture that prefers to believe that "all our children are above average," it has serious drawbacks. Many supervisors like to believe that their employees are all superior. Engineers who are doing effective work are hardly motivated by an "average" classification and may be encouraged to look elsewhere if placed in category 4 out of 5. Moen describes actions of companies such as General Motors and American Cyanamid Company in eliminating the use of forced distribution systems, and he paraphrases the appeal of quality guru Deming for more motivating appraisal systems:

> He suggests that the [old] systems of rewards nourish the win-lose philosophy and that they destroy people. Companies must adopt a win-win philosophy of cooperation, participation, and leadership directed at continuous improvement of quality

The primary emphasis in appraisal today, therefore, is on the contribution made toward achieving organizational objectives, which is the reason that personnel are employed to begin with. And with the increased emphasis on teamwork, there is greater emphasis on rewarding team members for team (or even total organization) performance rather than just individual performance. The engineering manager needs to find a happy median between team and individual recognition.

In fact, due to these challenges, some organizations have done away with annual performance reviews altogether, including Accenture, GE, and Microsoft. Those in favor of these types of change argue that annual reviews are a relic of the industrial age and recognize that 95 percent of employees are dissatisfied with their employer's appraisal process. Instead, some now argue for more regular and timely feedback systems including real-time 360-degree feedback approaches, where input on performance is sought from peers, subordinates, and higher levels of management.

AUTHORITY AND POWER

Other important human considerations in organizations, once they have been properly staffed, include the nature of authority and power and their effective delegation. These are considered in this section and the next.

Nature of Authority

Formal Authority.   The traditional view of authority is *legitimate power*, the *right,* based on one's position in an organization, to direct the work activities of subordinates. In the United States, formal authority over employees of corporations is thought to stem from society as a whole, through the guarantee of private property in the Constitution of the United States. Individuals invest their assets in corporate stock and elect a board of directors, delegating to them the right to manage their invested assets. The board, in turn, elects the executive officers of the corporation and they appoint subordinate managers, delegating authority to appoint lower-level managers. In this way the direction received by the lowest-level employee from their supervisor can be traced to the ownership authority of stockholders. Similarly, authority over government workers stems from national or state constitutional authority conferred on the legislative and executive branches of government, who in turn delegate authority and direction to the leaders of military and government agencies.

Acceptance Theory of Authority.   A different perspectie is provided by Chester Barnard who believed that authority originates when subordinates choose to accept the directives of superiors. According to Barnard the following is true:

> If a directive communication is accepted by one to whom it is addressed, its authority for [them] is confirmed or established. It is admitted as the basis for action. Disobedience of such a communication is a denial of authority for [them]. Therefore, under this definition the decision as to whether an order has authority or not lies with the persons to whom it is addressed, and does not reside in "persons of authority" or those who issue orders.

Despite this, we know that the overwhelming majority of requests or directives from superiors are, indeed, complied with. When a person enters employment with an organization, they are tacitly agreeing to accept any directives toward which the employee feels no strong objection.

Sources of Power

French and Raven have divided the sources of power and influence into five types:

1. **Legitimate** or **position power** (**authority**), stemming from one's appointment or election as leader
2. **Reward power**, the power to reward others for cooperation
3. **Coercive** or **punishment power**, stemming from fear of punishment
4. **Expert power**, stemming from a person's capability and reputation
5. **Referent power**, based on an attraction to or identification with another individual (or the program or cause that person is leading) that makes the follower want to behave or believe as the other does; it is similar to what is commonly called *charisma,* a special personal gift for inspiring others that is easier to give examples of than to define.

Thamhain bases his *System I* style of engineering program management on the first three of these five "bases of influence" (legitimate, reward, and coercive power), which derive *primarily* from one's formal position, and which are normally sufficient to obtain adequate (if not enthusiastic) response in traditional bureaucratic structures. In many jobs that engineers hold in modern "high-tech" organizations, and especially in project management, the formal authority granted is not enough to persuade others to get the job done. In this case, the combination of expert and referent powers Thamhain calls *System II* style, which stem primarily from one's personal capabilities and reputation, are necessary for effective leadership. Even when System I power is ample, the addition of System II influence makes the manager even more effective.

Pringle et al. list some sources of power in addition to sources listed above that have been suggested by others: (1) power through access to important individuals, (2) power obtained through ingratiation or praise, (3) manipulative power, (4) power of persistence or assertiveness, and (5) power gained through forming coalitions. Engineers may feel that they should automatically be granted enough power to get the job done and may find the "office politics" involved in acquiring power distasteful. Humphrey takes a more pragmatic approach:

> While power is the ability to cause action, politics is the art of obtaining power. Power and politics are important management concerns because they form the basis for all dealings between managers.

Status and Culture

Status refers to one's standing within a group or society in general, and it may lead to deference or special privileges. Two types may be distinguished. **Functional status** derives from one's type of work or profession; it explains the deference shown to the physician in a hospital or (sometimes) the professor in a nonacademic setting. The other is **scalar status**, due to one's level in the organization. In some companies an engineer may begin in small cubicle, move to a larger cubicle as a senior engineer, get a private office with a desk *and* table as a first level manager, and as an executive manager have both an adjoining conference room and an assistant.

In some organizations, such trappings are deliberately avoided to lessen the *social distance* between different levels of the organization and to promote close cooperation between all members of the *team*. One may describe the collectivity of such practices and habits as the *corporate culture*; corporate executives should try to foster in their organization the culture that will be most effective in achieving the goals of the organization.

DELEGATION

Assignment, Delegation, and Accountability

Three interrelated concepts of importance are the *assignment of duties, delegation of authority,* and *exaction of accountability,* as shown in Figure 7-2. Managers use their authority to *assign duties* to subordinates, making them *responsible for* carrying out the specified activities. This assignment proceeds in stages from top management down. A company president may assign responsibility for all technical matters to the vice president for research and engineering; the vice president may assign responsibility for all project matters to a chief project engineer, who in turn assigns the duty of carrying out a specific project to engineer X.

Once a subordinate has been assigned tasks to perform, it is important to provide them with the resources needed to carry out the assignment. This is called **delegation of authority** and can include

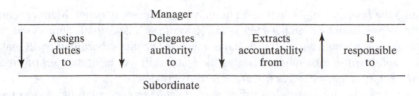

Figure 7-2 Assignment, delegation, and accountability

authority over people who will be needed to carry out the assignment as well as financial authority to acquire the equipment, perform the travel, or make other commitments of resources needed. Like assignment of duties, delegation of authority proceeds in stages from top management down. It is an essential management precept that "authority should be commensurate with responsibility," so that a subordinate has enough authority to carry out assignments effectively. Unfortunately, in many cases (especially in managing projects) the engineer is not given enough authority, and he or she must rely on personal influence, persuasion, or the threat (veiled or not) of appeal to higher authority.

When the manager has assigned duties to a subordinate and delegated the authority to carry them out, he or she is still not through. The manager must exact (insist on or require) **accountability** from the subordinate by making the subordinate *responsible to* the manager for carrying out the duties and reporting progress periodically. The manager has now made the subordinate "responsible for" the task and "responsible to" report progress, but the manager is still accountable (responsible) to the next higher level of executive to assure that the task is effectively carried out—hence the saying "you can't delegate responsibility."

Reasons for Delegation

Delegation relieves the manager of work the subordinate is capable of doing, substituting the need to assure that the work is actually done. The subordinate, on the other hand, is given a chance to develop their skills by being delegated more and more responsible problems. While some subordinates prefer the security associated with very detailed supervision, those with the most future potential will respond favorably to the delegation of increasing responsibility and initiative. Further, delegation tends to locate decision making closer to the work being performed, and this often results in more practical and prompt decisions.

Barriers to Delegation for Engineers

The engineer has been trained in a rigorous discipline and has been held responsible for every calculation and every decimal place through four or more years of college and subsequent years of engineering practice. When an engineer becomes a manager, however, they must now be responsible for the work of other people, and this can be especially threatening to the engineer. The engineer-manager has the responsibility to train new subordinates carefully (often with the help of their more experienced subordinates) and to assign jobs within the capability of the subordinate.

Just as a mother needs to cut the *apron strings* that limit the growth of a child's capability, the manager needs to give subordinates increasing room to grow in capability, which comes only through practice in carrying out increasingly difficult assignments. This requires the manager to let subordinates do their own work, even though the manager might do it more quickly or in some way better. Managers must recognize that the subordinate has more time, and as long as the subordinate's decision is appropriate, it need not be the same one the manager would make. The manager must realize that subordinates will make errors (just as managers do), and learn to trust subordinates as they gain skill, yet institute a set of broad controls to assure that those decisions that are truly critical are properly reviewed.

Insecure managers load themselves with their subordinates' problems through inadequate delegation. Oncken and Wass in their classic "Management Time: Who's Got the Monkey" give examples of how this can happen: subordinate meets manager in the corridor with "We've got a problem." The manager responds with "I'm in a rush—I'll get back to you" or "Send me a memo." In the first case the "monkey" (responsibility for the next action on the problem) has just jumped from the back of the subordinate to that of the manager; in the second case it will come riding in that afternoon on the requested memo. In these scenarios, if the manager is to become a more effective delegator, instead of either tossing the problem back and forth with the employee, they must work through it together. By collaborating together, the manager can help develop the employee's problem solving skills so that the next time something similar arises, the employee can solve it by themselves.

Decentralization

As organizations become larger, it no longer is effective (or even feasible) to make all decisions at the top. Alfred Sloan, Jr., recognized this when he introduced **decentralized management** to General Motors (GM) in about 1920. This concept, which permitted the tremendous early growth achieved by GM, essentially involves the widespread use of delegation throughout the organization. Lower-level decisions can usually be made more rapidly and can often be better than higher-level decisions because they are made closer to the problem.

In times of growth, when opportunities abound if seized promptly, decentralized management can be very effective. This is especially true where the *profit center* concept can be implemented, and the lower-level manager can be given responsibility for the major factors (usually both production and sales) that determine the profit contribution from a particular product, held accountable for results, and rewarded for success. Recent reshaping of corporate structures has resulted in elimination of several levels of middle management and concurrent increases in the number of people reporting directly to each surviving manager. Just to survive in the modern organization, today's managers must learn to delegate more and to coach rather than command subordinates.

The hazard inherent in decentralization is loss of control at the top, and Sloan's contribution was the effective balance of decentralized management with *centralized control* of key decisions (often the allocation of major financial resources). If top management does not retain this control, decisions made at lower levels can bankrupt the company. Especially in times of recession and financial losses, where expenses must be cut and hard decisions on reducing operations and personnel must be made, effective top management may have to institute some *recentralization,* taking back some decision-making authority that was earlier delegated in order to avert disaster.

COMMITTEES

A committee is created when two or more people are officially designated to meet to pursue some specific purpose. A **committee** is a type of small deliberative assembly that is usually intended to remain subordinate to another, larger deliberative assembly. Committees may be found in every type of organization:

large and small, public and private, profit-making, governmental, and volunteer. Some (standing) committees may have indefinite life and may be required in an organization's bylaws; others (ad hoc) may be appointed for a specific purpose and be discharged when the purpose is met.

Reasons for Using Committees

Committees provide some definite advantages over actions by single individuals. Some of the more important reasons for using committees follow.

Policy Making and Administration.   The highest level in most organizations is a policy-making committee, which may be called the board of directors, city council, or some other name. Such a group typically meets monthly or quarterly. Between such meetings, operating decisions are often made by a subset of this group, called an *executive committee* or by a general management committee consisting of the major executive officers.

Representation.   Organizations have many committees composed of representatives selected from each organizational unit affected by a particular class of problems. Universities abound in such committees, from the academic senate and graduate council to the tenure committee, publications committee, and many more, and they are present in all organizations of any size that make any pretense of participative management. Committee members are supposed to reflect the opinions and needs of the units that sent them in group deliberation. In engineering design, for example, the *configuration control board* needs to know the impact on cost and schedule of a proposed system design change from all affected areas. Representatives from production, training, documentation, scheduling, and subcontracting as well as affected design engineering groups may be part of that committee to assure that the effect on their functional responsibility is considered before a change is made.

Sharing Knowledge and Expertise.   Engineers meet many situations where no one person has the knowledge necessary to solve a complex problem or carry out a complex function. The *engineering design review,* for example, requires the participation of reliability, quality, safety, and manufacturing engineers and other specialists in addition to the original designers to assure that a complex new system design is ready for production.

Securing Cooperation in Execution.   Committees consisting of the leaders of affected groups or their appointed representatives can identify any problems created by a proposed change in operation. In the ensuing discussion their viewpoint is fully aired, and when the change takes place, they should at least feel that they had their "day in court." Japanese companies are famous for the (in Western eyes) interminable meetings used to achieve consensus; once consensus is achieved, however, implementation may be very rapid and trouble-free. The American model, on the other hand, values decisive executive decision making, but this speed is often at the expense of a lack of cooperation or even opposition when the executive tries to impose this solution on managers who had no part in the decision.

TEAMS

As discussed in Chapter 6, teams have become an integral part of the workforce and have largely taken the place of committees. Today teamwork is essential within modern industry. One of the principles for management of the modern enterprise is teaming. A **team** is defined as follows: a small number of people who are committed to a common goal, objectives, and approach to this goal that they are mutually accountable to reaching. A group in itself does not necessarily constitute a team. Teams normally have members with complementary skills and generate synergy through a coordinated effort, which allows each member to maximize his or her strengths and minimize his or her weaknesses. Team objectives, size, and composition affect the team processes and outcomes. The optimal size (and composition) of teams is debated and varies depending on the task at hand. There is no one model or ideal as far as team size goes. But team size is certainly a factor in team performance.

A critical factor to keep in mind is the importance of justifying the presence of each and every member on the team. The key deciders should be the individual roles, the complexity of the task, and the need for a certain number of people to execute the job effectively. Overall, research does seem to indicate that ultimately, small is the better way to go when forming a team. At least one study of problem solving in groups showed an optimal size of groups at four members. Other works estimate the optimal size to be between 5 and 12 members.

DISCUSSION QUESTIONS

7-1.   Outline the steps a large high-technology organization takes to identify its plan for personnel acquisition for the next year. Identify the uncertainties that apply to each step.

7-2.   Prepare a one page résumé of your qualifications, meeting the criteria described in this chapter. Make sure to focus on accomplishments from prior roles.

7-3.   Company representatives take a wide variety of approaches to campus interviews. Critique the approaches used by several such interviewers and your own preparation for and responses to them.

7-4.   You have been invited to a site (plant) visit as a result of a campus (or other) interview. Using the recommendations in "Interviewing Well," prepare two stories that highlight your key qualifications for this position.

7-5.   Describe a performance appraisal technique or form with which you are familiar, and assess its strengths and weaknesses.

7-6.   Give an example in which Barnard's "acceptance theory" of authority seems to apply especially well.

7-7.   You have just been offered a job with We Do Cool Stuff Inc. Your first two years will be spent in their management development program. The pay during this time is not great and you will be working many hours and moving often. However, you know that if you are successful during these two years, you will then have your choice of up to four different highly lucrative, challenging, and rewarding assignments. Using the steps of development and authority discussed here, outline the components that should be included in this program for you to be excited to accept the offer.

7-8.   It is a management dictum that authority should be equal to responsibility. Outline a situation you were in where these did not match. How did you manage it? How would you manage it now?

7-9.   Is it reasonable that managers from backgrounds other than engineering might find delegation easier? Support your conclusion.

SOURCES

Barnard, Chester I., *The Functions of the Executive* (Cambridge, MA: Harvard University Press, 1938), pp. 163–169.

Benne, K. D. and Sheats, P., "Functional Roles of Group Members," *Journal of Social Issues*, 4, Spring 1948, pp. 41–49.

Eichel, Evelyn and Bender, Harry E., *Performance Appraisal: A Study of Current Techniques* (New York: American Management Association, Research and Information Service, 1984), as summarized in Ronald D. Moen, "The Performance Appraisal System: Deming's Deadly Disease," *Quality Progress,* November 1989, p. 62.

Filley, A. C., "Committee Management: Guidelines from Social Science Research," *California Management Review,* 13:1, Fall 1960, p. 15.

French, John R. P., Jr., and Bertram Raven, "The Bases of Social Power," in D. Cartwright, ed., *Studies in Social Power* (Ann Arbor, MI: Research Center for Group Dynamics, 1959).

Hatlan, James T., "Managerial Appraisals—A Systems View," unpublished Master's thesis, University of Missouri–Rolla, 1972.

Horton, E. J., Jr., personal communication.

Humphrey, Watts S., *Managing for Innovation: Leading Technical People* (Englewood Cliffs, NJ: Prentice-Hall, Inc., 1987), p. 157.

Katzenbach, Jon R. and Smith, Douglas K., "The Discipline of Teams," *Harvard Business Review*, March–April 1993.

Kennedy, Marilyn Moats, "How to Talk Money in a Tough Market," *Graduating Engineer,* February 1993, pp. 51–56.

Koulopoulos, Thomas, "Performance Reviews Are Dead. Here's What You Should Do Instead," *Inc., February 2018.*

Lewis, Adele, *The Best Résumé for Scientists and Engineers* (New York: John Wiley & Sons, Inc., 1988).

Moen, Ronald D., "The Performance Appraisal System: Deming's Deadly Disease," *Quality Progress*, November 1989, p. 62.

Oncken, William, Jr. and Wass, Donald L., "Management Time: Who's Got the Monkey?" *Harvard Business Review*, 52:6, November–December 1974, pp. 76–80.

Pringle, Charles D., Jennings, Daniel F., and Longnecker, Justin G., *Managing Organizations: Functions and Behaviors* (Columbus, OH: Merrill Publishing Company, 1988), p. 250.

Sloan, Alfred P., Jr., *My Years with General Motors* (New York: Doubleday & Company, Inc., 1964).

Thamhain, Hans J., *Engineering Project Management* (New York: John Wiley & Sons, Inc., 1984), pp. 218–219.

Tillman, Rollie, Jr., "Problems in Review: Committees on Trial," *Harvard Business Review,* 38 May–June 1960, pp. 6–12, 162–172.

Wilke, Dana, "Is the Annual Performance Review Dead?," *Society for Human Resource Management, August 2015,* https://www.shrm.org/resourcesandtools/hr-topics/employee-relations/pages/performance-reviews-are-dead.aspx.

STATISTICAL SOURCEBOOK

Average Median Salary by Engineering Specialty and Degree, 2017, http://www.bls.gov/ooh/Architecture-and-Engineering/home.htm, May, 2017.

8

Controlling

PREVIEW

Controlling is a critical function because it ensures that all the management functions of leading, planning, and organizing are effective. Controlling includes establishing performance standards, which are aligned to the company's objectives, and also involves evaluation and reporting of actual job performance. A pivotal role of the manager is to control the progress made towards achieving the plans set by senior management. Monitoring the progress made entails identifying and correcting variances from the planned progress.

   This chapter begins by introducing the steps in the classical control process, three types of control, and the characteristics of effective control systems. Most of the chapter deals with financial controls since they are the most commonly utilized form of controls. Human resource controls such as management audits, human resource accounting, and social controls are discussed briefly. Finally, other nonfinancial controls that will be discussed in depth in later chapters are mentioned.

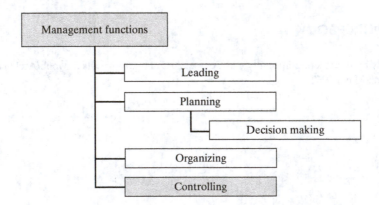

LEARNING OBJECTIVES

When you have finished studying this chapter, you should be able to do the following:

- Describe some of the important elements for establishing financial controls.
- Explain balance sheets, income statements, and ratios.
- Explain different nonfinancial control systems.

THE PROCESS OF CONTROL

Steps in the Control Process

Perhaps the simplest definition of controlling, attributed to B. E. Goetz, is "compelling events to conform to plans." Shannon in his management book states that "control techniques and actions are intended to insure, as far as possible, that the organization does what management wants it to do." Control is a *process* that pervades not only management, but technology and our everyday lives. Effective control must begin in planning; as shown in Figure 8-1, planning and control are inseparable.

The steps in the control process are simple.

- The first step, *establishing standards of performance*, is an essential part of effective planning. Standards should be measurable, verifiable, and tangible to the extent possible. Examples are:
  - standard rate of production established by work measurement;
  - budgeted cost of contracted activities;
  - targeted value for product reliability; or
  - desired room temperature.
- The second step (and the start of the actual control process) is *measurement of the actual* level of performance achieved.
- The third step is *comparison of the two*, measurement of the variance (deviation between them), and *communicating this deviation* promptly to the entity responsible for control of this performance, so that they might identify what changed to cause the deviation to occur and identify potential corrective actions.
- The final step is taking *corrective action* as required to "compel events to conform to plans."

Mechanical Process Control

Closed-loop control, also known as *automatic* control, monitors and manages a process by means of a self-regulating system. The essential feature of automatic control is a strong feedback system. The common home thermostat provides a simple example of an automatic control process. A desired (standard) temperature is set by adjusting a lever or wheel on the thermostat. A mechanism, such as a bellows,

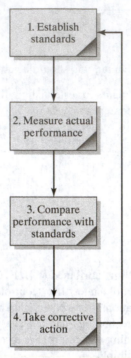

Figure 8-1 The control process.

converts the actual temperature surrounding the thermostat into physical movement. When the variance between desired and actual temperature exceeds some design maximum, sensor movement creates an electrical contact that communicates a signal to the correcting entity, in this case the control of a furnace or air conditioner, and the variance is automatically corrected. A more complex application is the automatic control of a nuclear reactor, designed to shut down the reactor under conditions of power surges that could become catastrophic long before a human operator could react.

Open-loop control requires an external monitoring system and/or an external agent to complete the control loop. Frequently, the automatic part of the control system provides a warning of a variance from planned values, but then human judgment is required to identify the reason for the variance and to determine corrective action. Even systems that are automated in the short run are ultimately open loop, because they permit an external agent to adjust the standard (or set point). Cruise control on an automobile, for example, operates automatically, but it may be turned off or set to a different speed by manual control.

In engineering management the last step in the control process, corrective action, usually requires human judgment. Consider the action required when a machining process fails to maintain a specified tolerance of ± 0.01 centimeter about some specified (planned) dimension. The problem (and its resolution) might include any of the following:

- The machine used is too worn to maintain such a tolerance (and should be fixed or replaced).
- The operator is not skilled enough to achieve the desired result (and needs training or a new process requiring less skill).

- The tolerance specified is more than can be reasonably achieved in the material being machined. (The designer should be asked to relax the specification or choose a more tolerant design or material.)

The choice among these solutions and others requires thought and decision making; the control system has done its job when it brings the problem and information surrounding it to the appropriate decision maker at the appropriate time.

Three Perspectives on the Timing of Control

Feedback Control. Engineers are usually comfortable with the idea of feedback systems, in which the output of a system can be measured and the variance between measured and desired output is used to adjust the system. Thus the rotational speed of a machine can be measured by the effect of centrifugal force on rotating balls (the traditional "governor"), and the difference between this physical movement and the desired (standard) value can be used to adjust the speed. The previous thermostat example is also a feedback system. Such feedback control (also called *post-action* or *output control*) is quite effective for continuing processes or for repetitive actions. For example, the lessons learned in building past McDonald's restaurants have certainly been used to make the next thousand restaurants more efficient. But for many applications, managers cannot afford to wait until an activity or product is complete before examining it, because the cost, risk, and schedule consequences of late discovery of failure are unacceptable. In addition, feedback controls generally lead to some level of defects reaching the customer or extensive internal rework.

Screening or Concurrent Control. Controls may also be applied concurrently with the effort being controlled. A new engineer may be given an unfamiliar assignment one step at a time, with review by the supervisor after each step. A production schedule may include several in-process inspection points so that further investment in defective parts can be avoided. A baseball coach will observe the effectiveness of a pitcher literally one pitch at a time, prepared at any point to start warming up a replacement in the bullpen. However, concurrent control can be expensive, stifle initiative, and can lead to inactivity while awaiting the next inspection.

Feedforward Control. The essence of **feedforward** control is a system that can predict the impact of current actions or events on future outcomes, so that current decisions can be adjusted to assure that future goals will be met. Engineers and managers have many applications where controls must be applied in the early phases of a project or program. A nuclear power reactor may take 10 years to produce, and the construction project manager needs management tools that will predict, as the project progresses, whether it is likely to be completed on time and within budget. As the project continues, control over the early tasks in this system gives us feed forward control over the total project duration. The *earned value* methods of Chapter 14 provide the same sort of feedforward control of costs. In the same chapter management tools, such as work breakdown structures and network systems (PERT or CPM), that enable us to identify the longest "critical path" of tasks that must be completed in sequence to complete the project are discussed.

Examples of feedforward control in manufacturing include careful screening of sequences for machine operations, inspection of raw materials, and *preventive maintenance* of machines, all in an attempt to reduce (control) later production problems. The prudent taxpayer does not wait until April 15 to discover his or her tax liability for the previous year; he or she tries to estimate it before the year ends in order to manage cash contributions, sales of stock, and other actions before December 31 to reduce or defer tax. Similarly, the comptroller of a corporation will try to forecast the next period's revenue and sales so that cash will be ready when needed (and effectively invested when not).

Characteristics of Effective Control Systems

An effective control system should satisfy most of the following criteria:

- *Accurate. Control* systems should measure what needs to be measured and controlled.
- *Efficient*. Control systems should be economical and worth their cost.
- *Timely*. Control systems should provide the manager with information in time to take corrective action. A tax accounting system is expected to show costs to the nearest dollar, but it does not need to do so for the year ending December 31 until the following April 15. A control system for monthly expenses, however, might be satisfied with ± 5 percent accuracy, but demand information within a week after the end of the month measured.
- *Flexible*. Systems should be tools, not straitjackets, and should be adjustable to changing conditions.
- *Understandable*. Control systems should be easy to understand and use, and they should provide information in the format desired by users.
- *Tailored*. Where possible, control systems should deliver to each level of manager the information needed for decisions, at the level of detail appropriate for that level of management.
- *Highlight deviations*. Good control systems will "flag" parameters that deviate from planned values by more than a specified percentage or amount for special management attention.
- *Lead to corrective action*. Systems should either incorporate automatic corrective action or communicate effectively to an agent who will take effective action; this is why the control system exists.

FINANCIAL CONTROLS

Engineers need to know about financial controls because their continued employment may be dependent upon how they support and contribute to their company's profitability. Many business owners do not realize that financial statements have a value that goes far beyond their use to prepare tax returns or loan applications. Financial controls include financial statements (especially the balance sheet and income statement), financial ratios used in ratio analysis, financial and operating budgets and the nature of the budgeting process, and financial audits. Financial statements provide the basic information for the control of cash and credit, which are essential to the survival of a company.

Budgets

Budgets are perhaps the most common and universally used control techniques. Creating a budget is the first step in the financial control process. Budgets are plans for the future allocation and use of resources (usually, but not always financial ones) over a fixed period of time. The budgeting process forces managers to think through future operations in quantitative terms and obtain approval of the planned scope of operations, and it provides a standard of comparison for judging actual performance in the control process.

Financial budgets describe where the firm intends to get its cash for the coming period and how it intends to use it. There are three common types. **Cash budgets** estimate future revenues and expenditures and their timing during the budgeting period, telling the manager when cash must be borrowed and when excess cash will be available for temporary investment. **Capital expenditure budgets** describe future investments in plant and equipment. Because expenditures for fixed assets require their use for an extended period to recover the investment, capital expenditures usually are scrutinized more carefully by upper management than are operating expenditures. Finally, a **balance sheet budget** uses the previous two estimates to predict what the balance sheet will look like at the end of the budgeting period.

For closer control, organizations are divided into *responsibility centers. Expense* or *cost centers* are those (such as manufacturing units or staff offices) where the manager's primary financial concern is control of costs. In a *revenue center*, such as sales or marketing, the manager has revenue targets to meet. Where an organization can be divided into business units containing both production and sales of a distinct product so that *profit centers* are created, the manager has more freedom to manipulate costs in order to increase profit.

Where one unit of a company has as its primary customer another unit of the same company, the transfer price credited to one profit center and debited to the other must be established with care, especially where no accurate market price for the product exists. Not only does this price establish which unit makes the most apparent profit but (where the units are in different states or countries) it also determines the amounts and beneficiaries of tax receipts on these profits.

Operating budgets can be created for each of these responsibility centers. These are three (corresponding) types: the *expense budget*; the *revenue budget*; and the *profit budget*, which is a combination of the expense and revenue budgets for an individual profit center.

Budgeting Process. Budgets can be prepared by a central staff group and imposed on everyone by top management (the "top-down approach"), but this approach is usually unwise. It does not take advantage of information from lower management levels that would improve the budget process, and it does not foster commitment from lower managers to conform to the budget. Alternatively, budgets could be prepared at the responsibility center level and then just added up, but such budgets tend to be inflated and often do not consider adequately upper management's goals and objectives for the coming period.

To combat these issues, many organizations employ a combination of these two approaches. Top management first provides guidelines for the budgeting process, including estimates of future sales and production levels and changes in priorities to meet new objectives. After middle management has provided more detail, the various responsibility centers prepare proposed revenue and expense budgets. These are merged, "massaged" (modified), and negotiated at each middle management level, approved at the top, and then passed back down as operating guidelines for the coming period.

Budgets are frequently proposed and approved as percentage increases or decreases in current levels, which makes it difficult to change priorities in resource use quickly to meet new priorities. The technique of **zero-base budgeting** was developed to overcome this problem. Each responsibility center develops a budget package with a core of resource expenditure that is absolutely necessary to meet next year's objectives, and one or more supplemental additions required to do the job more effectively or to carry out "nice-to-have" functions. Packages and supplements are then ranked on a cost–benefit basis at each management level, and top management allocates resources to meet organizational goals, which may require expansion of some units and shrinking or elimination of others.

Budgets should be tools, and management should be flexible in adapting them as conditions change, however, in government budgets are generally not flexible as they are authorized by Congress at a high level and the organizations must live with them. Many budgets are valid only for the level of production and sales on which they were based; thus, when the level of output can vary substantially, a **variable budget** is needed. In such a budget, costs for labor, materials, and certain overhead and sales costs are set up as functions of output, while others are kept fixed. For a given month, for example, budget expenditures might be authorized at the level corresponding to 60 percent of capacity.

Cost Accounting

The financial budgets just discussed are plans for the future in quantitative (dollar) terms. Before effective decisions for the future (plans) can be made, the costs of alternative decisions must be understood. Historical accounting systems that determine the profitability of past operations are needed to determine income tax liability and produce quarterly and annual reports for stockholders, but they are often not adequate for determining if particular products, whether produced in the past or proposed for the future, have been or will be profitable. To find that out, costs must be divided among (allocated to) specific products, and this is the arena of *cost accounting*.

Example

Assume that a plant produces 4,000 units of product A and 1,000 units of product B, and that each unit (whether A or B) requires one hour of direct labor at $10.00 per hour. Total labor cost is therefore $10.00 (4,000 + 1,000), or $50,000. Now if supervisory effort costing $5,000 is required to coordinate this production, it might be reasonably assumed that each hour of direct labor requires a proportional amount of supervision, resulting in an *overhead* or *burden* charge of $1.00 per direct labor hour, and a total cost for labor and supervision of $11.00 per unit (whether A or B).

Now, assume that costs of setting up the production line for products A and B total $8,000. If we allocate this overhead cost in proportion to direct labor, it will amount to $8,000/5,000, or $1.60, and we will now have a unit cost of $12.60 for both products A and B. However, this setup cost may represent one $4,000 setup activity for each of the products A and B, so that a fairer representation of setup cost would be $4,000/4,000, or $1.00 per unit of A, and $4,000/1,000, or $4.00 per unit of B. Now we find that the unit costs for direct labor, supervision, and setup total $12.00 for product A and $15.00 for product B. Knowing this, we may try to get a higher price for product B or we may want to quit making it.

Historically, direct labor formed the major part of manufacturing costs, and distribution of overhead costs in proportion to direct labor hours or direct labor dollars was often an acceptable estimate. With modern automation, direct labor costs are often reduced to less than 10 percent of total costs, and allocation of overhead costs by *activity-based costing*, as illustrated in our simple example of setup costs, becomes essential for making good decisions. Activity-based costing is an important control tool to support business decision making as discussed in the Activity-Based Costing to Drive Improvements vignette.

Activity-Based Costing to Drive Improvements

Even though the example provided above is a simple application of activity-based costing, one can see how this tool can provide insight to organizations about what it really costs to make their product or deliver their service. These insights can help organizations of all sizes in any industry be more profitable and competitive.

The key to activity-based costing is determining how any product or service *consumes* the resources of the organization. For items like direct materials (e.g., the steel in a car, or the food in a meal) and direct labor (e.g., the worker using a welding robot on an assembly line, or the cook preparing the food), capturing this consumption is straightforward. It becomes far more difficult with overhead expenses. This is especially pertinent to engineers and engineering managers as we are almost always overhead expense! For example, design engineers working for an automotive company must be paid, even though they did not directly manufacture the car you are driving. In addition, they must be provided with work space, a computer, and other tools to perform their design work. To understand how these expenses are consumed by the car in your garage, we must allocate them across all the cars the company manufactures. This allocation occurs through identification and measurement of *activities*. Examples of activities for design engineers might include conceptual design, modeling, wind tunnel testing, etc. We then measure the cost of all these activities (in this example, typically through the hours spent by the team and their respective hourly cost). Then these costs must be *allocated* to final products. This is done by determining *drivers* (also known as activity rates) of these activities. For the design engineering group, this is likely done by the percent of their total hours spent on any one model.

When all activities are defined, measured, and allocated, we combine them to understand the true cost to the organization of creating a given product or service. This information can be used in determining appropriate pricing (as in the example above) and in identifying areas for organizational improvement by understanding high cost activities and seeking ways to improve them. This later application is likely more valuable to the engineering manager seeking ways to improve organizational performance and productivity.

Financial Statements

The next step in the control process is to measure actual performance and this is what financial statements do. The **balance sheet** or **statement of financial position** is a summary of the financial balances of a sole proprietorship, a business partnership, a corporation or other business organization. The balance sheet shows the firm's financial position at a particular instant in time—a financial "snapshot," as it were. This snapshot is usually the financial status at the end of a calendar year or a financial year. The interval can be shorter, for example, at the end of a quarter. The balance sheet includes assets, liabilities, and equity.

Assets are what the company owns and consist principally of *current assets* (assets that can be converted into cash within a year) and *fixed assets* (property, plant, and equipment at original cost, less

Example

Table 8-1   Balance Sheet, Sterling B. Chemicals, Inc., December 31

ASSETS

Current assets		
Cash	$150,000	
Securities (at cost)	100,000	$250,000
Accounts receivable		400,000
Inventories (at lower cost or market)		
Raw materials and supplies	200,000	
Work in progress	180,000	
Finished goods	300,000	680,000
Prepaid expenses		30,000
Total current assets		$1,360,000
Property, plant, and equipment	4,500,000	
Less accumulated depreciation and depletion	2,400,000	
Net property, plant, and equipment		2,100,000
Total Assets		$3,460,000

LIABILITIES AND STOCKHOLDERS' EQUITY

Current liabilities		
Accounts payable	$100,000	
Installments due within one year on debt	30,000	
Federal income and other taxes	250,000	
Other accrued liabilities	120,000	
Total current liabilities		$500,000
Long-term debt		1,000,000
Total Liabilities		$1,500,000
Stockholders' equity		
Capital stock	500,000	
Retained earnings	1,460,000	
Total equity		1,960,000
Total Liabilities and Equity		$3,460,000

In this example the assets include current assets, inventories, prepaid expenses, and property. It gives total assets of $3.46 million. Liabilities include accounts payable, installments due within one year, tax, and other accrued liabilities. This gives $1.5 million for the total liabilities. The stockholders' equity is equivalent to the company's net worth or its assets after subtracting all of its liabilities. In this case, that is $3,460,000–$1,500,000 = $1,960,000 for total equity. For legal and accounting reasons, it is separated into $500,000 for capital stock and the retained earnings are $1.46 million. Thus, the two halves are always in balance.

the cumulative *depreciation* of plant and equipment [but not land] and *depletion* of natural resources since they were purchased).

Liabilities are what the firm *owes* and consist of *current liabilities* that must be paid within a year and long-term debt. The difference between assets and liabilities is the *net worth* or *equity* of the stockholders, and it consists of the original investment (what was paid in for common and preferred stock) plus the *retained earnings* (the cumulative profits over the years after dividends are paid).

Net worth is what is left over after liabilities have been subtracted from the assets of the business. In a sole proprietorship, it is also known as owner's equity. This equity is the investment by the owner plus any profits or minus any losses that have accumulated in the business.

Formally, **shareholders' equity** is part of the company's liabilities: they are funds "owing" to shareholders (after payment of all other liabilities); usually, however, "liabilities" is used in the more restrictive sense of liabilities excluding shareholders' equity. If liability exceeds assets, negative equity exists. Table 8-1 gives an example.

An **income statement** (see Table 8-2), also called a **profit and loss or revenue and expense** statement, shows the financial performance of the firm over a period of time (usually a year or a month).

Income Statement

While the balance sheet shows the fundamental soundness of a company by reflecting its financial position at a given date, the income statement may be of greater interest to investors. The reasons are twofold:

- The income statement shows the record of a company's operating results for the whole year.
- It also serves as a valuable guide in anticipating how the company may do in the future.

The **cash flow**, or *sources and uses of funds*, statement shows where funds come from (net profit plus depreciation, increased debt, sale of stock, and sale of assets) and what they are used for (plant and equipment, debt reduction, stock repurchase, and dividends). Like the income statement, it concerns financial activities over time. Note the special nature of depreciation and depletion. They represent an expense in that they permit recovery over time of earlier capital investment as a deduction from taxable income. Unlike other expenses, they are only *allocations* and do not represent money expended in the current period. The portion of revenue allocated as depreciation or depletion is therefore available without penalty of taxation for reinvestment in replacement assets or in entirely different assets.

Ratio Analysis

Financial ratios are ratios of two financial numbers taken from the balance sheet and/or the income statement. These ratios provide a framework for historical comparisons within the firm and for external benchmarking relative to industry performance. Such benchmarks are available through financial reporting firms such as Dun and Bradstreet. They can also be used to set financial targets or goals for the firm. The desirable levels of financial ratios vary with the industry, economy, culture, and recent company history. Used with care, however, they are invaluable tools for benchmarking within your industry.

Example

Sterling B. Chemicals had net sales of about $3.05 million for last year. Production costs (materials, labor, and production overhead costs) were $2 million, and the depreciation and depletion related to production were $250,000. Selling, advertising, and shipping cost $100,000, and "general and administrative" expenses ("G&A," the cost of general management, R&D, and miscellaneous activities not chargeable elsewhere) were $200,000. Subtracting the total expense of about $2.55 million from net sales leaves an *operating profit* of $500,000. After adjusting for interest and other nonoperating income and expense, the pretax income is found to be $540,000, and the net income (after taxes) is $280,000. The board of directors decided to return part of net income ($320,000) to the stockholders-owners as dividends and to reinvest the rest on their behalf as an addition to retained earnings.

Table 8-2 Income Statement, Sterling B. Chemicals, Inc., December 31

Gross sales	$3,200,000	
Less returns and allowances	150,000	
Net sales		$3,050,000
Less expenses and costs of goods sold		
Cost of goods sold	2,000,000	
Depreciation and depletion	250,000	
Selling expenses	100,000	
General and administrative expenses	200,000	2,550,000
Operating profit		$500,000
Plus interest and other income		60,000
Gross income		560,000
Less interest expense		20,000
Income before taxes		540,000
Provision for income taxes		260,000
Net income		280,000
Retained earnings January 1		1,500,000
		1,780,000
Dividends paid		320,000
Retained earnings December 31		1,460,000

Four types of ratios are ordinarily calculated: (1) liquidity, (2) leverage, (3) activity, and (4) profitability ratios. Each is discussed and calculated for the Sterling B. Chemical Company in Table 8-3. Values are taken from the balance sheet in Table 8-1, with exceptions noted.

Table 8-3 Financial Ratios for Sterling B. Chemicals, Inc.

Ratio

Liquidity ratios

Current ratio $\quad \dfrac{\text{Current assets}}{\text{Current liabilities}} \quad \dfrac{\$13,600,000}{\$500,000} = 2.72\%$

Acid test ratio $\quad \dfrac{\text{Current assets} - \text{inventory}}{\text{Current liabilities}} \quad \dfrac{\$680,000}{\$500,000} = 1.36\%$

Leverage ratios

Debt-to-assets ratio $\quad \dfrac{\text{Total debt}}{\text{Total assets}} \quad \dfrac{\$1,500,000}{\$3,460,000} = 0.434\%$

Activity ratios

Inventory turnover $\quad \dfrac{\text{Cost of goods sold}}{\text{Inventory}} \quad \dfrac{\$2,000,000}{\$680,000} = 2.94\%$

Asset turnover $\quad \dfrac{\text{Net sales}}{\text{Total assets}} \quad \dfrac{\$3,050,000}{\$3,460,000} = .88\%$

Accounts receivable turnover $\quad \dfrac{\text{Net sales}}{\text{Accounts receivables}} \quad \dfrac{\$3,050,000}{\$400,000} = 7.63\%$

Profitability ratio

Profit margin $\quad \dfrac{\text{Net income}}{\text{Net sales}} \quad \dfrac{\$280,000}{\$3,050,000} = 9.18\%$

Liquidity ratios measure the ability to meet short-term obligations. The most commonly used ratio is the **current ratio**, which measures a firm's current assets to current liabilities. The current ratio measures a firm's ability to pay its current obligations. Many analysts use a current ratio of 2.0 as a prudent minimum, but this ratio varies between industries. A high current ratio (such as 10.0) may simply indicate that assets are not being efficiently employed. A ratio lower than that of the industry average suggests that the company may have liquidity problems. Since quickly liquidating (converting to cash) the firm's inventories might prove difficult, analysts also use the **acid test ratio** or **quick assets ratio** of "quick assets" (current assets minus inventories) to current liabilities. An acid test ratio over 1.0 is prudent, and Sterling B. Chemical satisfies both liquidity tests.

Leverage ratios identify the relative importance of stockholders and outside creditors as a source of the enterprise's capital. A simple measure is the ratio of total debt to total assets (debt as a fraction of the sum of debt and stockholders' equity). A common alternative, which can be derived from this one, is the debt-to-equity ratio. Leverage ratios vary significantly by industry. For example, an electric utility might well have a debt/assets ratio of 0.5 (debt/equity ratio of 1.0), while retail firms might have much lower debt ratios.

Activity ratios (also known as *operating ratios*) show how effectively the firm is using its resources. One common measure is **inventory turnover**, measured in Table 8-3 by dividing the cost of goods sold (from *income statement*) by total inventory (both valued at the manufacturing cost invested in them). Another activity ratio is **asset turnover**, or sales/assets, a measure of how well the firm is using its assets to produce sales. A third is the **accounts receivable turnover**, the ratio of net sales (*income statement*) to accounts receivable. This ratio is often, in turn, divided into the traditional (but inaccurate) measure of 360 days per year to calculate the average collection period.

Profitability ratios describe the organization's profit. The **profit margin** measures the net income as a percentage of sales. Other measures are the profit as a percentage of total assets and the earnings per share of common stock (the net income less preferred stock dividends, divided by the shares of common stock outstanding), which the stock market investor can compare with the current market price. Remember that desirable levels of financial ratios vary with the industry, economy, culture, and recent company history.

Audits of Financial Data

The third step in the financial control process is the audit. Audits are investigations of an organization's activities to verify their correctness and identify any need for improvement. Audits of accounting and financial systems and records are the most common type, and these may be either internal or external. *External audits* are required at least annually for any publicly held organization and are performed by independent accounting firms. They determine if financial records are accurate and reflect generally accepted accounting practices, and provide stockholders and creditors with greater confidence in the firm's financial statements. Most large firms also have *internal auditing* staffs, who spend their time auditing the several subunits of the organization. These staffs often have a more intimate knowledge of the firm's accounting systems, and they may be charged by management to evaluate organizational efficiency as well as just the accuracy of financial data; this does not replace the legal need for periodic external audits. The fourth step to take corrective action would usually follow the audit.

Financial Control Process

Financial control refers to the running of a firm's costs and expenses in relation to budgeted amounts. It is a measure of how well a corporation or department controls its costs and it is sometimes articulated as how far over or under a budget it is. The steps in the financial control process are:

1. Budgets—Establishing standards of performance.
2. Balance Sheet—Measure actual performance.
3. Audits—Compare performance with standards.
4. Take corrective action.

HUMAN RESOURCE CONTROLS

Just as essential as financial performance conforming to plans is assuring that human and organizational performance conform to expectations. On an individual basis this is accomplished with the tools of performance appraisal discussed in Chapter 7, especially management by objectives (MBO), which is, by its very nature, a control system. Two tools used to evaluate collective human and organizational performance are the management audit and human resource accounting. Finally, one should consider social controls through group values and self-control.

Management Audits.   The definition of an audit provided under the prior subheading "Audits of Financial Data" can be applied equally well to other areas. One area of increasing importance is the audit of the entire system of managing an enterprise. A number of the major accounting firms have developed management services staffs that are prepared to conduct management audits, or firms may seek many of the same objectives through an **enterprise self-audit**. Some of the questions on administrative effectiveness that might be asked in such an audit can be found by using web searches for best practices in your industry.

Human Resource Accounting.   Conventional financial accounting deals with the prudent handling of revenue and expenses and with investments in tangible items that appear as assets on the balance sheet. Increasingly, however, the biggest assets of an enterprise are its people. Investments in acquiring out-standing people and in extensive training programs for them represent capital investments in the future as much as does the purchase of new machinery. Quantifying the value of human resource investment is difficult, but a number of approaches are being tested. (Similarly, costs for R&D and in-process improvement are written off as current expense, but these might more appropriately be recognized as capital investments.)

Social Controls.   No organization that relies on formal controls only will be truly effective. Peters and Waterman found the central importance of the underlying values imbued in their corporate cultures and inculcated into all employees, as explained in this excerpt from *In Search of Excellence*:

> The excellent companies live their commitment to people, as they do their preference for action—any action— over countless standing committees and endless 500-page studies, their fetish about quality and service standards that others, using optimization techniques, would consider pipe dreams, and their insistence on regular initiative (practical autonomy) from tens of thousands, not just 200 designated $75,000-a-year thinkers....

> The excellent companies seem to have developed cultures that have incorporated the values and practices of the great leaders and thus those shared values can be seen to survive for decades after the passing of the original guru. Second, ... it appears that the real role of the chief executive is to manage the values of the organization.

For values imbued in the corporate culture to be effective requires that employees in general exercise self-control over their actions. Like other control systems, self-control requires:

- The existence of standards (knowledge by the general worker of the organization's objectives and values).
- Comparison with actual outcomes (which implies feedback of performance to the individual, not just to management or a "quality control" group).
- Corrective action (which requires that the individual have the tools, the autonomy, and the motivation to make corrections).

Obviously, an emphasis on self-control is a "Theory Y" approach to leadership (see Chapter 3). It will not work with every person, and it requires careful selection and training of personnel, but carries with it handsome payoffs for success.

Other Nonfinancial Controls

The control process pervades all the functions and applications of management, and it is addressed in a number of later chapters. In Chapter 9, methods of evaluating the effectiveness of research activities are considered. In Chapter 10, control systems for drawing release and for engineering design changes (configuration management) are discussed. Effective production management (Chapters 11 and 12) requires inventory control and quality control, among other control systems. Project management (Chapters 14 and 15) requires control systems monitoring all three of its key variables: schedule, cost, and the performance of the resulting product.

The Other Management Function—Coordination

In Chapter 1, the fundamental management functions were introduced. These are leading, planning, organizing, and controlling. But Fayol initially included five "elements." Fayol believed that management had five principle roles: to forecast and plan, to organize, to command, to coordinate, and to control. Coordination is the alignment and harmonization of a groups' efforts, but few authors treat coordination as a separate management function. Among the major coordination problems in any large organization is that between central office and field units. In many cases, coordination boils down to two conditions: 1) that people and units know what they are to do and 2) when they are to do it. Thus communication is the prime coordinating mechanism for this management function.

Effective communication is significant for managers in the organization to perform the basic functions of management. Leaders and managers must communicate effectively with their subordinates so as to achieve the organization goals. Communication helps managers to perform their jobs and responsibilities. Communication serves as a foundation for planning. All the essential information must be communicated to the managers who in turn must communicate the plans so as to implement them. Organizing also requires effective communication with others about their job task. Controlling is not possible without written and oral communication. Communication and coordination are vital in streamlining goals and ensuring unified visions are achieved in a timely manner.

Coordination is the framework used to ensure that otherwise fundamentally different forces will all pull together. All the functions of management are affected by coordination. Hence coordination is essential for achieving the objectives of the organization. It is also required for the survival, growth and profitability of the organization. Coordination encourages team spirit, gives direction, motivates employees, and makes proper utilization of resources.

Source: Henri Fayol, *Administration Industrielle et Générale*, Constance Storrs, trans. (London: Sir Isaac Pitman & Sons Ltd., 1949), and Samuel C. Florman, "Engineering and the Concept of the Elite," *THE BENT of Tau Beta Pi*, Fall 1992, p. 19.

DISCUSSION QUESTIONS

8-1.   Provide two additional examples of (**a**) feedback, (**b**) screening (concurrent), and (**c**) feedforward control. In each example identify the four steps of the control process.

8-2. Corporate control systems try to have a mix of all three control types. What feedforward controls exist in your personal life? For an area where you only have feedback controls, how might you develop a feed-forward control?

8-3. Suggest some characteristics that distinguish an effective budgeting system from an ineffective one.

8-4. How does the existence of *profit centers* assist top executives in doing their job?

8-5. Discuss how allocation of overhead costs on the basis of direct labor might distort product pricing where some products are produced by automated machining centers and others by more labor-intensive methods.

8-6. In cases of distortion, what allocation approaches might work better?

8-7. Recalling what you learned about motivation in Chapter 3, how might you encourage your technical employees to support corporate goals and values through self-control?

8-8. Explain why Activity Based Costing might better enable engineers to gain support for engineering improvements than traditional budgeting approaches?

PROBLEMS

8-1. Hytek Corporation ended last year with cash of $50,000, accounts receivable of $100,000, and inventory of $300,000. Property, plant, and equipment were valued at their original cost of $470,000, less accumulated depreciation of $170,000. Current liabilities other than income taxes owed (see details that follow) were $120,000, and long-term debt was $250,000. Stockholders' equity consisted of (**a**) $90,000 capital stock investment and (**b**) accumulated retained earnings, which had totaled $130,000 at the end of 2017. Net sales for 2018 were $900,000. Expenses included $500,000 as cost of goods sold, $50,000 as allowance for depreciation, $85,000 as selling expense, and $65,000 as G&A expense. Interest income and expense were $5,000 and $25,000, respectively, and income taxes for the year (unpaid at year's end) were $80,000. Dividends of $20,000 were paid. Prepare a balance sheet and an income statement reflecting these figures.

8-2. Use the output of Problem 8-1 to calculate the current ratio, acid test ratio, leverage ratio, and profit margin (formulas found in Table 8-3). Comment on the values you obtain.

8-3. Excelsior Corporation reported the following status (in thousands of dollars) as of December 31 of last year: accounts payable of $150; accounts receivable of $250; cash of $150; inventory of $200; long-term debt of $260; net plant and equipment of $500; notes payable during 2013 of $250; and stockholders' equity of $440. (**a**) Prepare a balance sheet as of 12/31, and (**b**) calculate as many financial ratios as you can with the information provided.

8-4. Last year a company reported (in millions of dollars) net sales of $10.0, cost of goods sold of $4.4, other (sales, G&A, and interest) expense of $1.2, and income taxes of $1.6. As of December 31, the company had $1.0 cash and securities, $1.4 accounts receivable, and $2.0 inventory; it owed $2.0 in current liabilities (including unpaid taxes) and $2.5 in long-term debt. Calculate as many financial ratios as you can with the information provided.

8-5. ABC Corporation produces 50,000 units of product X and 5,000 units of product Y at a direct materials cost of $3.00 per unit. Product X requires 3 minutes and product Y 30 minutes direct labor per unit (at $20.00 per hour). Other costs (tooling, setup, and equipment depreciation and maintenance) for this period amount to $60,000. (**a**) If these "other costs" are allocated on the basis of direct labor hours, what is the apparent unit cost of each product? (**b**) Production of product X is highly auto-mated to reduce direct labor cost; it is responsible for $55,000 of this "other cost," and product Y only $5,000. Using *activity-based costing*, what do the unit costs now become? (**c**) What difference might this make in ABC Corporation's actions?

SOURCES

Peters, Thomas J. and Waterman, Robert H., Jr., *In Search of Excellence: Lessons from America's Best-Run Companies* (New York: Harper & Row Publishers, Inc., 1982).

Shannon, Robert E., *Engineering Management* (New York: John Wiley & Sons, Inc., 1980), p. 261.

How to Read a Financial Report. 2003. Merrill, Lynch, Pierce, Fenner & Smith Incorporated, http://www.ml.com/media/14069.pdf.

Part III

Managing Technology

9

Managing Research and Development

PREVIEW

Now that the four basic management functions of leading, planning, organizing, and controlling have been defined, the next few chapters will deal with the management of technology. The first topic is research and development (R&D)—examining new product strategies, organization for research, and the sequential process of winnowing the many ideas for product research and development to an affordable level, according to technical, market, and organizational considerations. Next follows a contributed section on the important topic of protecting ideas through patents, trade secrets, and other means. Finally, creativity, which is essential to effective research, is considered carefully.

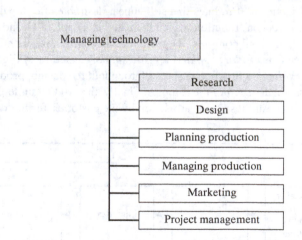

PRODUCT AND TECHNOLOGY LIFE CYCLES

A new product begins as an idea for the solution of a problem or the satisfaction of a need. In nature only a few out of a hundred tadpoles survive to become frogs; in research only a few out of many research ideas will be vigorous enough to survive and will reach the right environment to mature into a successful product. Like the 8-track, cassette tape, or CD, our product will have its day and will then be replaced by newer ideas that satisfy new needs. This cradle-to-grave sequence is known as the **product life cycle** (Figure 9-1).

This product life cycle begins with an *identification of need* or suggestion of a product opportunity, which might come from the customer, researchers, observation of a competitor, or fear of a potential enemy. The product idea must then be subjected to a screening process to select from the many ideas available that are technically and economically feasible. Then a program is proposed for their successful design and development. These preliminary steps (the *product planning and research* functions in Figure 9-1) are the subject of this chapter.

Proposed products that appear attractive at this point are approved for the *product design* function, itself a process of several steps discussed as *systems engineering phases* or *engineering stages of new product development* in Chapter 10. Products that still appear desirable after the design process then go to the *production* (and/or construction) function, which is treated in Chapters 11 and 12. Finally, the products are put into use, and if they are at all complex, they will require continuing technical effort to support their operation and maintenance (the *product use and logistic support* function in Figure 9-1), as discussed in Chapter 13. The *product evaluation* function is spread throughout the design, production, and early system use phases and is discussed under each of these topics. Finally, the product undergoes phase-out, disposal, reclamation, and/or recycling. All these steps are driven by the customer further down the line.

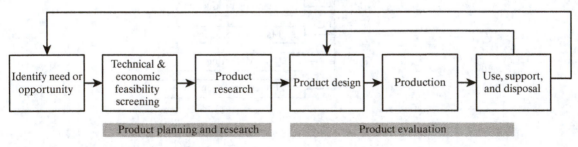

Figure 9-1 An example of product life cycle.

The preceding model of the product life cycle fits the construction of a building or a ship or the design and development of an aerospace system well. For a product line (or family of products) based on a technology that is developed and improved over a period of years of product manufacture, the model of the *technology life cycle* portrayed by Betz (Figure 9-2) is more appropriate. Betz illustrates this model using the automobile as an example:

> When a new industry (based on new technology) is begun, there will come a point in time that one can mark as the inception point of the technology. In the case of the automobile, that was 1896, when Duryea made and sold those first 13 cars from the same design.
>
> Then the first technological phase of the industry will be one of rapid development of the new technology—technology development. For the automobile this lasted from 1896 to 1902, as experiments in steam-, electric-, and gasoline-engine-powered vehicles were tried....
>
> In any new technology, the early new products are created in a wild variety of configurations and with differing features....Finally, when enough experimentation has occurred to map out the general boundaries of possibilities of the product line, some managerial genius usually puts all the best features together in one design and creates the model which then becomes the standard design for the industry. Thereafter all product models generally follow the standard design. This makes possible large market volume growth. For the automobile, this occurred [in 1908] with Ford's Model T design.
>
> After the applications launch, there occurs a rapid growth in the penetration of technology into markets (or in creating new markets). After some time, however, the innovation rate slows and market creation will peak. This is the phase of technology maturity. Finally,...when competing or substituting technologies emerge, the mature technology begins to degrade in competition with the competing technologies.

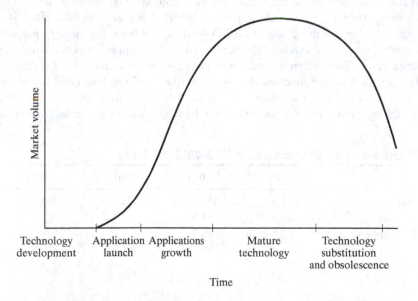

Figure 9-2 Technology life cycle. (Source: Based on Frederick Betz, *Managing Technology: Competing Through New Ventures, Innovation, and Corporate Research*, (Englewood Cliffs, NJ: Prentice-Hall, Inc., 1987) pp. 72–74.

NATURE OF RESEARCH AND DEVELOPMENT

R&D Defined

Research and development are commonly lumped together under the catchall term "R&D." To distinguish between them, let us adopt the definitions commonly used by the National Science Foundation:

Research, both basic and applied, is systematic, intensive study directed toward fuller scientific knowledge of the subject studied.

Basic research is ... research devoted to achieving a fuller knowledge or understanding, rather than a practical application, of the subject under study ... [although when funded by commercial firms, it] may be in fields of present or potential interest to the company.

Applied research is directed toward the *practical application* of knowledge, which for industry means the discovery of *new* scientific knowledge that has specific commercial objectives with respect to either products or processes.

Development is the systematic *use* of scientific knowledge directed toward the production of useful materials, devices, systems, or methods, including design and development of prototypes and processes.

Distribution by Expenditure and Performance

U.S. investment in research and development amounted to approximately $495.1 billion in 2015. This spend over time and a comparison with the spend of other countries is shown in Tables 9-1 and 9-2.

Although R&D expenditures in the United States, Japan, and Germany had been comparable in the late 1980s as a percentage of gross domestic product (about 3 percent), in the United States about one-third of the total (and two-thirds of federal) R&D funding had been for military purposes, so that the United States spent less proportionately than these two major competitors on nonmilitary products for the global marketplace. This difference has only grown over time. In today's political climate, budget cuts continue to slash federal investment in science. Alan Leshner, past chief executive officer of the American Association for the Advancement of Science, has stated that not only is this bad for science, but it is bad for the economy whose growth is driven by advances in science and technology.

Table 9-1   R&D Expenditures in 2012–2015, $ Billions

Year	2012	2013	2014	2015
Total	433.6	454.0	475.4	495.1
Basic	73.3	78.5	82.1	83.5
Applied	87.1	88.3	91.9	97.2
Development	273.3	287.1	301.5	314.5

Source: National Science Board, 2018. *Science and Engineering Indicators 2018 Table 4-4*, Arlington, VA: National Science Foundation (NSB 10-01).

Table 9-2   International Comparisons of R&D Expenditures as Percentage of Gross Domestic Product, 2018

Year	United States	Japan	Germany	France	United Kingdom	India	South Korea	China
2001	2.63	2.97	2.39	2.13	1.63	0.72	2.34	0.94
2008	2.75	3.34	2.60	2.06	1.64	0.87	3.12	1.44
2015	2.74	3.29	2.93	2.22	1.70	0.63	4.23	2.07

Source: National Science Board, 2018. *Science and Engineering Indicators 2018 Table 4-5,* Arlington, VA: National Science Foundation (NSB 10-01).

RESEARCH STRATEGY AND ORGANIZATION

New Product Strategies

Within a specific industry, deciding the relative investment a company should make in R&D is a part of strategic planning and should be based on the organization's concept of its fundamental mission and objectives. Ansoff and Stewart suggest four alternative new product strategies:

First-to-market. This strategy demands major expenditures for research before there is any guarantee of a successful product. It also demands heavy development expenditures and perhaps a large marketing effort to introduce an innovative product. The possibilities of reward from the R&D, however, are tremendous. Apple is an example of a company that follows this strategy.

Follow-the-leader. This strategy does not require a massive research effort, but it demands strong development engineering. As soon as a competitor is found to have had research success that could lead to a product, the firm playing follow-the-leader joins the race and tries to introduce a product to market almost as soon as the innovator. Samsung is a company that historically followed this strategy, but is shifting more to a first-to-market approach.

Me-too. A me-too strategy differs from follow-the-leader in that there is no research or development. In its purest form this strategy means copying designs from others, buying or leasing the necessary technology, and then concentrating on being the absolute minimum-cost producer. The firm following this strategy will try to maintain the lowest possible overhead expenses. ZTE is a smart phone company that has followed this strategy.

Application engineering. This role involves taking an established product and producing it in forms particularly well suited to customers' needs. It requires no research and little development, but a good deal of understanding of customers' needs and flexibility in production. Companies that do after-market conversions of vehicles are typical examples of application engineering strategy.

Corporate Research Organizations

Through the end of the nineteenth century, industrial support of research was unknown. The first corporate research laboratory (more commonly research and development facility today) in the United States began when General Electric Company observed that newer inventions were making its principal product, the carbon filament lamp, technically obsolete, and hired MIT Professor Willis R. Whitney to organize what became the GE R&D Center in 1900. Other early corporate research laboratories that were very successful were those of AT&T (Bell Labs), DuPont, Dow Chemical, and General Motors. Today, most large corporations consider some level of corporate research essential. Although some companies are extremely successful in creating profitable new products from research, others are not.

Large corporations normally have two kinds of research activities: applied research staffs attached to each of the major business units, and a central laboratory with a broader scope of scientific expertise and a long-range outlook. In General Electric, for example, the central laboratory represents only about 10 percent of the research effort, but it plays an essential role. Central corporate laboratories also make their special expertise available to the business units to solve current problems, but they must be careful that this does not cripple their basic function.

SELECTING R&D PROJECTS

Need for Selection

Any successful technology-based manufacturing firm will have many more ideas for research projects than it has resources to invest in them. Booz, Allen, and Hamilton, Inc. has suggested approximately the following ratio of raw new product ideas to profitable products (also illustrated in Figure 9-3):

- Sixty ideas (from researchers, other employees, customers, and suppliers) need to be screened quickly down to
- Twelve ideas worthy of preliminary technical evaluation and analysis of profitability, to produce
- Six defined potential products worth further development, to obtain
- Three prototypes for detailed physical and market testing, resulting in
- Two products committed to full-scale production and marketing, of which
- One product should be a real market success.

Initial Screening

To slash 60 crude ideas into 12 worthy of any significant evaluation requires a method that is quick and inexpensive. A common method is use of a simple *checklist,* in which the proposed product is given a simple judgmental rating (poor/fair/good/excellent or $-2/-1/+1/+2$, for example) for each of a number of characteristics. Seiler suggests, for example, scoring 10 items:

1. *Technical factors* (availability of needed skills and facilities; probability of technical success)
2. *Research direction and balance* (compatibility with research goals and desired research balance)
3. *Timing* (of R&D and market development relative to the competition)
4. *Stability* (of the potential market to economic changes and difficulty of substitution)
5. *Position factor* (relative to other product lines and raw materials)
6. *Market growth factors* for the product
7. *Marketability* and compatibility with current marketing goals, distribution methods, and customer makeup
8. *Producibility* with current production facilities and manpower

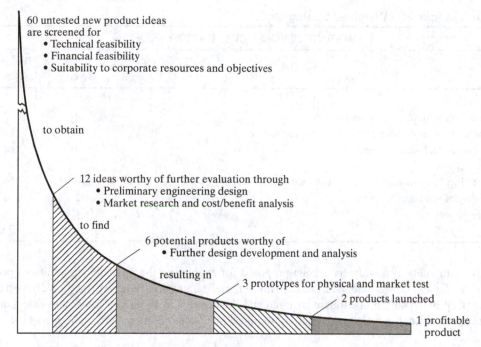

Figure 9-3 Screening of research project ideas.

 9. *Financial factors* (expected investment need and rate of return from it)
 10. *Patentability* and the need for continuing defensive research

Only slightly more sophisticated is the use of a *weighted checklist* or *scoring model* in which each factor is scored on a scale, often from 0.0 to 1.0. A relative *weight* representing the importance of that factor is then used as a multiplier, and the weighted scores for all factors are added. Table 9-3 provides an example of such a scoring model. In this example, a potential new product has been given a raw score of 36 (60 percent of the maximum 60) and a weighted score of 75 (only 50 percent of the maximum 150). The product was judged very favorably on technical factors and could be developed with some confidence of technical success. However, it was rated poorly on its marketing factors (which had been assigned greater weight in the model) and therefore probably would not be developed.

Quantitative Approaches

Once the large number of ideas for research projects has been screened to a more manageable number, the remaining proposals justify more detailed consideration of their technical and financial merits. The technical evaluation can take place in several stages increasing in depth and detail (such as the conceptual, technical feasibility, development, and commercial validation stages of new product development discussed in Chapter 10), with a decision point at the end of each phase. Hand in hand with evolution of the technology should come an increasingly detailed analysis of costs of producing the proposed product and market estimates of potential sales and profits.

Table 9-3 Example of a Weighted Scoring Model

PRODUCT CONCEPT EVALUATION SHEET

Criteria	Weight	Score	Weighted Score
Technical factors			
Compatibility with research objectives	1	9	9
Compatibility with production facilities and capabilities	2	8	16
Probability of technical success	2	9	18
Marketing factors			
Compatibility with marketing goals, distribution, customers	4	4	16
Probability of marketing success	4	2	8
Potential profitability	2	4	8
Totals		36	75

    Many mathematical models have been proposed for evaluating the financial suitability of proposed projects (see, for example, Balderston et al., Dhillon, and Shannon). Typically, they involve estimating the relationship between the investment required and the benefits to be gained. Easiest to calculate is the **simple payback time** T_{pb}, which is the ratio of required investment I and mean annual gross profit A:

$$T_{pb} = \frac{I}{A} \qquad\qquad (9\text{-}1)$$

Simple payback time is often used to justify investments that need to be recovered quickly because of uncertainties, but it is unsuitable for longer-term investments because it ignores profits expected beyond the point of payback and does not consider the time value of money (the fact that a dollar profit returned at some future time has less value than a dollar available today). Many engineers learn these valuable methods of justifying investment in a new project or purchase of new equipment in a course in engineering economy and return to tell their instructors that it was one of the most useful, practical courses they took in college. Using the standard engineering economy nomenclature for **time value of money**, consider the following:

P = present worth of future cash flow
A_j = cash flow (revenue less expense) in the jth year
i = discount rate (minimum attractive rate of return) required by the organization to justify investment, expressed as a decimal
n = number of years of future cash flow

Any sum P today, placed at an (annually compounded) interest i would compound to $A_1 = P(1 + i)$ in one year, $A_2 = P(1 + i)^2$ in two years, and $A_j = P(1 + i)^j$ in j years. Therefore, the present worth of any future sum A_j can be calculated as

$$P = \frac{A_j}{(1 + i)^j} \qquad\qquad (9\text{-}2)$$

The present worth of n years of such cash flow would be

$$P = \sum_{j=1}^{n} \frac{A_j}{(1 + i)^j}$$ (9-3)

Example

Suppose it was proposed to invest ($I=$) $400,000 today with the certainty of a return of ($A_1=$) $209,000 in one year and ($A_2=$) $242,000 in two years. On the surface, the return of $451,000 for an investment of only $400,000 seems attractive, and the payback time is under two years. However, if the rate of return on corporate investment must be at least 10 percent ($i = 0.1$), the overall net present worth (NPW) of the proposal would be

$$\begin{aligned} \text{NPW} &= -\$400,000 + \frac{\$209,000}{1.1} + \frac{\$242,000}{(1.1)^2} \\ &= -\$400,000 + \$190,000 + \$200,000 \\ &= -\$10,000 \end{aligned}$$

This negative net present worth shows that the project would not earn the required return on investment, and the research proposal would be rejected.

Even if the net present worth were positive, there would normally be no certainty that the projected future earnings would be realized. For this reason, it is prudent to calculate a maximum expenditure justified E_{mj} based on the estimated probabilities of success:

$$E_{mj} = F_c \times F_t \times P$$

Here, F_c is the probability of commercial success, F_t is the probability of technical success, and P is the NPW assuming certainty of success (from the previous equation, excluding initial investment). Other quantitative methods that calculate the expected return on investment or the benefit-to-cost ratio achieve much the same result.

MAKING R&D ORGANIZATIONS SUCCESSFUL

Three topics are treated herein: the relation of R&D strategy to business strategy, evaluating the effectiveness of R&D (both at the organizational and individual levels), and providing effective support for researchers.

R&D and Business Strategy

Chapter 4 discussed the importance of strategic planning to the success of any enterprise. In the technology-driven organization, a carefully planned technology strategy must be thought through to support the overall strategy of the enterprise. This strategy should encompass research, product and process development, and manufacturing engineering. Erickson et al. identify *three broad classes of technologies* a typical firm must consider:

- *Base technologies.* These are the technologies that a firm must master to be an effective competitor in its chosen product-market mix. They are necessary, but not sufficient....The trick for R&D management is to invest enough—but only enough—to maintain competence in these technologies.
- *Key technologies.* These technologies provide competitive advantage. They may permit the producer to embed differentiating features or functions in the product or to attain greater efficiencies in the production process.
- *Pacing technologies.* These technologies could become tomorrow's key technologies. Not every participant in an industry can afford to invest in pacing technologies; this is typically what differentiates the leaders (who do) from the followers (who do not). The critical issue in technology management is balancing support of key technologies to sustain current competitive position and support of pacing technologies to create future vitality.

Evaluating R&D Effectiveness

Organizational Effectiveness.   Balderston et al. suggest the following 11 criteria for business enterprise R&D:

1. Ratio of research costs to profits
2. Percentage of total earnings due to new products
3. Share of market due to new products (usually computed as the volume of sales from a firm's new products in a specific product market to the total sales available from that market, which confounds the measure by including marketing proficiency as well)
4. Research costs related to increases in sales
5. Research costs to ratio of new and old sales
6. Research costs per employee
7. Ratio of research costs to overhead expenses such as administrative and selling costs
8. Cash flows (continuing evaluation of the pattern of outflows for research expense and actual and projected inflows from resulting revenue)
9. Research audits, including indicators of administrative and technical objectives such as costs, time, completion dates, probability of technical success, probability of commercial success, expected market share, expected profits, expected return on investment, design, and development. Blake provides a checklist of questions to ask in such an audit.
10. Weighted averages of costs and objectives (a measure of the extent the average R&D dollar contributed toward objectives with weights on a scale, such as 0.0 equals *project badly missed objectives* to 3.0 equals *project far exceeded objectives*)
11. Project profiles (a more complex weighted scoring of each project, using criteria such as those in the research audits, item 9).

A number of these measures (such as items 1, 4, and 5) are obscured by the lag between research expenditures and the sales and profits that result from them, as well as the contribution of production and marketing to sales and profits. Others (items 6 and 7) are measures of the intensity of research expenditures

Table 9-4 Patents Granted by the U.S. Patent Office, 2015

Rank	Company Name	Total Patents in 2015
1	INTERNATIONAL BUSINESS MACHINES CORP	7,309
2	SAMSUNG ELECTRONICS	5,059
3	CANON KABUSHIKI KAISHA	4,127
4	QUALCOMM, INC.	2,900
5	GOOGLE, INC.	2,835
6	TOSHIBA CORPORATION	2,582
7	SONY CORPORATION	2,448
8	LG ELECTRONICS INC.	2,241
9	INTEL CORPORATION	2,046
10	MICROSOFT TECHNOLOGY LICENSING, LLC.	1,955

Source: https://www.uspto.gov/web/offices/ac/ido/oeip/taf/topo_15.htm, December 2018

rather than research effectiveness. The last three are more time-consuming and require subjective opinion, but they also may be more effective.

A measure of R&D effectiveness is the number of patents a company receives in a given year as shown in Table 9-4. The U.S. Patent and Trademark Office (USPTO) had issued a total of 298,407 utility patents in 2015. IBM is in the top slot, as it has been for the previous 28 years, with 7,309 patents.

Individual Effectiveness. The effectiveness of individual researchers can be evaluated by the normal techniques of performance appraisal introduced in Chapter 7, especially management by objectives (MBO), emphasizing research goals. A few quantitative measures such as the number of patents and publications, and citations by others of those publications, give limited insight into research effectiveness.

Support for R&D

Quality supporting services need to be supplied to make the work of the highly trained scientist and engineer more efficient and productive. A few special types of assistance that are needed in research and engineering are listed as follows:

1. Technician support to carry out repetitive testing and other functions not requiring a graduate engineer or scientist
2. Shop support of mechanics, glassblowers, and carpenters to produce test and research equipment based on researchers' sketches
3. A technical library with technical information specialists conversant in the fields of the company's interest and willing and able to suggest sources to researchers, and structure and run searches in the appropriate databases for them
4. Technical publication support, including typing, editing, and graphical support to simplify researchers' production of reports, technical papers, and presentations
5. A flexible, responsive system for approving and acquiring equipment as needed by researchers

6. Ample computer facilities conveniently available to researchers, and programming assistance to provide consultation and programming to those researchers not wishing to do it themselves

7. A strong internal commercialization process in place to take research to product

Protection of Ideas

By Dr. Donald D. Myers
Professor of Engineering Management
Missouri University of Science and Technology

Strategic planning for competition implies searching for means of capturing a sustainable advantage. R&D is conducted to develop and improve technological products and processes that provide the organization a competitive advantage. Likewise, development of organizational goodwill through marketing and other means is used to gain a competitive advantage. If these advantages can be readily duplicated by others, then there are often insufficient reasons for expending the initial resources for a short-term advantage. As the more advanced nations develop products and services that have high creative value-added content, it is vital to the economic well-being of the creative organizations (and countries) that there be some means of protection of these intellectual properties. Fortunately, there are means for protection of ideas in all industrialized nations.

There are generally four legal means to protect an organization's (or individual's) ideas and right to benefit from those ideas. They are patents, copyrights, trade secrets, and trademarks or other marks. This area of law is generally referred to as *intellectual property law.* Through the efforts of the World Trade Organization, intellectual property law is becoming more uniform across national boundaries, although it is important to recognize that there are still significant differences and conflicts.

Each of the legal protection means is discussed in a subsequent section based on U.S. law. It should be noted that intellectual property law is the most dynamic area of law in terms of the number of precedential cases. Rapidly advancing technology has pushed the boundaries of legal precedents and principles. Accordingly, although the concepts presented here may appear to be noncontroversial, be assured that there have been major legal battles all the way to the U.S. Supreme Court over the interpretation of those concepts, and it is likely to continue as rapidly as new technology emerges.

Patents

A U.S. patent is an exclusive property right to an invention issued by the Commissioner of Patents and Trademarks, U.S. Department of Commerce. The rights granted are limited to the "claims" of the patent. There are three classifications of patents: (1) utility, (2) design, and (3) plant. A *utility* patent may be obtained by the inventor(s) for a process, machine, article of manufacture, composition of material, or any new and useful improvement. The life of the utility patent is, generally, 20 years from the date of application. Utility patents cannot be obtained on laws of nature, scientific principles, or printed matters.

To be patentable, the invention must be (1) new or novel, (2) useful or have utility, and (3) nonobvious. If the invention has been used, sold, or known by others in the United States or patented or disclosed in a printed publication in the United States or a foreign country before the invention was made by the inventor, a patent is barred. It is also barred if the invention was patented or described in a publication or in public use or on sale in the United States more than one year prior to the application for the patent. An applicant would also be barred if it was made before the date of the invention by others not concealing it. Useful inventions must advance the useful arts and benefit the public. The test of obviousness is whether it is obvious to those "with ordinary skill in the art involved."

A design patent is granted to the inventor on the new, original, and ornamental design of an article of manufacture for a term of 14 years from the date the design patent is granted. *In contrast to the* utility patent, the design patent is not concerned with how the article of manufacture was made and how it was constituted, but with how it looks. The design must be primarily ornamental rather than primarily functional to be valid. *Plant patents* are granted for 20 years from date of application for anyone who invents or discovers and asexually reproduces any distinct and new variety of plant, with the exception of tuber-propagated plants or plants found in the uncultivated state.

Establishing Patent Rights   The invention process includes (1) conception and (2) reduction to practice. In the United States, if the first to conceive makes a reasonable, diligent effort to reduce the invention to practice, he or she will receive the patent, even if someone else actually reduces it to practice earlier. Accordingly, it has been essential for the American inventor to maintain good records to establish the date of conception and diligence in reduction to practice in case of any later interference. The filing of the patent application satisfies reduction to practice if, from the patent specification, one skilled in the art to which it relates is capable of constructing or carrying out the invention.

A written disclosure of the invention should be made as soon after conception as possible. There is no specific requirement about the form a written disclosure must take to document the conception of an invention. A disclosure's primary purpose is to prove the date of conception where there is question of invention. The disclosure should include sufficient description and sketches to describe fully what has been conceived. The disclosure should be witnessed by at least two persons who fully understand its content.

To demonstrate diligence to "reduce to practice," a written record of developmental activities should be maintained in a bound notebook. Daily entries are encouraged. Each page should be signed and witnessed in proximity to the entries on that page. Each entry should be made in chronological order. Notebook pages should be consecutively numbered, with all entries made in ink. If an error is made in an entry, it should not be erased; it should be crossed out. All entries should be made by the inventor in their own handwriting. Although it is permissible for an inventor to file their own application, it is strongly advised that a patent attorney or patent agent be used to make and prosecute the application.

In almost all other countries, patents are awarded to the first person to file, rather than the first to conceive. There continues to be considerable pressure for the United States to harmonize with other countries by awarding patents to the first person to file. However, Congress has not chosen to modify the existing patent law. Private inventors, small businesses, and universities are opposed to such a change. The 2005 Patent Reform Act included provisions to change the United States to a first-to-file country, but was not passed. This change was finally passed in 2013 with the *America Invents Act*, making the United States a first to file country.

Just over half of U.S. utility patents have been awarded to Americans in recent years; the first 10 companies that were awarded the most U.S. patents in 2015 included six foreign owned companies. See Table 9-4 for details.

Trademarks and Other Marks

The Lanham Act defines a mark as "any word, name, symbol, or device, or any combination thereof." The U.S. Patent and Trademark Office recognizes four types of marks: trademarks, service marks, certification marks, and collective marks. A trademark is "used by a manufacturer or merchant to identify his goods and distinguish them from those manufactured or sold by others." A trademark differs from a trade name. *Intel* may be both a trademark and a trade name, but only the trademark attached to a product is protected by federal statutes and registered with the Patent and Trademark Office. The potential of a sustainable competitive advantage of the mark for technological products is readily recognized by recalling Intel's strategic decision to distinguish its memory chip in PCs from competitors by implementing the "Intel Inside" mark.

A service mark is associated with services rather than goods. A *certification mark* indicates that the marked goods or services meet standards established by the mark's owner—for example, Good Housekeeping. A *collective mark* identifies members of a group such as an organization, union, or association.

The rights to a mark can be lost, especially if a mark is abandoned or allowed to become a generic word. To avoid losing a mark, vigilance must be exercised even to the point of suing infringers. Under the Trademark Law Revision Act of 1988, beginning November 16, 1989, application for a mark can be made before any use has taken place. Previously, a mark had to be used and products bearing the mark sold and shipped to a commercial customer before the mark could be registered. Now the applicant need only indicate a bona fide intent to use the mark within the next three years.

Almost all states have their own trademark law. If a mark is to be used entirely within one state, the only protection it has, other than common law, is registration under the state's trademark law. Federal trademark law applies only to marks used in interstate commerce.

A mark does not have to be registered, but the symbol "®" or the notice "Reg. U.S. Pat. and TM Off." should be used with registered trademarks and, "TM" or "Trademark" with nonregistered marks. A nonregistered mark has common-law rights. Official registration, however, provides distinct advantages.

Copyrights

A copyright is a bundle of rights to reproduce, derive, distribute, perform, and display an original creative work in a tangible form for the life of the author, plus 70 more years thereafter. Exceptions to this term include work for hire, where the copyright lasts for 120 years from the date of creation or 95 years from the year of first publication. Copyright owners can sue anyone who infringes on their rights to stop illegal reproduction; impound infringing articles; collect lost profits, court costs, and attorney's fees; and in extreme cases, invoke criminal penalties.

Copyrights can be given for literary works; musical works, including any accompanying music; dramatic works, including any accompanying music; pantomimes and choreographic works; pictorial, graphic, and sculptural works; motion pictures and other audiovisual works; sound recordings; and architectural works. A copyright protects expressions, not ideas. A potentially patentable idea expressed in a copyrighted text may be used by others.

As a result of the United States joining the Berne Convention in 1988, a copyright is secured automatically when the work is first created. The fundamental tenet of the Berne Convention is that the enjoyment of copyright protection shall not be subject to any formalities. However, there are distinct advantages in registration and imprinting proper notice on copies, such as the right to bring suit for domestic works (not required for international works), proof of copyright validity if registration is within five years of publication, rights to statutory damages, and rights to attorneys' fees and costs. A copyright notice has the following three elements: (1) the copyright symbol ©, the word "copyright," or the abbreviation "copr."; (2) the year of first publication; and (3) the name of the copyright owner. A copyright notice can appear anywhere in or on the work as long as it can be readily seen, but in a book such as this the notice is usually on the back of the title page. Copyright registration is not a condition for protection, but is a prerequisite for an infringement suit. Copyrighted material is registered with the copyright office at the Library of Congress, which requires one copy of unpublished work and two copies of published work, plus a $30.00 fee for the processing of registration forms.

There are a number of exceptions to the rights of a copyright. The most notable and highly publicized is the "fair use exception." One may, without permission, make a fair use of a copyrighted work for purposes such as criticism, comment, news reporting, teaching, scholarship, or research. Fair use is determined by consideration of such factors as the purpose of the use, the nature of the work, the amount and substantiality used, and market effect.

Trade Secrets

Trade secrets, or confidential technological and commercial information, are the most important assets of many businesses. The law protects trade secrets as alternatives to patents and copyrights. Trade secrets have no precise definition, but to be protected by the courts, they must be secret, substantial, and valuable. The secret can be almost anything as long as it is not generally known in the trade or industry to which it applies. A trade secret provides its owner with a competitive advantage. It may be a formula, process, know-how, specifications, pricing information, customer lists, supply sources, merchandising methods, or other business information. It may or may not be protected by other means.

Unlike patents or copyrights, trade secrets have no time limitations and there is no registration with any government agency. A trade secret, however, has value only while it remains secret. For instance, a trade secret may lose its privileged status when it is ascertained through "reverse engineering" or when it is discovered independently. A trade secret revealed in these ways can be used without any obligation to the trade secret's originator or owner. If a trade secret is unlawfully obtained—for example, by breach of trust or violation of a confidential relationship—the courts could award the trade secret's owner compensation for damages suffered and forbid the infringer use and further disclosure of the trade secret.

It should be recognized that, although trade secrets have no direct cost in obtaining any property right, they in fact are generally expensive to establish adequate protection systems. These would include establishing security systems and confidentiality agreements, identifying confidential information with physical restrictions, limiting plant tours, making covenants not to compete, etc.

Comparison of Means of Protecting Ideas

Table 9-5 compares the various means of protecting ideas just discussed. Any innovator or author should be familiar with these options so that an intelligent decision can be made on the proper protection needed for each idea. Different options offer very different kinds of protection. For example, the Coca-Cola Company has elected to protect the ingredients, mixing, and brewing of its principal product, Coca-Cola, as trade secrets. This decision does not prevent another company that claims to have discovered these secrets from marketing a similar product. The trade-secret approach, however, protects the Coca-Cola Company's information for as long as it remains secret. Had the company patented these formulas, the knowledge would have been dedicated to the public 20 years after the patent application.

Many ideas that are protected as trade secrets cannot be patented. On the other hand, an item that is patentable can theoretically be protected as a trade secret. If the idea can be easily discovered through reverse engineering, however, a patent is the only practical choice for protection.

Computer software may be protected by copyright as literary works. It may be that a utility patent could be used to protect it. A utility patent protects the idea, whereas the copyright would only protect the expression. The distinction of what constitutes the idea and what constitutes the expression is one that is often decided by the courts. Recent practice has been to seek protection of software by utility patents to ensure the strongest protection.

Databases that consist of facts are not protectable by copyright. That leaves only the means of trade secret. However, if the value of the database is in making it available to the public, it cannot be protected. The European Union provides protection for databases, while debate has continued about a means of protection in the United States since the 1991 Feist case.

In theory, a design may be protected not only by a design patent but also a copyright under the category of pictorial, graphic, or sculptural works. However, the design of a useful article may be considered a pictorial, graphic, or sculptural work only if the design features can be identified separately from, and are capable of existing independently of, the utilitarian aspects of the article.

In summary, intellectual property law is a rapidly changing environment with many nuances. The engineering manager must understand the fundamentals sufficiently to be able to know when and how to interact with the legal experts. Failure to do so can be costly in terms of lost sustainable strategic competitive advantages.

Table 9-5 Comparison of Means of Protecting Ideas

Category	Utility Patents	Design Patents	Trademarks	Copyrights	Trade Secrets
Idea or subject matter	New and useful processes, machines, articles of manufacture, and compositions of matter	New ornamental designs for articles of manufacture	Words, names, symbols, or other devices that serve to distinguish goods or services	Writings, music, works of art, and the like that have been reduced to a tangible medium of expression	Almost anything that is secret, substantial, and valuable
Sources of protection	U.S. Patent and Trademark Office patent	U.S. Patent and Trademark Office patent	Registration with the U.S. Patent and Trademark Office Registration with the secretary of state Common-law protection through courts as long as proper use continues	Federal law protects only a tangible medium of expression Enforceable only when registered with the copyright office	Primarily common-law protection through courts
Terms of protection	20 years from application filing date	14 years from issue date of patent	10 years from registration with federal office; renewable for additional 10-year terms	Life of author, plus 70 years	For as long as it remains a secret
Tests for infringement	Making, using, or selling invention described in patent claim	Making, using, or selling design shown in patent claim	Likelihood of confusion, mistake, or deception	Copying of protected subject matter	Taking of trade secret by breach of trust or violation of a confidential relationship

CREATIVITY, INNOVATION, ENTREPRENEURSHIP

Nature of Creativity

Creativity is the ability to produce new and useful ideas through the combination of known principles and components in novel and nonobvious ways. Another definition for creativity given by Lumsdaine is "playing with imagination and possibilities, leading to new and meaningful connections and outcomes while interacting with ideas, people, and the environment." Creativity exists throughout the population, largely independent of age, gender, and education. Yet in any group a few individuals will display creativity completely out of proportion to their number. To have an effective research organization requires understanding the creative process, identifying and acquiring creative people, and maintaining an environment that supports rather than inhibits creativity.

The Creative Process

There are a number of models for problem solving. One method, often inefficient, is simple trial and error. A second is the planning/decision-making process introduced in Chapter 4 (see Figure 4-1), which involves problem definition, identification of alternatives, and evaluating alternatives against objectives. Its major thrust is analytical reasoning, although its success is enhanced by some creativity in selection of alternatives to be evaluated. The creative process uses some of the same steps, but it emphasizes the insight that can occur subconsciously when a perplexing problem is not resolved through the analytical process and is temporarily set aside. Following are the steps usually identified in describing this process.

1. *Preparation.* Shannon describes this step as "a period of conscious, direct, mental effort devoted to the accumulation of information pertinent to the problem.... Quite often the problem is solved at this stage as one submerges oneself in the problem while trying to (a) structure the problem, (b) collect all available information, (c) understand relations and effects, (d) solve subproblems, and (e) explore all possible solutions and combinations that may lead to a satisfactory solution."
2. *Frustration and incubation.* Failure to solve the problem satisfactorily by the analytical process leads to frustration and the decision to set it aside and get on with something else. However, the problem, fortified with all the facts gathered about it, "stews" or incubates in the subconscious mind.
3. *Inspiration or illumination.* A possible solution to the problem may occur as a spontaneous insight, often when the conscious mind is at rest during relaxation or sleep. Many creative individuals are never without a notepad and pen on their person or bedside table, to write down these flashes of insight.
4. *Verification.* Intuition or insight is not always correct, and the solution revealed in a flash of insight must now be tested and evaluated to assure it is, indeed, a satisfactory solution to the problem.

Shannon defends this model:

> How do we know this process is true? Because thousands of creative people have described exactly this process when discussing their work. Over and over again we see this interplay between the conscious and the subconscious. For creative work we have this wondrously competent coupling where each part

(conscious and subconscious) is indispensable in its own way, but each is helpless without the other. When applied to problem solving, the human mind has two aspects: (1) a judicial, logical, conscious mind that analyzes, compares, and chooses; and (2) an imaginative, creative, subconscious mind that visualizes, foresees, and generates ideas from stored knowledge and experience.

Brainstorming and Other Techniques for Creativity

Dhillon describes eight creativity techniques designed for one, two, or up to a dozen people. Best known is *brainstorming,* a modern method for "organized ideation" first employed in the West by Alex Osborne in 1938, although he reports that a similar procedure had "been used in India for more than 400 years as part of the technique of Hindu teachers" under the name *Prai-Barshana,* literally "outside yourself-question." The essence of brainstorming is a creative conference, ideally of 8 to 12 people meeting for less than an hour to develop a long list of 50 or more ideas. Suggestions are listed, without criticism, on a whiteboard or flip chart as they are offered; one visible idea leads to others. At the end of this session participants are asked how the ideas could be combined or improved. Organizing, weeding, and prioritizing the ideas produced is a separate, subsequent step and make take place immediately following when these sessions are scheduled for several hours.

The preceding description is of unstructured brainstorming. For a more structured brainstorming, the Nominal Group Technique is used. In this case, the problem is presented and participants write down their ideas quietly for a short period of time (5 to 10 minutes). Then each participant in an organized manner with no repetitions presents one idea at a time. When one pass is finished, another is begun until all the ideas are presented. Then the process continues as with the unstructured brainstorming. The advantage of this process is that everyone participates, and the quiet time often leads to ideas that otherwise would not have been considered.

Dhillon next lists two brainstorming techniques that can be used by two people. In one, known as the *tear-down* approach, the first person (person A) must disagree with the existing solution to a problem and suggest another approach; next, person B must disagree with both ideas and suggest a third; then person A must suggest yet another solution; this cycle continues until a useful idea clicks. In a variant, known as the *and-also* method, person A suggests an improvement on the subject under study; person B agrees, but suggests a further improvement; this sequential improvement continues until a sound solution is reached.

In a somewhat different group technique developed by W. J. Gordon, a team explores the underlying concept of the problem. For example, if a new can opener is desired the team would first discuss . . . the meanings of the word opening and examples of opening in life things. The method encourages finding unusual approaches by preventing early closure on the problem. Gordon used a team of six meeting for about a day on a problem.

Dhillon describes two approaches in which individuals are given a description of a problem and required to list solutions in advance of group effort. In the simpler method, each participant has to have a certain number of solution ideas, say 17, to the problem before they are allowed to attend the meeting. In a more complex version known as the collective notebook method each member of a team is given a notebook with a problem statement and supporting material a month in advance. Each day during that month, the team member writes one or more ideas in the notebook, and at the end of the month selects the best idea along with suggestions for further exploration. A problem coordinator collects and studies notebooks and prepares a detailed summary for distribution; if necessary, all team members then participate in a final meeting.

Finally, Dhillon includes two methods that individuals may use. In an *attribute-listing* approach, a person lists attributes of an idea or item, then concentrates on one attribute at a time to make improvements in the original idea or item. The other method tries to generate new ideas by creating a forced relationship between two or more usually unrelated ideas or items. For example, an office equipment manufacturer might consider the relationship between a chair and a desk, start up a line of free associations, and end up with a combined unit consisting of both desk and chair.

Mindmapping combines aspects of brainstorming, sketching, and diagramming. A mindmap consists of a central word or concept with 5 to 10 main ideas that relate to that word, similar to creating a spiderweb. Tony Buzan, a British researcher, invented mindmaps in the 1970s, and they can be applied to a variety of situations including note taking, creative and report writing, studying, meetings, and think tanks. A procedure adapted from Lumsdaine for drawing a mindmap follows:

1. Start your mindmap (in a team or individually) by writing the main topic in the center of a large piece of blank paper [or whiteboard].
2. Think about what main factors, ideas, concepts, or components are directly related to your topic. Write down the most important factors as main branches off the central concept. Connect them to the main topic.
3. Now concentrate on one of these headings or main ideas. Identify the factors or issues related to this particular idea. Additional branches and details can be added if needed. Use key words, not phrases, if at all possible, to keep the map uncluttered.
4. Repeat the process for each of the main ideas. During this process, associations and ideas will not always come to mind in an orderly arrangement—soon you will be making extensions all over the mindmap. Continue the process for at least 10 minutes until you can no longer add ideas to the map.
5. Next comes the organization and analysis phase of mindmapping. Connect the related ideas and concepts. Review, annotate, organize, and revise. Edit and redraw the mindmap until you are satisfied with the logic of the relationships among all the ideas.
6. Finally, you are ready to begin writing. The time spent thinking up and organizing the mindmap will make the writing task easier. The result will be a well-organized and well-understood product.

Characteristics of Creative People

There have been many studies comparing more creative with less creative people. Characteristics of creative people can be grouped into the following categories:

Self-confidence and independence. Creative people seem to be self-confident, self-sufficient, emotionally stable, and able to tolerate ambiguity. They are independent in thought and action and tend to reduce group pressures for conformity and rules and regulations that do not make sense.

Curiosity. They have a drive for knowledge about how or why things work, are good observers with good memories, and build a broad knowledge about a wide range of subjects.

Approach to problems. Creative people are open-minded and uncritical in the early stages of problem solving, generating many ideas. They enjoy abstract thinking and employ method, precision, and exactness in their work. They concentrate intensively on problems that interest them and resent interruptions to their concentration.

Some personal attributes. Creative people may be more comfortable with things than people, have fewer close friends, and are not joiners. They have broad intellectual interests: They enjoy intellectual games, practical jokes, creative writing, and are almost always attracted by complexity.

Providing a Creative Environment

Creative people tend to be independent, nonconformist, and to work intensively for long periods, but with a disregard for conventional work hours. They are most effective in an organization that accepts differences, eliminates as much routine regulation and reporting as feasible, provides support personnel and equipment as required, and recognizes and rewards successes.

Creativity and Innovation

Invention (the creative process) only produces ideas. Ideas are not useful until they are reduced to practice and use, which is the process of innovation. Roberts and Wainer have identified five kinds of people who are needed for technological innovation:

> *Idea generator—the creative individual*
> *Entrepreneur—the person who carries the ball*
> *Gatekeeper—discussed below*
> *Program manager—the person who manages without inhibiting*
> *Sponsor or champion—the person, often in senior management, who provides financial and moral support*

Technological Gatekeepers in R&D Organizations

Allen and Cohen found that only about 15 percent of the scientific and technical ideas being worked on in industrial laboratories came directly from the scientific and technical literature—most of it reached lab members in a two-step process involving **gatekeepers** (although the role might be more accurately defined as enablers). These are research staff members who, through their professional work habits, bring essential information into the organization. Gatekeepers (1) are more likely to read the more sophisticated (refereed) journals, (2) are in contact with outside specialists, and (3) form a network with other gatekeepers. They often are high technical performers, usually produce more than their share of conference papers and refereed articles, and are likely to be promoted to first- and second-line supervision ahead of their peers. Gatekeepers are not appointed, but the wise research manager recognizes them and their function. Professional staff who are hired away from other organizations or who transfer in from other parts of a corporation provide another important source of new ideas and ways of doing things.

Entrepreneurship

It takes a special kind of person to lead the innovation task successfully—the entrepreneur is one who undertakes the effort to transform innovations into economic goods. Betz extends this:

> The entrepreneur is a kind of business hero; and like all heroes, they have qualities to be admired: initiative, daring, courage, commitment. These values are especially admired in turbulent business conditions, when initiative is required for survival.

The Entrepreneurial Mindset

As noted in the section on entrepreneurship, entrepreneurs are often lionized in Western culture as a hero and the idea of intrapreneurship is relatively new in business contexts. At the core of this type of creativity and risk taking is mindset—both of individuals and the organization. By approaching new problems and opportunities with a mindset that is curious about the world, open to new ideas, and seeks holistic solutions; engineers can be true difference makers in their organizations.

The KEEN Network states that: "Engineers with an Entrepreneurial Mindset Transform the World" because they "understand the bigger picture, can recognize opportunities, evaluate markets, and learn from mistakes to create value for themselves and others." They define this mindset to have three components: curiosity, connections, and creating value. These three components are critical because:

- Only through curiosity can we make sufficient discoveries to support a rapidly changing world.
- Discovery alone is not enough; only through connecting these discoveries with information on the broader systems can we develop innovative solutions.
- These innovative solutions won't be impactful if they are not matched to needs; when unmet needs are filled in an innovative way, real value is created.

What does this mean for the established or aspiring engineering manager? First, an engineering skillset is no longer enough for a long-standing successful career, much less to change the world. Technical skills represent the foundation of engineering practice, but they are made far more impactful with the mindset described here. How do you develop this mindset in yourself and your employees? A key step is approaching every design problem with a focus on the opportunity by thinking through the following questions. Why is this design needed? How can we better meet the needs of those we are designing for? Who else might need this design? How can we make it more impactful?

When we approach our engineering problems in this manner, we are well positioned to make a difference. Whether that is as the "heroic" entrepreneur or the successful intrapreneur driving value in an existing business.

Source: The Kern Family Foundation

Kern Family Foundation. "What Is Entrepreneurial Mindset?" https://engineeringunleashed.com/mindset-matters.aspx. December 2018.

While the initial concept of an entrepreneur is of a person who creates a new business for personal profit, established corporations need continuing entrepreneurial activity to create the new products and new businesses that will assure future growth of the organization; the term **intrapreneurship** has been coined to describe this activity. The challenge in managing technology is to provide a climate where intrapreneurs are encouraged to take risks, are given needed resources and time, and are permitted early failures, while shifting to closer control of resources and costs as products become mature.

Creativity is the ability to produce new and useful ideas and there are a number of techniques for creativity. The creative process only produces ideas and the ideas are not useful until they are reduced to practice and use. This is called innovation. Some ideas might lead to being an entrepreneur, some might lead to a patent, some might lead to both, and other ideas might fall by the wayside.

DISCUSSION QUESTIONS

9-1. Figure 9-1 presents a generic product lifecycle flow. Draw a more detailed version and state why the steps you have added are important.

9-2. Use the full-range leaderhsip model (Chapter 3) to outline the kind of leader who would be suitable at each stage of Betz's technology life cycle? Do they differ by stage? Why or why not?

9-3. Summarize the principal contributions to U.S. R&D activity by each of (**a**) the federal government; (**b**) industry; and (**c**) universities.

9-4. Explain the roles of entrapreneurship and intrapreneurship in the U.S. economy? Is it easier to see one of these than another? Why?

9-5. Discuss the relationship between the central corporate research laboratory and divisional research in a corporation you know or have found described in the literature.

9-6. Why are simple checklists used as a first screening of ideas in research projects by many companies?

9-7. As an R&D manager, what actions might you take or programs might you implement to assure your organization got maximum benefit from patentable ideas?

9-8. How do the kinds of ideas best protected by patent differ from those best protected by keeping them a trade secret?

9-9. What are some of the steps a manager can take to encourage creativity in his or her technical employees?

9-10. What is the role of mindset in creativity and driving innovation? According to KEEN, how might this mindset be developed in engineering students?

9-11. How would you try to evaluate the effectiveness of researchers if you were their research manager? Outline specific controls you might want to use.

9-12. What are some of the support services an organization might provide to make the work of researchers and design engineers more effective?

9-13. Discuss how the management functions of planning, organizing, leading, and controlling relate to research and development.

PROBLEMS

9-1.   An engineer proposes to buy a machine for $100,000 today that will save $60,000 in labor costs at the end of each of the next two years. If the company demands a 15 percent return on investments such as this, what is the net present worth (NPW) of the proposal? Should it be funded?

9-2.   Your company has two alternative opportunities, each requiring your entire capital investment budget of $325,000. Alternative A will return $390,000 at the end of one year; alternative B will return $216,000 at the end of each of the first two years. Which (if either) alternative should you recommend on the basis of (**a**) simple payback time? (**b**) net present worth using a 15% rate of return?

9-3.   If you have been exposed to capital investment analysis and/or engineering economy, comment on the proposal to invest $1 million in a new product now that is projected to generate $200,000 profit at the end of each year for eight years, assuming that your company requires 15 percent return on investment before taxes.

SOURCES

Allen, T. J. and Cohen, D. I., "Information Flow in Research and Development Laboratories," *Administrative Science Quarterly*, 14, 1969, pp. 12–19.

Allen, Thomas J., *Managing the Flow of Technology: Technology Transfer and the Dissemination of Technological Information Within the R&D Organization* (Cambridge, MA: The MIT Press, 1977), pp. 144–149, 163–173.

Ansoff, H. Igor and Stewart, John M., "Strategies for a Technology-Based Business," *Harvard Business Review*, 45:6, November–December 1967, pp. 71–83.

Balderston, J., Birnbaum, P., Goodman, R., and Stahl, M., *Modern Management Techniques in Engineering and R&D* (New York: Van Nostrand Reinhold Company, Inc., 1984), pp. 34–58.

Betz, Frederick, *Managing Technology: Competing Through New Ventures, Innovation, and Corporate Research* (Englewood Cliffs, NJ: Prentice-Hall, Inc., 1987).

Blake, Stewart P., *Managing for Responsive Research and Development* (San Francisco: W.H. Freeman and Company, 1978), pp. 250–261.

Dhillon, B. S., *Engineering Management* (Lancaster, PA: Technomic Publishing Company, Inc., 1987), pp. 79–104.

Erickson, Tamara J., Magee, J. F., Roussel, P. A., and Saad, K. N., "Managing Technology as a Business Strategy," *Sloan Management Review*, 31:3, Spring 1990, pp. 73–78.

Frosch, Robert A., "GM's Healthy Pain," *Mechanical Engineering*, December 1987, pp. 23–25.

GE Corporate Research Laboratories, descriptive material (Fairfield, CT: General Electric, 1980), p. 22.

Kidder, Tracy, *The Soul of a New Machine* (Boston: Little, Brown, 1981).

Leshner, A. I. and Kresa, K., "Will Science Get the Ax?" *The Gainesville Sun*, Gainesville, FL, September 30, 2012, p. 7F.

Lumsdaine, Edward and Lumsdaine, Monika, *Creative Problem Solving: Thinking Skills for a Changing World* (New York: McGraw-Hill, 1995) pp. 14–55.

Myers, Don, some of Dr. Myers's remarks have appeared in John M. Amos and Bernard R. Sarchet, *Management for Engineers* (Englewood Cliffs, NJ: Prentice-Hall, 1981).

National Science Foundation, *Patterns of R&D Resources*, Report 74–304 (Washington, DC: U.S. Government Printing Office, 1974), p. 17.

Osborn, Alex F., Applied Imagination, 3d ed. (New York: Charles Scribner's Sons, 1963), p. 151.

"Reinventing America 1992," *BusinessWeek*, October 23, 1992, p. 169.

Roberts, E. B. and Wainer, H. A., *IEEE Trans. Engineering Management*, 18:3, 1971, pp. 100–109, summarized in George E. Dieter, *Engineering Design: A Materials and Processing Approach* (New York: McGraw-Hill Book Company, 1983), p. 25.

Seiler, Robert E., *Improving the Effectiveness of Research and Development: Special Report to Management* (New York: McGraw-Hill Book Company, 1965).

Shannon, Robert E., *Engineering Management* (New York: John Wiley & Sons, Inc., 1980), pp. 235–257.

"USPTO Releases Annual List of Top 10 Organizations Receiving Most U.S. Patents: American Innovation Continues to Top the Field," United States Patent and Trademark Office Press Release #06-03, January 10, 2006.

Whelan, J. M., "Project Profile Reports Measure R&D Effectiveness," *Research Management*, 14, September 1976, pp. 14–16.

10

Managing Engineering Design

PREVIEW

This chapter begins by considering the nature of engineering design and the tasks or stages in the systems engineering and new product development processes, and the modern emphases on concurrent (simultaneous) engineering. Special control systems in engineering design—drawing/design release, configuration management, and design review—are considered. Then design criteria are introduced, which require special precautions. These criteria, which are important to design, are discussed: liability, reliability, maintainability, availability, human factors engineering, standardization, producibility, and value engineering.

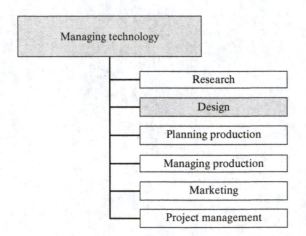

NATURE OF ENGINEERING DESIGN

Design is the activity that best describes the engineer. To design is to create something that has never existed before, either as a solution to a new problem or as a better solution to a problem solved previously. J. B. Reswick summarizes the process of design well:

> Design is the central purpose of engineering. It begins with the recognition of a need and the conception of an idea to meet that need. It proceeds with the definition of the problem, continues with a program of directed research and development, and leads to the construction and evaluation of a prototype.

Essentially, design is the process of creating a **model**, usually described in terms of drawings and specifications (whether on paper or in computer memory), of a system that will meet an identified need of the customer. The model can then be reproduced by some suitable manufacturing process and distributed for use, as described in Chapters 11 and 12.

Engineering design is a process of transforming information, as illustrated in Figure 10-1. Information provides the input to the process: a statement of the problem to be solved, design standards, design methods, and the methods of engineering science. Through the activity represented by the box labeled "Engineering design process," the engineer performs some logical sequence of activities, decisions, and analyses to develop a solution to the problem. However, this solution is of little use until the engineer communicates the solution in the form of drawings, specifications, financial estimates, written reports, and oral presentations to explain and promote the solution. Unfortunately, many engineers do not realize the importance of this vital last step of communication and do not give it the focus it deserves. Without successfully completing this last step, the rest of the work will be wasted.

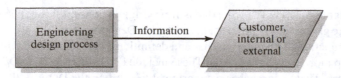

Figure 10-1 The engineering design process.

SYSTEMS ENGINEERING/NEW PRODUCT DEVELOPMENT

The design of a complex engineered system, from the realization of a need (for a new system or improvement of an existing system) through production to engineering support in use is known as **systems engineering** (especially within military or space systems) or as **new product development**. The (U.S.) National Aeronautics and Space Administration (NASA) offered a good definition of the first in their *NASA Systems Engineering Handbook*:

> **Systems engineering** is a robust approach to the design, creation, and operation of systems. In simple terms, the approach consists of identification and quantification of system goals, creation of alternative system design concepts, performance of design trades, selection and implementation of the best design, verification that the design is actually built and properly integrated, and post-implementation assessment of how well the system meets (or met) the goals. The approach is usually applied repeatedly and recursively, with several increases in the resolution of the system baselines (which contain requirements, design details, verification procedures and standards, cost and performance estimates, and so on).

Whether called systems engineering or new product development, the engineering of complex systems is carried out in a series of sequential phases or stages. Systems engineering as defined by the International Council on Systems Engineering (INCOSE) is an engineering discipline whose responsibility is ensuring that the customer and stakeholder's needs are satisfied in a high quality, trustworthy, cost efficient, and schedule compliant manner throughout a system's entire life cycle. INCOSE was formed in 1990 and is leading the development of the systems engineering discipline, which has been evolving over the last 30 years.

This process is usually comprised of the following seven tasks:

- State the problem
- Investigate alternatives
- Model the system
- Integrate
- Launch the system
- Assess performance
- Re-evaluate.

It is important to note that the systems engineering process is not sequential. The functions may be performed in a parallel and are almost always done in an iterative manner.

Tasks/Stages in Systems Engineering

The tasks of the **system life cycle** (which extends from original concept through systems engineering to product disposal) were given slightly different names by the Department of Defense (DOD) and NASA, but they cover the same functions. Other approaches and definitions were developed within different fields as organizations develop more complex systems. These included ANSI/EIA-632, ISO/IEC 15288, and IEEE Std 1220 among others. These approaches were originally based on the DOD, MIL-STD-499B. INCOSE was created to address the need for improvements in systems engineering practices and education.

Each stage begins with approval to expend the resources that phase will require and agreement on the work to be accomplished in that phase. Next comes accomplishment of the work of the stage, which may be modest to enormous. The results of that stage are then compiled: designs and specifications, analyses and reports, and a proposed plan for conducting the next phase if one is recommended. At this point there should be a conscious, and often formal, review to decide whether the expense of the next stage, which will usually represent a substantial increase in resource commitment, is justified. Typically, one of three types of decisions should come out of this review: (1) to cancel the development, if study to that point has shown that further development cannot be justified; (2) to go back (recycle) and do more work in the present stage if too many uncertainties still exist; or (3) to proceed with the next stage and its increased resource expenditure with confidence.

The stages are defined by INCOSE as follows:

State the problem

The problem statement starts with a description of the top-level functions that the system must perform: This might be in the form of a mission statement, a concept of operations or a description of the deficiency that must be ameliorated. Most mandatory and preference requirements should be traceable to this problem statement. Acceptable systems must satisfy all the mandatory requirements. The preference requirements are traded-off to find the preferred alternatives. The problem statement should be in terms of *what* must be done, not *how* to do it. The problem statement should express the customer requirements in functional or behavioral terms. It might be composed in words or as a model. Inputs come from end users, operators, maintainers, suppliers, acquirers, owners, regulatory agencies, victims, sponsors, manufacturers and other stakeholders.

Investigate alternatives

Alternative designs are created and are evaluated based on performance, schedule, cost and risk figures of merit. No design is likely to be best on all figures of merit, so multicriteria decision-aiding techniques should be used to reveal the preferred alternatives. This analysis should be redone whenever more data are available. For example, figures of merit should be computed initially based on estimates by the design engineers. Then, concurrently, models should be constructed and evaluated; simulation data should be derived; and prototypes should be built and measured. Finally, tests should be run on the real system. Alternatives should be judged for compliance of capability against requirements. For the design of complex systems, alternative designs reduce project risk. Investigating innovative alternatives helps clarify the problem statement.

Model the system

Models will be developed for most alternative designs. The model for the preferred alternative will be expanded and used to help manage the system throughout its entire life cycle. Many types of system models are used, such as physical analogs, analytic equations, state machines, block diagrams, functional flow diagrams, object-oriented models, computer simulations and mental models. Systems Engineering is responsible for creating a product and also a process for producing it. So, models should be constructed for both the product and the process. Process models allow us, for example, to study scheduling changes, create dynamic PERT charts and perform sensitivity analyses to show the effects of delaying or accelerating certain subprojects. Running the process models reveals bottlenecks and fragmented activities, reduces cost and exposes duplication of effort. Product models help explain the system. These models are also used in tradeoff studies and risk management.

As previously stated, the Systems Engineering Process is not sequential: it is parallel and iterative. This is another example: models must be created before alternatives can be investigated.

Integrate

No man is an island. Systems, businesses, and people must be integrated so that they interact with one another. Integration means bringing things together so they work as a whole. Interfaces between subsystems must be designed. Subsystems should be defined along natural boundaries. Subsystems should be defined to minimize the amount of information to be exchanged between the subsystems. Well-designed subsystems send finished products to other subsystems. Feedback loops around individual subsystems are easier to manage than feedback loops around interconnected subsystems. Processes of co-evolving systems also need to be integrated. The consequence of integration is a system that is built and operated using efficient processes.

Launch the system

Launching the system means running the system and producing outputs. In a manufacturing environment this might mean buying commercial off the shelf hardware or software, or it might mean actually making things. Launching the system means allowing the system do what it was intended to do. This also includes the systems engineering of deploying multi-site, multi-cultural systems.

This is the phase where the preferred alternative is designed in detail; the parts are built or bought (COTS), the parts are integrated and tested at various levels leading to the certified product. In parallel, the processes necessary for this are developed – where necessary – and applied so that the product can be produced. In designing and producing the product, due consideration is given to its interfaces with operators (humans, who will need to be trained) and other systems with which the product will interface. In some instances, this will cause interfaced systems to co-evolve. The process of designing and producing the system is iterative as new knowledge developed along the way can cause a re-consideration and modification of earlier steps.

The systems engineers' products are a mission statement, a requirements document including verification and validation, a description of functions and objects, figures of merit, a test plan, a drawing of system boundaries, an interface control document, a listing of deliverables, models, a sensitivity analysis, a tradeoff study, a risk analysis, a life cycle analysis and a description of the physical architecture. The requirements should be validated (Are we building the right system?) and verified (Are we building the system right?). The system functions should be mapped to the physical components. The mapping of functions to physical components can be one to one or many to one. But if one function is assigned to two or more physical components, then a mistake might have been made and it should be investigated. One valid reason for assigning a function to more than one component would be that the function is performed by one component in a certain mode and by another component in another mode. Another would be deliberate redundancy to enhance reliability, allowing one portion of the system to take on a function if another portion fails to do so.

Assess performance

Figures of merit, technical performance measures, and metrics are all used to assess performance. Figures of merit are used to quantify requirements in the tradeoff studies. They usually focus on the product. Technical performance measures are used to mitigate risk during design and manufacturing. Metrics (including customer satisfaction comments, productivity, number of problem reports, or whatever you feel is critical to your business) are used to help manage a company's processes. Measurement is the key. If you cannot measure it, you cannot control it. If you cannot control it, you cannot improve it. Important resources such as weight, volume, price, and communications bandwidth and power consumption should be managed. Each subsystem is allocated a portion of the total budget and the project manager is allocated a reserve. These resource budgets are managed throughout the system life cycle.

Re-evaluate

Re-evaluate is arguably the most important of these functions. For a century, engineers have used feedback to help control systems and improve performance. It is one of the most fundamental engineering tools. Re-evaluation should be a continual process with many parallel loops. Re-evaluate means observing outputs and using this information to modify the system, the inputs, the product or the process.

Variations

Like all processes, the Systems Engineering process at any company should be documented, measurable, stable, of low variability, used the same way by all, adaptive, and tailor-able! This may seem like a contradiction. And perhaps it is. But one size does not fit all. The above description of the Systems Engineering process is just one of many that have been proposed. Some are bigger, some are smaller. But most are similar to this one.

Disposal Stage.   Although this is not listed as a separate stage, every product causes waste—during manufacture, while in use, and at the end of its useful life—that can create disposal problems. The time to begin asking, "How do we get rid of this?" and "How do we protect the environment?" is in the early stages of product or process design. The U.S. nuclear weapons program leaves us with many billions of dollars in costs to mitigate radioactive waste, much of which could have been saved had disposal been considered from the beginning. Chemical, petroleum, steel, and other *smokestack* industries send enormous amounts of waste into our air and water, much of which can be eliminated (sometimes even at a profit in material recovered) by improving production processes.

The simplest example of waste problems created during product use is the automobile. Used tires, discarded lead acid batteries, fluorocarbons from air conditioners, and gasoline tank fumes each present an environmental hazard, and require a complex recovery network. Packaging gives us mountains of glass, plastic, steel cans, and aluminum, each of which requires a different process to recycle.

CONCURRENT ENGINEERING

Concurrent (Simultaneous) Engineering

In traditional engineering, a relatively short time was spent defining the product. A relatively long time is spent designing the product, and a surprisingly longer time is often spent redesigning the product. The key to shortening the overall design time is to better define the product and better document the design process. To do this, a new approach is applied to engineering design that creates products that are better, less expensive, and more quickly brought to market. This trend unites technical and nontechnical disciplines such as engineering, marketing, and accounting. While always focusing on satisfying the customer, these representatives work together in defining the product to be manufactured.

This approach to reduce time-to-market has become widely adopted under the name *concurrent engineering* for development of both industrial and military systems.

Benefits of concurrent engineering (CE) include 30 to 70 percent less development time, 65 to 90 percent fewer engineering changes, 20 to 90 percent less time-to-market, 200 to 600 percent higher quality, and 20 to 110 percent higher white-collar productivity. [As reported by the National Institute of

Standards and Technology, Thomas Group Inc., and Institute for Defense Analyses in *Business Week*, April 30, 1990.]

Most manufacturing firms have targeted at least a 50 percent reduction in the time it takes to launch a new product from idea to production. Yet, few organizations have pushed this concept to the point of having a corporate design strategy or a way of projecting the design and full-range planning of all their products five years into the future, a direction many companies are now seeking. Good ideas that are novel have a unique motivating quality. People become excited about them, and eventually there will be competition and disagreement about their origin. Nonetheless, most ideas, good or bad, are never acted upon either by individuals, groups, or, enterprises.

To summarize CE in practice, the following are used for faster product development and fewer changes:

- Colocate key functional disciplines.
- Organize cross-functional teams.
- Use computer-aided design (CAD) and prototyping software.
- Conduct thorough design reviews at design concept and definition stages.
- Involve key disciplines, especially manufacturing, early in development.
- Prepare properly for CE implementation.
- Allow for a CE learning curve.
- Implement CE in small, manageable bites.

How, then, can the sequential systems engineering and new product development multistage process be retained while gaining the benefits of concurrent engineering? Figure 10-2 illustrates the interplay of

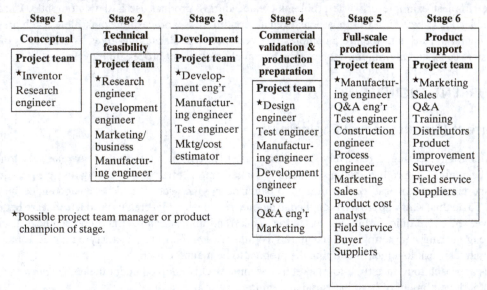

Figure 10-2   Formulation of a multispeciality project team designed to accomplish concurrent engineering in the stages of new product development. (Source: Adapted from National Society of Professional Engineers, *Engineering Stages of New Product Development*, NSPE Publication #3018, Figure 22, p. 10 Reprinted by permission of the National Society of Professional Engineers (NSPE) www.nspe.org.)

technical specialties and changes in team leadership that take place as a CE team carries a system from its conception through to postproduction product support.

CONTROL SYSTEMS IN DESIGN

In creating a complex system (e.g., an automobile or rocket), hundreds or thousands of engineers, technicians, and other workers may be involved in creating designs, reviewing them, manufacturing or constructing in accordance with them, or inspecting to assure that what has been made agrees with what was specified. However, no complex system is specified perfectly to begin with, and the needs of the user may change during the design phase; so design changes are inevitable. Control systems for drawing/design release and configuration management are essential to assure that everyone knows what the official design (configuration) is at any instant, while change can be managed effectively. Design review, introduced earlier in this chapter, is also an essential control system in design.

Drawing/Design Release

Drawing release is the process of identifying when a particular design drawing or change has been officially accepted. When a designer finishes with a drawing, it must go through a series of checks and/or analyses by appropriate specialists. At some point, the drawing or design (which might consist of a number of drawings) needs to be officially *released*, so that others may work with it. Depending on the design phase, this release might be to further development, to validation testing, or to production. In the past, a drawing made the rounds through designated analysts and checkers; was signed off by the appropriate supervisor(s) and/or project engineer(s); and then was released by rubber-stamping the reproducible master at an official *release desk*, from which copies were sent to an official distribution list with instructions to destroy any previous version—those not on the list were left in the dark. In team management, the needed specialists are brought together in integrated product teams empowered to create, review, and approve designs concurrently; design release is then affected by changing the design detail recorded in a common electronic database accessible to all who need it (including, increasingly, suppliers and even customers).

In team management, business as usual is ineffective because of the complexity of political, regulatory, and technological changes. Globalization (Chapter 18) drives a greater need for communication and coordination across all departments and different time zones requires organizations to react faster. Flexibility in information technology (IT) enables faster adaptation to these constant and rapid changes. IT plays a fundamental role in supporting critical changes efficiently and effectively.

In modern organizations, IT is seen as the critical force in the transformation of competition, firm structures, and firm boundaries. The latest innovations in information system (IS) of the goods and services and changes in drawing and design can be made at any time, in any place, and in any variety. The virtual organization for the customer is cost-effective, produces instantaneous results, and is customized to the customer's request. Management that is linked by IT can simultaneously share skills, costs, and potential changes, as well as access and assess each other's markets. Today flexibility and responsiveness of the decentralized organization are important elements for success.

Configuration Management

Designing a new system or product is a very complex undertaking. With global corporations and organizations all racing to bring new ideas and products quickly to market, the ability to control and organize diverse teams working on a joint design project is critical. Along the way, design specialists often find some need for or advantage in changing the design in one way or another. However, the designer often cannot tell what the impact of the change will be on the other parts of the complex system being developed. It is critical for all of this information to be conveyed to the rest of the system design team. A design change in one area may require engineers in a second discipline to provide more electrical energy, those in another group to find a way to carry off more heat, test engineers to modify their test equipment, and training-manual writers to discard what they have written and start over. Moreover, unless everyone knows the design criteria their design neighbors are working toward, they may be investing substantial effort in designing a system that no longer exists. Therefore, there must be agreement as to the current design criteria (or configuration), and there must be a control system to define this configuration and rigorously control changes to it. It is a very important part of the management process to ensure that the communication lines are kept open between the designers and the workers in the field, and that changes are relayed and the correction is followed through and completed.

In systems design programs for the U.S. DOD, there are several points where the current design criteria are to be specified. These are their baseline documents. Examples of these are as follows:

- At the end of the concept exploration and definition phase, a functional baseline is developed to identify the functional characteristics and design constraints that must be satisfied by the design.
- At the end of the demonstration and validation phase, an allocated baseline is prepared, which describes the performance characteristics each subsystem and component must meet.
- At the end of the engineering and manufacturing development phase, a product baseline is established consisting of all the detailed specifications required for production.

Baselines are therefore part of the material submitted at the end of a phase for approval in the design review (described subsequently), and they form the basis for beginning work in the next phase.

Changes to these baselines during a phase of the design process are governed by a system known as configuration management (or control). This system usually involves a committee known as a configuration control board (CCB). The CCB is made up of members from the major design branches and other functions (reliability, production planning, training, etc.) that are affected by change. If, during the design process, a need or desire arises to change a significant part of the design, the exact change proposed is identified to all CCB members. The CCB then analyzes the impact in dollars and the time delays of the proposed changes to their project. The CCB then discusses the total impact and compares it with the benefit afforded by the change. After review, the decision to change or not to change a baseline must be made. The executive appointing the CCB or—where the system is being created under contract to someone else—the customer or client then makes a decision based on the CCB estimate. Up to that decision point, no work is done on the proposed new configuration. If it is adopted, all parts of the organization are immediately notified and everyone begins incorporating the change. Where change in a configuration item affects its interface with other systems (its space envelope or the physical energy or other interactions that pass through it), an interface control working group of some sort is needed to coordinate these changes. Currently, configuration management is making inroads into all aspects of product and systems development. With a globalization of ideas and engineers working in diverse locations, projects are now required to be accessed from anywhere and at any time.

In many businesses, automation is becoming one of the principle means of achieving greater productivity and higher product quality. To resolve the problems of more work with less, computer-assisted software and other engineering tools for the development and integral support processes are now available from a large number of vendors worldwide. These include tools to aid in performing the software configuration management (SCM) process.

One definition of SCM is the identification, control, status accounting, and audit of a software product as it evolves from a conceptual stage through delivery and into maintenance. More information on these definitions can be found in MIL STD-973 (EIA/IS649). SCM is an active discipline that is integral to the software engineering process and must support the definition and implementation of the software process itself. It must manage changes to all project components as they move through their approval life cycle and meet the needs of both engineers and management. SCM must also manage the evolving process model applied to the objects under control. The definition of SCM carries the implicit requirement that it must be automated; otherwise, it will fail its mission by neither providing a current view of what is taking place nor supporting the rapid pace of the engineering process.

An SCM system represents the system structure because it captures a collection of files as a composite that can have a series of versions that may contain source objects, files, build rules, documents, and a software product hierarchy. By capturing all of these artifacts, the system structure is represented and placed under configuration control. There are several different classifications of tools that are available for evaluation and selection. Some systems may aid the developers in selecting previous versions to recreate or compose a new version, whereas other systems will provide for a series of transactions by a team of developers working on a selected version to create a new version. Other systems may include the use of change sets or deltas that enable the developer to make a choice of adding or subtracting from an established version or a baseline to create a desired version or update the baseline.

One way to address the problem of the present system of configuration management is with the use of automated version control and configuration management. The benefits of automated version control and configuration management include the following:

- *Improves communication among extranet partners.* By automating the communication process, a version control system enables the webmaster to establish a single, consistent channel for communication and processing change requests for all parties, ensuring that none fall through the cracks.
- *Protects shared web source files under rapid development.* A version control system helps you store and track changes to web source files.
- *Enhances development work flow.* It encourages the establishment of good work flow practices, develops work flow by enabling the webmaster to quickly prioritize and assign web content requests, runs reports to determine the status of any request, determines whether project files are still checked out, or views a summary of the modifications made to project files.
- *Saves time.* Integrated system enhancements, new features, and content can be added much more quickly and at less expense.
- *Reduces the number of defects introduced into the system.* Defects caused by accidental overwrites, lack of communication, and manual merging of changes can be prevented by an effective version control system.

- *Reduces the costs and time to find defects that are introduced.* Most systems feature a severity rating system that enables team members to specify the priority level of their change requests.
- *Reduces maintenance costs.* The ability to recreate an earlier revision or build of the system.
- *Improves productivity of the development team.* When communication is streamlined and everyone has visibility into all aspects of a project, true team collaboration is possible and productivity skyrockets.
- *Reduces the costs of content and application development.* Eliminates unproductive meeting time and redundant communication while eliminating rework and unnecessary changes, and eliminates time spent preparing manual reports.
- *Improves the quality of extranet applications.* Ensures that outstanding issues are resolved, enables early and ongoing participation by nontechnical staff, and encourages software component reuse.

Secondary benefits include better corporate image, improved team morale, less overtime and fewer working weekends required from the development staff, increased respect for the extranet development team from organizations external to the effort, more competitive stance in the marketplace, increased customer satisfaction, and improved communications among all staff at all levels and between levels.

Design Review

Design reviews are generally scheduled prior to each evolution step in the design process. In some instances, this may entail a single review toward the end of each phase (i.e., conceptual, preliminary system design, detail design, development). For other projects, where a large system is involved and the amount of new design is extensive, a series of formal reviews may be conducted on designated elements of the system. This may be desirable to allow for early processing of some items while concentrating on more complex high-risk items. All projects start with a kick-off meeting where a project manager prepares a project plan that is distributed to all project team players. The plan describes drawing and/or design release procedures, drawing change procedures, and how and when design reviews will be conducted. Many companies have policies, procedures, and standards already in place that cover these areas.

Although the quality and type of design reviews scheduled may vary from program to program, four basic types are readily identifiable and common to most programs. They include the conceptual design review (i.e., systems requirement review), the system design review, the system software design review, and the critical design review.

- *Conceptual design review.* The conceptual design review may be scheduled during the early part of a program (preferably four to eight weeks after program start) when operational requirements and the maintenance concept have been defined. Feasibility studies justifying preliminary design concepts should be reviewed. Logistics support requirements at this point are generally included in the specification of supportability constraints and goals, and in the maintenance concept definition generally contained in the system specification.
- *System design review.* System design reviews are generally scheduled during the preliminary design phase when preliminary system layouts and specifications have been prepared (before their formal release). These reviews are oriented to the overall system configuration in lieu of individual equipment items. Supporting data may include functional analysis and allocations, preliminary supportability analysis, and the trade-off study reports. There may be one or more formal reviews scheduled,

depending on the size of the system and the extent and complexity of the new design. The purpose of the review is to determine whether the design is compatible with all system requirements and whether the documentation supports the design.

- *System/software design review.* System/software design reviews are scheduled during the detail design and development phase when layouts, preliminary mechanical and electrical drawings, functional and logical diagrams, design databases, and component part list are available. In addition, these reviews cover engineering and hardware, software models or mock-ups, and prototypes. Supporting the design are reliability analyses and predictions, maintainability analyses and predictions, human factor analyses, and logistics support analyses. The design process at this point has identified specific design constraints, additional or new requirements, and major problem areas. Such reviews are conducted prior to proceeding with finalization of the detail design.
- *Critical design review.* The critical design is scheduled after detail design has been completed, but prior to the release of firm design data to production. Such a review is conducted to verify the adequacy and producibility of the design. Design is essentially frozen at this point, and manufacturing methods, schedules, and costs are reevaluated for final approval.

The critical design review covers all design efforts accomplished subsequent to the completion of the system/software review. This includes changes resulting from recommendation for corrective action stemming from the equipment/software design review. Data requirements include manufacturing drawings and material list, a production management plan, final reliability and maintainability predictions, engineering test reports, a firm supportability analysis, and a formal logistics support plan.

DESIGN CRITERIA

Product Liability

Through the centuries, a relationship of buyer and seller stemming from the Roman philosophy of *caveat emptor* (let the buyer beware) persisted in Western thinking. Although English law permitted recovery by a plaintiff based on the negligence of a defendant, this was possible only where a direct contractual relationship (*privity of contract*) existed. In the famous 1842 case *Winterbottom* v. *Wright*, for example, the injured driver of a defective mail coach could not sue the maker of the coach, because there was no such privity. This changed in U.S. law in New York in 1916, when a man named MacPherson was awarded damages from Buick Motor Company for harm done by a defect in his car, even though his contract for the car was with a dealer (who in turn had purchased the car from Buick).

Thus began the era of *product liability*, which has had far-reaching effects on how companies make and describe their products, and which requires great care on the part of their engineers and managers. Initially, the plaintiff (the injured party) had to prove *negligence*; that is, to show that the manufacturer omitted doing something that a *reasonable person* guided by the *ordinary considerations* that regulate human affairs, would have done, or did something that a "reasonable and prudent person" would not have done. The manufacturer, on the other hand, could defend itself by showing that the plaintiff did not use the product as a *reasonable person* would (and was therefore guilty of *contributory negligence*). However, in the 1960 California case of *Hennington* v. *Bloomfield Motors*, another auto manufacturer was found liable when the steering mechanism on a new car failed at only 20 mph, causing the car to swerve and hit a wall. Although there was no privity and negligence could not be proven, the court concluded that there had been a breach of an *implied warranty* of merchantability and fitness for use.

More constraining was the case of *Greenman* v. *Yuba Power Products*, in which Mr. Greenman was injured when a piece of wood he was turning on his combination lathe/saw/drill press flew out and struck his head. The California Supreme Court ruled that "a manufacturer is *strictly liable* in tort when an article he places on the market, knowing it is to be used without inspection for defects, proves to have a defect that causes injury to a human being." Still worse is the threat of *absolute liability*, where "a manufacturer could be held strictly liable for failure to warn of a product hazard, even if the hazard was scientifically unknowable at the time of the manufacture and sale of the product."

Liability problems must be attacked prudently and rapidly, for they can destroy even large companies. Manville Corporation was forced into bankruptcy in 1982 because of the claims of tens of thousands of workers exposed to asbestos. A. H. Robins Co. faced 200,000 claims of injuries (about 20 resulting in death) because of a plastic intrauterine device (IUD) the size of a nickel; the company was forced into bankruptcy in 1985 and established, with their insurers, a $2.38 billion trust fund to meet claims. In December 1999, an appeals judge upheld $259 million in damages against DaimlerChrysler in a case in which a six-year-old child was thrown out of the rear of a 1985 Dodge Caravan. At issue was a door-latch design that the rest of the auto industry had abandoned. General Motors lost a case in California in July 1999 concerning fuel tank fires. The company was ordered to pay $4.9 billion to six people who were burned when their 1979 Chevrolet Malibu exploded after a rear-end collision. Other cases of large punitive damage awards are the McDonald's coffee case and the tobacco controversy. See the Toyota and Takata story for current issues in this area. Johnson and Johnson, on the other hand, withdrew their profitable Tylenol product from retail shelves everywhere immediately on learning that poison had been inserted into some bottles of product; they retained consumer confidence and regained most of their market share when they resumed deliveries in new tamper-resistant containers.

Reducing Liability.   To protect against product liability, designers must foresee even unlikely conditions, taking active responsibility for how their products might be used. A manufacturer producing wooden doors with a window at the top packaged the doors in a stack with windows aligned and a cardboard cover over the stack to protect the windows. A stevedore walked across a stack of these doors in a ship hold, fell through the glass windows, and sued for injuries. The manufacturer was held liable on the theory that they should have known that this is the way stevedores behave.

How can the designer reduce the threat of product liability? The simplest way to support this outcome is for designers to take active responsibility for how their product might be used or misused. Some specific actions that designers and managers can take to increase safety and reduce future liability problems are:

- Include safety as a primary specification for product design.
- Use standard, proven materials and components.
- Subject the design to thorough analysis and testing.
- Employ a formal design review process in which safety is emphasized.
- Specify proven manufacturing methods.
- Assure an effective, independent quality control and inspection process.
- Be sure that there are warning labels on the product where necessary.

Product Liability: Toyota and Takata

Product liability is a risk for even the most respected brands. Companies that are considered to have the highest quality levels (quality management is discussed in Chapter 12). Perhaps the most surprising of these recent cases were the challenges faced by Toyota Motor beginning in 2009. The issues started with reports of drivers experiencing sudden, unexpected, acceleration leading to crashes and even deaths. Initially the company sought to deflect claims of liability for the problem, pointing to lack of information and driver error. In the end, Toyota was found to be at fault for two mechanical defects: "sticking" accelerator pedals and a design flaw that allowed the accelerator to become trapped under floor mats, while being exonerated of electrical defects that lead to acceleration (following a 10-month study led by NASA engineers). Initially, the defects led to the recall of nearly 8 million vehicles and $48.8 million in civil penalties. It also led to Toyota's President, Akio Toyoda, making a very public apology. Over time, the initial settlements expanded greatly. After over four years of investigation and legal actions, the initial cover-up of the issues led to a $1.2 billion deferred prosecution agreement.

As if their own manufacturing issues were not trouble enough, almost immediately following the accelerator issues, issues with supplier parts to Toyota (and 18 other manufacturers) created the largest and most complex recall in U.S. automotive history. The source of the recall was a defect in the inflators of Takata manufactured airbags installed between 2002 and 2015. At the time of this writing, the full scope of the recall is still being investigated but it is expected to impact 37 million vehicles and take years to resolve. It is also a very real possibility that Takata will no longer be a viable business once the recall is complete.

Sources

Consumer Reports. (2018). Takata Airbag Recall: Everything You Need to Know. *Consumer Reports*. Retrieved from https://www.consumerreports.org/car-recalls-defects/takata-airbag-recall-everything-you-need-to-know/

Ross, B., Rhee, J., Hill, A. M., Chuchmach, M., & Katersky, A. (2014). Toyota to Pay $1.2B for Hiding Deadly "Unintended Acceleration." Retrieved from https://abcnews.go.com/Blotter/toyota-pay-12b-hiding-deadly-unintended-acceleration/story?id=22972214

U.S. Department of Transportation. (2015). U.S. Department of Transportation Releases Results From NHTSA-NASA Study of Unintended Acceleration in Toyota Vehicles [Press release]. Retrieved from https://www.transportation.gov/briefing-room/us-department-transportation-releases-results-nhtsa-nasa-study-unintended-acceleration

- Supply clear and unambiguous instructions for installation and use.
- Establish a traceable system of distribution, with warranty cards, against the possibility of product recall.
- Institute an effective failure reporting and analysis system, with timely redesign and retrofit as appropriate.
- Document all product safety precautions, actions, and decisions through the product life cycle.

An increasing number of engineers will be involved in product liability work, supported by organizations such as the System Safety Society, and seeking certification as Certified Safety Professionals. Some will work for manufacturers, some for governments or consumer groups, and some will serve as *expert witnesses* before courts of law. Ethical codes of the engineering professional require placing the public interest and safety paramount, and careful attention to product safety is one way engineers can meet their professional obligations.

Reliability

Significance of Reliability.   A television is present in almost every American home. Television originated in the West (United States and Europe), and Western manufacturers maintained market leadership by emphasizing picture quality and, through frequent model changes, various innovations and gadgetry. Quality control guru J. M. Juran contrasted the approach of the Japanese, beginning with the early 1970s:

> Consumer emphasis has been on reliability and function....In response to this emphasis, Japanese manufacturers attained a clear leadership in reliability—a leadership which they hold to this day....During the middle 1970s the Western color TV sets were failing in service at a rate of about five times that prevailing in Japanese sets.

Western automobiles experienced a similar problem. *Consumer Reports* annually publishes frequency of repair statistics for automobiles, taken from surveys of the magazine's many readers. Over the past nearly two decades, readers have reported least repairs needed with Toyota, Nissan, Honda, and other Japanese cars—only recently have American brands started to regain positions in these lists. Consumers bought millions of imported cars because they have the reputation of reliability.

Reliability and Risk Defined.   To this point we have used the term *reliability* without defining it. The rigorous definition of reliability has four parts:

1. Reliability is the *probability* that a system
2. will demonstrate specified performance
3. for a stated period of time
4. when operated under *specified conditions*.

If the required function, the duration, or the environment in which a system operates changes, so does the probability of success (reliability). As an example, the *Challenger* space shuttle solid rocket motor was designed and qualified to operate in the range of 50°F to 90°F, and it could not be expected to have the same reliability after the cold night of January 27 and 28, 1986, for which temperatures at the launch site of 18°F were predicted. The political decision to launch anyway cost seven lives and a delay of over 30 months in the U.S. space program. (The ethical considerations of this decision are discussed in Chapter 16.)

Several different measures of reliability and its complement, failure, are in common usage. Four of the more common ones, which are used in the rest of this section, are defined in Table 10-1.

Risk may be defined as the chance (i.e., the probability) of injury, damage, or loss. We need a basic feeling for probability and statistics to be able to make good decisions in our daily lives as citizens of a democracy in a technological age. For example, many people have a fear of flying, but

Table 10-1   Some Reliability Measures

Measure	Symbol	Definition
Reliability (four-part definition)	$R(t)$	Number surviving at time t ÷ (number existing at $t = 0$)
Failure CDF (cumulative distribution function)	$F(t)$	Cumulative failures by time t ÷ (number existing at $t = 0$)
Failure PDF (probability density function)	$f(t)$	Number failing/unit time at time t ÷ (number existing at $t = 0$)
Failure or hazard rate	$\lambda(t)$	Number failing/unit time at time t ÷ (number existing at time $t = 0$)

it has been calculated that the same increased chance of death (one in a million) produced by traveling 1,000 miles by jet is produced by traveling 300 miles by car, 10 miles by bicycle, or six minutes by canoe!

Others feel that nuclear power plants present too great a radiation hazard to their surroundings, but this same source reports that the same 0.000001 increased chance of death from cancer caused by radiation from living five years in the open air at the boundary of a nuclear reactor plant site can be caused by (1) one chest X-ray (even in a good hospital), (2) living two months in an average stone or brick building, (3) vacationing two months in the mile-high city of Denver (if you normally live at sea level), or (4) flying 6,000 miles in a jet.

Simple Reliability Models.   When we are designing systems, we can often obtain a good estimate of the reliability of the individual components we plan to use under approximately the conditions of our application. We need to combine these known reliabilities to estimate the overall reliability of our system, and we use reliability models for this estimation.

For example, consider a system whose purpose is to turn on an electric light on demand over a period of a year under household conditions. Our components are two lamps (incandescent lightbulbs) with a reliability over that period of $R_L = 0.8$, and two switches with a reliability $R_S = 0.9$. (We are assuming that failures of the power source and of the wire and connections themselves may be ignored.) There are several ways in which we might connect these components, as shown in Figure 10-3.

Simple Series Model.   If we place one switch and one lamp in series, such that *both must work* for the system to work, the total reliability R_T of the system is the product of the reliabilities of the components. In our example (Figure 10-3a),

$$R_T = R_L \times R_S = 0.8 \times 0.9 = 0.72$$

Even if components are reasonably reliable individually, when a large number are placed in series in a complex system, the system reliability can be unacceptable. For example, a system consisting of 14 components in series, each 95 percent reliable, will have a systems reliability of 0.95^{14}, or 0.488. Modern complex systems have hundreds or thousands of components.

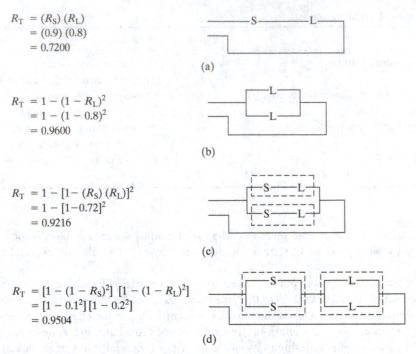

$$R_T = (R_S)(R_L)$$
$$= (0.9)(0.8)$$
$$= 0.7200$$

(a)

$$R_T = 1 - (1 - R_L)^2$$
$$= 1 - (1 - 0.8)^2$$
$$= 0.9600$$

(b)

$$R_T = 1 - [1 - (R_S)(R_L)]^2$$
$$= 1 - [1 - 0.72]^2$$
$$= 0.9216$$

(c)

$$R_T = [1 - (1 - R_S)^2][1 - (1 - R_L)^2]$$
$$= [1 - 0.1^2][1 - 0.2^2]$$
$$= 0.9504$$

(d)

Figure 10-3 Simple reliability models: (a) series; (b) simple parallel; (c) series in parallel; (d) parallel in series. S, switch; L, lamp; R_T, total system reliability; dashed "boxes" indicate subsystems analyzed within [] in calculation.

Simple Parallel Model. If we place two components in parallel so that *both must fail* for the system to fail (providing *redundancy*), the probability of failure F_T and the reliability R_T of a system consisting of two switches in parallel, of which only one must work, are as follows:

$$F_T = F_S \times F_S$$

$$R_T = 1 - F_T = 1 - F_S^2 = 1 - (1 - R_S)^2$$

$$R_T = 1 - (1 - 0.9)^2 = 1 - 0.01 = 0.99$$

Similarly, the reliability of a pair of lamps in parallel (Figure 10-3b) is 0.96.

Series–Parallel Models. Systems of any complexity consist of a combination of series and parallel arrangements of components. Consider the use of two switches and two lamps, with the requirement that one switch and one lamp in series must work for the system to work. Figure 10-3c calculates a reliability of 92.16 percent for two series systems placed in parallel; Figure 10-3d calculates a reliability of 95.04 percent for two parallel systems placed in series.

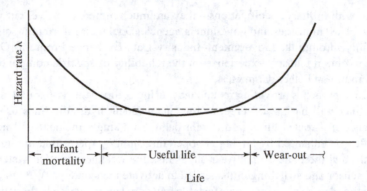

Figure 10-4 The *bathtub curve*.

All of the preceding assumes that failures of components are *statistically independent* of each other—that is, that the failure of one component has no effect on the probability of failure of another component. Further, only one type of failure was considered for each component. If failure of the switch to open (and therefore turn off the light when desired) was considered as well as the failure to close, our reliability calculations would have been more complex.

Bathtub Curve Model.   Figure 10-4 shows the pattern of hazard rate (instantaneous failure rate, assuming no previous failure) versus time, looking somewhat like the cross section of a bathtub, which is true of many components and systems. During the early *infant mortality* period, numerous failures due to substandard or defective parts or assembly take place—a phenomenon all too familiar to buyers of electronics. Following this early period on many systems is a constant failure rate period, where only a low level of random failures occurs. Finally comes the *wear-out* period, when important parts of the system come to the end of their useful life. A vast array of things exhibit this pattern of behavior, including humans and most other living creatures.

The (approximately) constant failure rate period is the preferred useful life of the system. During this period, the system can be modeled as having a constant hazard (instantaneous failure) rate lambda $(1/\lambda)$ in failures per unit time. The inverse of this rate $(1/1/\lambda)$ is the *mean time between failures (MTBF)*, a common figure of merit for reliability.

Developing Reliability over the Product Life Cycle.   Reliability is a continuing concern throughout design, manufacture, and use of a complex system. The first step in planning for reliability is establishing a *reliability goal* (the desired probability of successful operation) and its complement, the acceptable failure rate, for the system. This system failure rate is then divided into acceptable failure rates for each subsystem and component (*reliability apportionment*). The component failure rates, in turn, become the design targets for component designers.

Designing for Reliability.   A number of techniques are used in designing components and subsystems to meet reliability objectives. One is to "start with the best"—to specify and use parts of known high quality. In electronics this may mean specifying HI-REL (high reliability) parts that have been produced and tested in accordance with military specifications—they are much more expensive, but worth it where the cost of failure is high. U.S. industries and government agencies pool their information on (principally electronic) parts reliability through the Government–Industry Data Exchange Program (GIDEP). Designers can use this source to obtain industry experience on the reliability of specific components under specific conditions of temperature and other parameters.

A second design approach is providing redundancy, using components in parallel, as previously modeled. Thus a jet airliner will have two, three, or four independent hydraulic lines or electrical wires to control a critical function—routed through different paths, so a single incident will not affect them all. Redundancy can often be enhanced by having nonoperating *standby* spares that are not turned on unless the primary unit fails, and therefore do not wear out as fast; the weakness in such systems is often unreliability in the added sensor and switching units needed to activate the standby.

Reliability is enhanced by assuring a comfortable *factor of safety*, which is the ratio of the minimum strength provided by the design to the maximum stress anticipated in use. An electrical analog is *derating*, in which electrical components of higher-than-necessary strength (or "rating") are specified to assure high reliability and durability in the actual service expected. Another approach is *fail-safe* design, in which if failure does occur it leaves the system in a safe (although perhaps inoperable) condition.

Flattening the Bathtub Curve.   Figure 10-4 shows that component and system failures are often higher during the early *infant mortality* period due to defective parts or assembly. These failures can be reduced by careful quality control. Useful life may be extended by replacing those parts that wear out quickly (such as the brake linings on a car), but sooner or later there comes a point where it is cheaper to replace a system than to maintain it.

Reliability Growth.   Reliability is evaluated and improved throughout the system life cycle. Early breadboard tests of critical systems are used to give a first indication. Subsystem tests may be run at several points as parts and components are defined and prototypes become available. Subsystems, and then the system itself, will be qualified by conducting rigorous qualification tests under a range of expected environmental conditions (temperature, shock, vibration, and others). At each step, failure modes exhibit themselves, and reliability is enhanced by redesign to eliminate them. In manufacture, each unit or batch is tested to provide further reliability information. In field use, still another set of problems arises, and systems may have to be retrofitted in the field to correct problems not found before. Throughout this process, system reliability improves. Still, the best and most economical place to minimize failures is in the design phase, using such techniques of reliability engineering as failure modes and effects analysis (FMEA).

The Reliability Profession.   Reliability engineering has become an established profession, involving a number of professional societies. The Reliability Division of the American Society for Quality (ASQ) publishes the quarterly *Reliability Review*; ASQ also has a Certified Reliability Engineer (CRE) designation awarded after a rigorous, day-long examination. The Institute of Electrical and Electronic Engineers (IEEE) has a Reliability Division, and it publishes the *IEEE Transactions on Reliability*. The Society of Reliability Engineers (SRE) also has a journal. These societies and a half dozen others jointly sponsor an annual Reliability and Maintainability Symposium that is extremely well attended.

OTHER CRITERIA IN DESIGN

Maintainability

Blanchard states that **maintainability** "is an inherent design characteristic of a system or product [and] it pertains to the ease, accuracy, safety, and economy in the performance of maintenance actions." One can create a four-part definition for maintainability by adding and striking out words in the definition already given for reliability:

1. Maintainability is the probability that a failed system
2. will demonstrate it can be restored to specified performance
3. within a stated period of time
4. when operated maintained under specified conditions.

Maintenance downtime has three components:

1. *Administrative and preparation time.* Processing the repair request, waiting for an available worker, travel, and obtaining tools and test equipment
2. *Logistics time.* Delay to obtain parts (or test equipment or transportation) after maintenance personnel are available
3. *Active maintenance time.* Actually doing the job (including studying repair charts before repair and verifying and documenting the repair afterward)

Maintenance may be divided into *corrective maintenance*, made necessary by failures, and *preventive maintenance*, designed to prevent failures. The average time between maintenance actions (regardless of type) is the *mean time between maintenance (MTBM)*, and the average total time for the three components of maintenance is the *mean downtime (MDT)*. Maintainability may alternatively be defined by just the active maintenance time for corrective maintenance *mean time to repair (MTTR)*, since only the preceding item 3 is substantially influenced by the designer (although specification of standard parts, tools, and test equipment can shorten the others). The reliability measure of *MTBF*, the inverse of the hazard rate $1/\lambda$, is often used with the MTTR.

The designer can reduce active maintenance time by providing easy access to the system, dividing the system into modules that can be replaced as units, specifying preventive maintenance that will delay deterioration and identifying worn parts, and providing clear, comprehensive maintenance manuals. Maintainability can be enhanced by creating realistic system models—physical mockups on which maintenance actions can be at least simulated by typical repair people or, using the output of a CAD process, three-dimensional computer simulations that can be rotated and enlarged to provide visibility and understanding of potential maintenance difficulties.

Another aid to maintenance, especially of electronic equipment, is provision for built-in test (BIT). BIT may consist of the simple provision of test points to facilitate a mechanic's diagnosis, or it may include an extensive system of sensors, a computer, and software that periodically checks the condition of avionic systems and provides an automatic printout of potential defects as soon as an aircraft lands. (BIT systems can themselves become so complex that a significant number of the problems they identify are false indications, due instead to defects in the BIT.)

Availability

Many users are more concerned that a system operates satisfactorily when called upon, a condition called *availability*, than they are in pursuing some ultimate reliability by making the system so complex that it defies repair. Norman Augustine has observed that, as military aircraft become more costly and more complex, the maintenance crew hours per flight hour increase. He therefore proposes (with tongue in cheek) the Augustine-Morrison Law of Unidirectional Flight: "Aircraft flight in the 21st century will always be in a westerly direction, preferably supersonic, to provide the additional hours needed each day to maintain all the broken parts."

Two definitions of availability, based on the four measures just identified under maintainability, merit mention. The *inherent availability* A_i of a system considers only corrective maintenance in an ideal support environment (with neither administrative nor logistic delays):

$$A_i = \frac{MTBF}{MTBF + MTTR} \tag{10-1}$$

Operational availability A_o, on the other hand, considers both preventive and corrective maintenance, conducted in the actual support environment:

$$A_o = \frac{MTBM}{MTBM + MDT} \tag{10-2}$$

Human Factors

Human factors engineering, also known as **ergonomics**, is concerned with ways of designing machines, operations, and work environments to match human capacities and limitations. Its origin can be traced to early scientific management studies, such as the tailoring of shovel size to material density by Frederick Taylor in 1898 and the efficient arrangement of bricks on the scaffold by Frank Gilbreth in 1911. Human factors engineering did not really emerge as a discipline, however, until the mid-twentieth century, as a result of World War II experience. For example, the location of three critical controls in three military aircrafts in common use in 1945 were as shown in Table 10-2.

The hazard created when a pilot is making a difficult landing in an unfamiliar aircraft is obvious. A similar hazard exists when the typical business traveler arrives at an airport on a rainy night and jumps

Table 10-2   Control Placement in Military Aircraft

| Aircraft | Control Placement on the Throttle Quadrant | | |
	Left	Center	Right
B-25	Throttle	Propeller	Mixture
C-47 (DC3)	Propeller	Throttle	Mixture
C-82	Mixture	Throttle	Propeller

Source: Adapted from Alphonse Chapanis, *Man-Machine Engineering*, Wadsworth Publishing Company, Inc., Belmont, CA, 1966, p. 95.

in an unfamiliar rental car to traverse an unfamiliar city street to find a hotel. Automobile manufacturers, using ergonomics, have made great strides in locating critical controls in consistent locations that drivers can reach without taking their hands off the steering wheel, in providing climate control and comfortable seats, and in locating and lighting displays for easy visibility.

Human factors engineers have developed a wide variety of standards for illumination, sound, accessibility, controls, displays, and other factors affecting work. As an example, by the time the Apollo spacecraft was being designed in the 1960s, many of these standards were well into development. Babcock, who then had design responsibility for the Launch Escape System in the Apollo command module, recalled extensive design reviews on the placement of critical control switches and displays so that the astronauts could perform essential functions despite the gravity, vibration, and other forces of launch.

Standardization

A **standard** is defined as a set of specifications for parts, materials, or processes intended to achieve uniformity, efficiency, and a specified quality. One of the important purposes of a standard is to place a limit on the number of items in the specifications so as to provide a reasonable inventory of tooling, sizes, shapes, and varieties.

At one time there were no standards for bolts, nuts, and screw threads, and a $\frac{1}{2}$-inch nut removed from one bolt would not fit another. The same applied to lamp bases—manufacturers once offered 175 different ones, and now there are only about a half dozen. Any large design organization has a standards manual that identifies fasteners, tolerances, processes, and the like that are considered acceptable in that organization; specification of nonstandard alternatives requires strong justification. Standardization can be important for reliability, too. In the Apollo program, essentially all devices in which an electrical signal was translated into a pyrotechnic or explosive pulse, whether on a launch stage or command, service, or lunar excursion module, had to use the same initiator; thousands of these initiators were fired in development, and an outstanding reliability record resulted.

The National Institute of Standards and Technology (NIST) is only one of many government agencies involved in standardization. Trade associations, industries, professional societies, and government organizations work together in the American National Standards Institute (ANSI) to coordinate standardizing activities. Unfortunately, the United States is alone among developed countries in not standardizing on the metric system of measurement, although increasing numbers of individual companies and industries are doing so to remove this barrier to international trade in American goods.

Producibility

As a product is being designed, careful attention should be paid to ensure that it can be produced economically, using available processes and equipment where possible. Manufacturing engineers familiar with production capabilities should be involved in reviewing parts as they are designed, suggesting tolerances, materials, and shapes that are more producible. Two-way understanding is developed—understanding by designers of manufacturing preferences, and by manufacturing engineers of the performance consequences if certain critical specifications are relaxed. Furthermore, the transfer from design to manufacturing is greatly simplified, and the ultimate product is produced not only at lower cost, but also with less transition time. These considerations are an essential part of the modern thrust of *concurrent engineering* discussed earlier in this chapter.

Value Engineering/Analysis

Value engineering or value analysis (VE/A) is a methodical study of all components of a product in order to discover and eliminate unnecessary costs over the product life cycle without interfering with the effectiveness of the product. Fasal would use the term **value engineering** in developing new products and value analysis in reviewing old products, but most people use the terms interchangeably. One of the techniques of VE/A is asking a series of penetrating questions about a product, system, process, or component. Weiss provides a typical set of questions to ask about each item:

1. What is it?
2. What does it do?
3. What does it cost?
4. What is it worth?
5. What else might do the job?
6. What do the alternatives cost?
7. Which alternative is least expensive?
8. Will the alternative meet the requirements?
9. What is needed to implement the alternative?

Value engineering activities are encouraged—often required—of contractors by the U.S. DOD and by NASA. Practitioners share their experiences in the Society of American Value Engineers (with the appropriate acronym SAVE) and can earn the SAVE title Certified Value Specialist, by examination.

Value Engineering

Value analysis, value management, and value control are considered synonymous with value engineering (VE). VE is an effective technique for reducing costs, increasing productivity, and improving quality. It can be applied to hardware and software; development, production, and manufacturing; specifications, standards, contract requirements, and other acquisition program documentation; and facilities design and construction. It may be successfully introduced at any point in the life cycle of products, systems, or procedures. VE is a technique directed toward analyzing the functions of an item or process to determine best value, or the best relationship between worth and cost. In other words, best value is represented by an item or process that consistently performs the required basic function and has the lowest total cost. VE originated in the industrial community, and it has spread to the federal government due to its potential for yielding a large return on investment. VE has long been recognized as an effective technique to lower the government's cost while maintaining necessary quality levels. Its most extensive use has been in federal acquisition programs.

VE contributes to the overall management objectives of streamlining operations, improving quality, and reducing costs and can result in the increased use of environmentally sound and energy-efficient practices and materials. The complementary relationship between VE and other management techniques increases the likelihood that overall management objectives will be achieved.

Source: https://www.whitehouse.gov/sites/whitehouse.gov/files/omb/circulars/A131/a131-122013.pdf, May 2019.

DISCUSSION QUESTIONS

10-1. Outline how the design approaches outlined in this chapter might have led to the failures experienced by Toyota and Takata.

10-2. How does the design work done in the technical feasibility stage of new product development differ from that done in the later design stages?

10-3. Are there products that should not consider human factors in their design? Why or why not?

10-4. For an engineering design or project management system you are familiar with, describe the drawing release and design review processes.

10-5. Summarize the history of gradually increasing liability of industry for damage caused by their products. What is driving this change?

10-6. Research the results of two recent product liability cases. Based on these examples, are companies being held accountable for their actions? Why or why not?

10-7. Select a product line in a specific industry, and list actions that can be taken to reduce product liability. At what stage in design should each of these actions be taken?

10-8. Identify a company and product (aside from televisions and automobiles), and tell how good or poor reliability has significantly affected company success. How has reliability changed on the product selected?

10-9. Describe some mechanisms a designer can use to improve maintainability. How do these steps impact the overall product lifecycle outlined in Figure 9-1?

10-10. Give an example of a consumer product with which you are familiar that, through recent redesign, seems to be a greater value (a better ratio of utility to apparent cost).

10-11. Discuss how the management functions of planning, organizing, leading, and controlling relate to the engineering design process.

10-12. For a component to which the bathtub curve model of reliability applies, describe the provisions you would make to assure a low hazard rate in use of the component.

PROBLEMS

10-1. Given four components, each with a reliability of 0.9, calculate the reliability of a total system in which the four are arranged in: (**a**) four in series, (**b**) four in parallel, and (**c**), (**d**) two different series/parallel designs each using a *total* of only three components, (**e**) given the performance and cost differences (assume all components cost the same) of each design, which would you recommend?

10-2. An engineered system consists of one each of three components X, Y, and Z with reliabilities RX, RY, and RZ of 0.94, 0.80, and 0.95. (**a**) What is the system reliability, assuming that one component of each type must work? (**b**) If required system reliability is 0.85, show how you can meet this goal by replacing one of these components with two of that same component in parallel.

10-3. An engineered system has a hazard rate of 0.01 failure per hour. (**a**) What is its MTBF? If the same system has an MTBM of 60 hours, an MDT of 20 hours, and an MTTR of six hours, what are its (**b**) inherent availability and (**c**) operational availability?

SOURCES

Anstett, Patricia, "Dow Corning Takes Final Step Toward Settlement of Breast-Implant Cases," *Detroit Free Press*, June 28, 1999.

Augustine, Norman R., *Augustine's Laws, Revised and Enlarged* (Washington, DC: American Institute of Aeronautics and Astronautics, 1983), p. 74.

Ball, Jeffrey and Geyelin, Milo, "GM Ordered by Jury to Pay $4.9 Billion—Auto Maker Plans to Appeal Huge California Verdict in Fuel-Tank-Fire Case," *The Wall Street Journal*, July, 1999, East. Ed.: A3C.

Bass, Lewis, "Designing for the Jury," *System Safety Society Newsletter*, 2:5, October 1986, p. 1.

Berlack, H. Ronald, "Evaluation & Selection of Automated Configuration Management Tools," *Crosstalk—The Journal of Defense Software Engineering*, Nov/Dec 1995. http://www.stsc.hill.af.mil/crosstalk/1995/nov/Evaluati.asp National Aeronautics and Space Agency, *NASA Systems Engineering Handbook*, August 2012.

Blanchard, Benjamin S. and Fabrycky, Wolter J., *Systems Engineering and Analysis*, 2d ed. (Englewood Cliffs, NJ: Prentice Hall, Inc., 1990), pp. 389–390.

Bussey, John and Sease, Douglas R., "Speeding Up: Manufacturers Strive to Slice Time Needed to Develop Products," *Wall Street Journal*, February 23, 1988, p. 1.

Crosstalk—The Journal of Defense Software Engineering.

Fasal, John H., *Practical Value Analysis Methods* (Hasbrouck Heights, NJ: Hayden Book Company, Inc., 1972), pp. 8–9.

Geyelin, Milo, "Judge Upholds $259 Million in Damages In DaimlerChrysler Minivan-Latch Case," *Wall Street Journal East*. Ed.: B12:4.

Juran, J. M., "Japanese and Western Quality: A Contrast," *Quality Progress*, December 1978, pp. 10–18.

Miller, Roger Leroy, "Drawing Limits on Liability," *Wall Street Journal*, April 4, 1984, editorial page.

National Society of Professional Engineers, *Engineering Stages of New Product Development*, NSPE Publication #3018 (Alexandria VA: NSPE, 1990).

"Reliability Forecast for New Cars," *Consumer Reports*, April 2000, p. 32.

Reswick, J. B., foreword to Morris Asimow, *Introduction to Design* (Englewood Cliffs, NJ: Prentice-Hall, Inc., 1962), p. iii.

Shigley, Joseph and Mitchell, Larry, *Mechanical Engineering Design* (New York: McGraw-Hill Book Company, 1983), p. 14.

University of Michigan Transportation Research Institute data graphed in *Newsweek*, March 30, 1992.

Weiss, Gorden E., "Value Engineering/Analysis, Part III," *UMR Extension Division Continuing Education Series* (Rolla, MO: University of Missouri–Rolla, 1973), p. 1.

Wilson, Richard, "Analyzing the Daily Risks of Life," *Technology Review*, February 1979, pp. 40–46.

11

Planning Production Activity

PREVIEW

The next topic applying the management functions to the management of technology is planning engineering production activity. This chapter begins by emphasizing the importance of competitive production processes to any economy, and the central position of the engineer in the production organization. Next, the importance of plant location, design, and layout in planning manufacturing facilities is considered. Three quantitative production planning tools are then discussed: economic order quantity, break-even charts, and learning curves. A summary of production planning tools is presented, and finally, different production systems are presented.

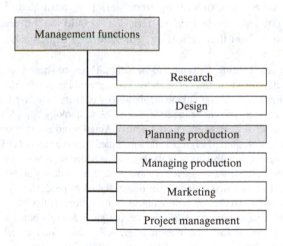

LEARNING OBJECTIVES

When you have finished studying this chapter, you should be able to do the following:

- Describe the position of the engineer in the production process.
- Describe considerations in planning manufacturing facilities.
- Be able to use production planning tools of economic order quantity, break-even analysis, and learning curves.
- Describe the major methods used for production planning and control.

INTRODUCTION

Vital Nature of Production

Alexander Hamilton, the founding father featured in the hit Broadway musical, is reported to have said the following: "Not only the wealth, but the independence and security of the country appear to be materially connected to the prosperity of manufacturers." For much of the last century, U.S. manufacturers were the envy of the world. In the 1980s and 1990s, this changed dramatically. Manufactured goods produced in the United States as a percentage of those consumed varied in the comfortable range of 100 to 105 percent from 1966 until 1982, when it crossed the 100 percent line; it then plummeted to only 85 percent by 1986, and has remained around 90 percent since that time.

Some authors might view a manufacturing trade deficit as natural and acceptable for the United States. After all, are we not a *postindustrial society*, and should we not expect to emphasize service industries while "lesser nations" get their hands dirty in factories? Cohen and Zysman refute this:

> Mastery and control of manufacturing is…critical to the nation. This…has been obscured by the popular myth that sees economic development as a process of sectoral succession. Economies develop as they shift out of sunset industries into sunrise sectors. Agriculture is followed by industry, which in turn is sloughed off to less developed places as the economy moves on to services and high technology. Simply put, this is incorrect. It is incorrect as history and it is incorrect as policy prescription. America did not shift out of agriculture or move it offshore. We automated it; we shifted labor out and substituted massive amounts of capital, technology, and education to increase output. Critically, many of the high value added service jobs we are told will substitute for industrial activity are not substitutes, they are complements. Lose industry and you will lose, not develop, those service activities. These service activities are tightly linked to production just as the crop duster (in employment statistics a service worker) is tightly linked to agriculture. If the farm moves offshore the crop duster does too, as does the large-animal vet. Similar sets of tight linkages—but at a vastly greater scale—tie "service" jobs to mastery and control of production. Many high value added service activities are functional extensions of an ever more elaborate division of labor in production. The shift we are experiencing is not from an industrial economy to a postindustrial economy, but rather to a new kind of industrial economy.

This chapter looks not only at existing production methods, but also at the nature of the "factory of the future" and the importance of the engineer's contribution to it.

The Engineer in Production Activity

Types of Positions. Production organizations vary tremendously with the industry involved, with the size of the organization, with the type of production (mass production of standard items or small-quantity production of specialty items), and with many other factors. However, it is helpful to create a model of the *typical* manufacturing plant from which we can generalize the functions needed and the way in which engineers and engineering managers might fit into such an organization. Our hypothetical plant is assumed to be one of several at different locations, producing products that are researched, designed, financed, and marketed at a corporate headquarter separate from the plant. The plant organization, all reporting to a single *plant manager,* might look something like Figure 11-1.

An obvious job for an engineer in this organization is the position of **plant engineer** at the right of the diagram. The engineering design function under the plant engineer will normally be responsible for designing small changes to the plant and its production equipment. For more extensive changes, plant engineers would just specify what is needed and monitor construction; the detailed design and construction would then be carried out by an architect/engineering (A/E) firm specializing in that type of manufacturing plant or by the central engineering group at corporate headquarters, since it is not efficient to maintain that level of design capability at each plant.

Maintenance engineering (the design and specification of the criteria for maintenance tasks) commonly might appear under the plant engineer; the routine conduct of maintenance is often under the general superintendent (or plant production manager). The plant engineer is commonly responsible for the utilities (heat, power, steam, water, telephone) throughout the plant and for certain other functions, such as housekeeping and security, that relate to the facility rather than the product.

Another important class of engineering functions in the plant comes under the heading of **industrial engineering** (IE). Traditional IE functions of plant layout, time-and-motion analysis, and standards setting are performed here. In many metal cutting and chemical processing plants the **process engineer**, often in a separate organizational unit, makes a major contribution. The manager of quality assurance, or quality control, will be an engineer or scientist in most plants of significant size or product complexity. Two functions are shown under the quality manager: quality control (or quality engineering), responsible for the analysis of quality problems and their prevention (commonly headed by an engineer); and inspection (which often is not).

First-line production management positions, typically supervisors, provide employees an excellent opportunity to learn production problems and to test their wings as leaders. Most such positions will be filled by nonengineers, but they provide a good starting place for the engineer interested in manufacturing management. Positions as shift manager are natural steps up the promotional ladder. Several other positions may call for engineers in larger plants: the safety engineer, whether under human resources department or elsewhere; materials control (where automated storage, retrieval, and transport are extensive); and even purchasing, where the technology of the item being purchased is paramount. More often, such posts and others (technical employment, for example) will be staffed by technical people at the corporate level, with nonengineers implementing their recommendations at the plant level.

Use of Engineers. During the 2015–2016 academic year, U.S. colleges and universities granted 112,721 bachelor's degrees in engineering and 64,602 master's degrees. These numbers represent a substantial growth in engineering graduates over the past decade. Since the recession of 2008 made degrees in STEM (Science, Technology, Engineering and Math) more attractive to those seeking work, undergraduate

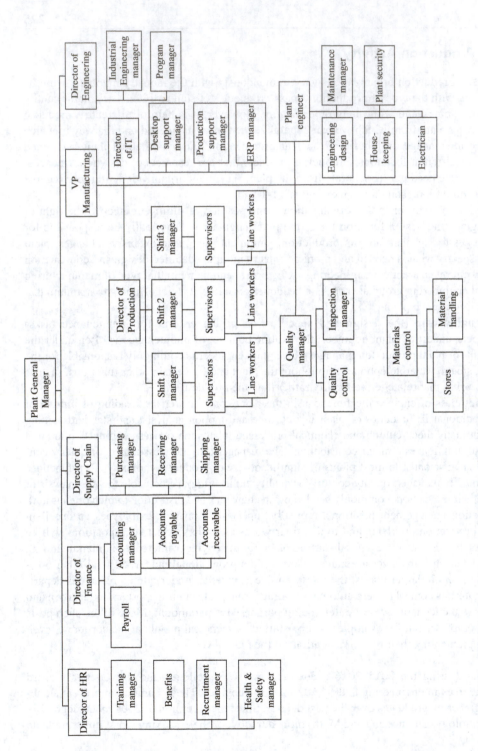

Figure 11-1 Organization chart for a plant in a larger company.

engineering enrollment has grown by 54 percent. Overall engineering employment is expected to grow by 7 percent from 2016 to 2026 decade, creating about 194,300 new jobs.

There appears to be a shift of engineers from large to medium-sized or small companies, as large companies downsize and spin off business units. Many large companies are outsourcing to specialized engineering and consulting firms. Some companies are also moving from a philosophy of employing specialized engineers to a team of flexible "systems" engineers who have broad all-around knowledge. In conclusion, although fewer engineers may be working in manufacturing because of the increased use of computers and automated machines, there is still a growing need for engineers in other areas.

Future Demands on Manufacturing Engineers. The Society of Manufacturing Engineers commissioned the study, *The Manufacturing Engineer in the 21st Century,* which summarized more than 10,000 opinions from manufacturing practitioners. This report (by A. T. Kearney, Inc.) concluded the following:

The manufacturing engineer of the future will be faced with new challenges in the form of:

- an environment of exploding scope [increasing product sophistication and variation, a global manufacturing environment, and extensive social and economic changes];
- multiple roles [with the manufacturing engineer acting as an operations integrator and manufacturing strategist as well as a technical specialist];
- advanced tools [including more powerful computer hardware, more and larger databases, a greater choice of software and expert systems, and advanced CAD/CAM (computer-aided design/manufacturing) systems]; and
- changed work emphasis [focusing on teams, not individuals, with a more human, less technical orientation and the use of more outside services].

Recent advances in additive manufacturing (3D printing) are likely to accelerate these changes and further transform the role of the manufacturing engineer. Jack Welch, former CEO of General Electric Company, has emphasized the need to give higher priority to manufacturing engineering.

Today manufacturing is undergoing a change that is every bit as significant as the introduction of interchangeable parts or the production line. Digital technology is transforming manufacturing, making it leaner and smarter, and raising the prospect of an industrial renaissance.

To whom then, is it left to see that American innovation is dynamic enough, and American productivity growth sufficiently rapid, to win in world markets? In large measure it is the engineer, and in that context America needs to see the profession as the bodyguard of its standard of living. If it does; if the country perceives the nexus between a powerful engineering base and our way of life, educational and motivational programs that will preserve and nurture that base will be more forthcoming.

PLANNING MANUFACTURING FACILITIES

Plant Location

Whether it is Amazon seeking a location for a second headquarters, or Foxconn looking to identify a location for a major U.S. plant, the question of location includes a very important decision on the region of the country to select as a location. Many factors, such as transportation, labor supply and attitude, resource availability, and political climate, had to be considered before selecting either Washington, DC or New York (in the case of Amazon) or Wisconsin for Foxconn. Amrine et al. outline "seven basic steps in locating and building every new plant" followed by one large company:

1. Establish the need for a new plant.
2. Determine the best geographical area for the plant on the basis of the company's business needs.
3. Establish the requirements (e.g., product to be made, equipment and buildings needed, utilities and transportation necessary, number of employees, etc.).
4. Screen many communities within the general area decided upon.
5. Pinpoint a few communities for detailed studies.
6. Select the best location.
7. Build the plant.

Some of the factors affecting the choice of region, community, and site are as follows:

- Transportation (highway, rail, air, water)
- Labor (supply, skill level, local wage rates, union membership and attitudes)
- Geographical location (relative to raw materials, customers, or other company activities)
- Utilities (supply and cost of water, electric power, and fossil fuels)
- Business climate (taxes, pollution controls, community attitudes)
- Amenities (climate, educational facilities, nearby recreation)
- Plant sites (land availability and cost, zoning, space for expansion)

The most important factors for plant location will vary with the industry and its critical factors, as in the following examples:

- Kilns used to create charcoal for briquettes from hardwood will be close to the raw material supply to reduce transportation cost, since four-fifths of the mass disappears in charring.
- Aluminum production and data centers have traditionally sought a source of low cost electricity, since they are energy intensive.
- High-technology electronic firms have tended to cluster together where technical professionals and educational institutions are available.
- U.S. clothing manufacturers have moved from high-labor-cost areas to lower-labor-cost areas in the United States, and then, increasingly, overseas.

Plant Location

There are a number of references for determining plant location and the information affecting the choice of region, community, and site. Economic development offices within a community are one. This website http://factfinder2.census.gov provides surveys about communities, as well as demographics, economic census, and housing, business, and government statistics.

Plant Design

Once the site is selected, engineers must decide on the nature of the plant and its arrangement on the site. *Multistory* plants conserve land area, permit use of gravity flow in moving product along the production line, and are cheaper to heat. However, *single-story* construction is more flexible, permits lighter foundations and columns, and allows higher floor loadings. Most new U.S. plants are now built near major highways on the edge of a city, where available and economical land provides room not only for single-story construction, but also for parking—often larger in area than the plant itself. Materials for plant construction may be steel-reinforced concrete (most expensive, but lowest in maintenance cost and most fire resistant), exposed steel beams and trusses, or wood (for low buildings and light loads where fire is not a hazard).

The arrangement of the building on the site will depend on such things as the contours of the site, railroad and truck access, parking-lot provisions, and appearance. Some large companies have their own corporate engineering staffs for plant design, but most companies will call on an architect/engineering (A/E) firm for this specialized service.

Plant Layout

Plant layout attempts to achieve the most effective arrangement of the physical facilities and personnel for making a product. The three principal methods of moving the product through the manufacturing steps are product layout, process layout, and group technology. (In a fourth method, *fixed-position layout,* the product remains stationary and the processes are brought to it. This method is largely confined to shipbuilding and other massive construction.)

In **product layout**, machines and personnel are arranged in the sequence of product manufacture so that the product can be moved along the production (assembly) line with a minimum of travel between steps, as shown in Figure 11-2a. This method is especially useful when a large quantity of standardized products are to be produced over a long period of time, and it is the basis for mass production of most automobiles, major household appliances, and the like. Ideally, the assembly line and the plant structure are designed in parallel, since adapting an existing building to a new mass production need can involve undesirable compromises.

In **process layout**, all machines or activity of a particular type are located together. Thus, a plant may have separate departments for turning, planning, grinding, milling, drilling, and painting, as shown in Figure 11-2b. Individual products are transported from department to department in the sequence needed for their production. This layout is particularly useful for the job-shop environment, in which a large number of different products are to be produced by the same equipment and workers. It provides great flexibility in the use of expensive equipment and skilled personnel at the expense of substantial in-plant transportation.

Operations Management

Operations management is an area of business that refers to the management of the people, processes, and equipment that transform inputs into delivered goods and services that meet customer expectations.

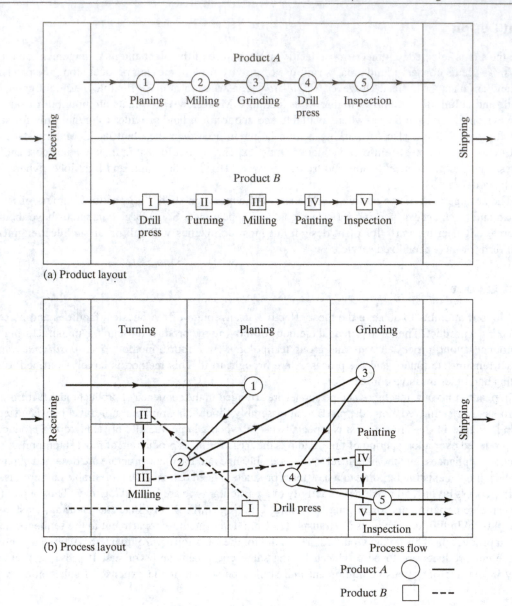

Figure 11-2   Schematic representations of (a) product and (b) process layouts. (From Arthur C. Laufer, *Production and Operations Management*, 3rd ed., South-Western Publishing Company, Cincinnati, OH, 1984, pp. 232–233).

Case Study: Operations Management

Dunn and Savastano Orthodontics is a local successful small business in Florida that implements many tools and techniques of operations management. The business mission is to provide the finest orthodontic treatment possible in a sensitive, patient-oriented, and fun environment. Their goal is to create patient relationships, motivation, and self-esteem for their patients. This, along with a caring attitude is an integral part of the treatment philosophy of Dunn & Savastano Orthodontics. To achieve that goal Drs. Rick Dunn and Nick Savastano spent a great deal of time and effort with their patients and transformed the experience into reality. Looking at the business mission and goal we can identify the factors affecting their mission: First, the orthodontists' philosophy and values, second their emphasis on keeping an excellent public image, targeting certain customers that search for the quality of work and last, maintaining a certain rate of growth.

THE BUSINESS STRATEGY

In the competitive market of orthodontists, Dunn and Savastano Orthodontics adopted a differentiation strategy to be better than, or at least different from other practices. The unique services provided for the patients and their families are the key for such differentiation. Flat TVs at the workstation where every patient is being treated, a small arcade room for the kids, a small business office for the parents to check their e-mail messages while waiting for the service, and prize incentives for their patients to encourage them to follow the required treatment are some of the characteristics that differentiate their practice from other similar practices. This office went beyond both the physical characteristics and service attributes to encompass everything that impacts customers' perception of value.

LOCATION STRATEGY

Location decision is one of the top 10 important decisions an operations manager has to deal with. Dunn and Savastano Orthodontics chose their location to provide the market with high-quality services and the convenience of proximity. With two convenient orthodontist offices in Longwood and Lake Mary, both just north of Orlando, the practice can service up to nine different areas, Orlando, Sanford, Heathrow, Casselberry, Oviedo, Deltona, and Apopka. Both Dr. Dunn and Dr. Savastano understood that once committed to a location, many resources and cost issues are difficult to change.

LAYOUT STRATEGY

When entering Dunn and Savastano Orthodontics office one immediately notices the great utilization of space, equipment, and people. The practice studied the requirements of work cells and provided the right resources to fulfill customer satisfaction. Positioning the workers and their equipment provided the flexibility of movement for doctors, workers, and patients. One can also experience the distinguished customer interaction with workers and doctors due to the comforting environment available at each station.

Source: Dr. Nabeel Yousef, Daytona State College, December 2012.

In **group technology**, a set of products requiring similar processing equipment is identified, and a small group of the machines needed to make this set of similar products is placed together. Transportation between steps in the manufacturing process is therefore minimized, inventory accumulating between steps can be almost eliminated, and products are produced much faster. For example, a General Electric plant reported that productivity in making motor frames increased 240 percent, floor space needed was reduced 30 percent, and the manufacturing cycle was reduced from 16 days to 16 hours! By adding computer control, automated pallet handling of the workpiece, and automatic tool changing, one can create a **flexible manufacturing cell** (FMC) capable of producing this group of related parts with a minimum of human intervention. *Flexible manufacturing systems* (FMS) are discussed in a later section.

QUANTITATIVE TOOLS IN PRODUCTION PLANNING

Three specific tools are discussed here: the economic order quantity (EOQ) approach to inventory control, break-even charts, and learning curves.

Inventory Control

Types of Inventory.   Most types of manufacturing processes begin with some type of *raw material* (sheet steel, lumber, leather) that requires processing. They add *purchased parts* (valves, switches, hinges), and consume *supplies* (cutting oils, time cards, drill bits). As work progresses, there will be a considerable investment in *work-in-process* before the *finished goods* are delivered to the warehouse to await sale and shipment. Each of these types of inventory represents an investment of capital, requires storage space, and is subject to loss, so it would seem desirable to make or purchase very small quantities at a time. However, each time a lot of product is made there is a *setup cost,* consisting of the clerical cost of processing and tracking the order and the cost of finding tooling and adjusting machines to make the item; these costs are less when lots are larger. Inventory control is the process of identifying and implementing inventory levels that result in a minimum total cost.

Economic Order Quantity (EOQ).   Consider an inventory item for which the annual requirement is R units. Storing each unit of the item in inventory will cost I dollars per year. These storage costs include interest on the working capital invested in the unit, warehouse expense, and threat of deterioration, theft, and obsolescence while the unit is in storage. If, every time the last item is used, you renew the inventory with a batch of Q units, your average inventory will be $Q/2$ units and you will need R/Q batches per year. Each such batch involves an ordering or setup cost of S dollars. The total annual cost C_T of that inventory item is, therefore,

$$C_T = I\left(\frac{Q}{2}\right) + S\left(\frac{R}{Q}\right).$$

$\qquad$ **(11-1)**

Setting the differential of total cost (with respect to Q) to zero and solving for Q yields the economic order quantity:

$$EOQ = \sqrt{\frac{2RS}{I}}$$

$\qquad$ **(11-2)**

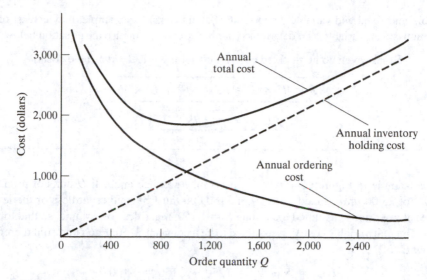

Figure 11-3 Economic order quantity.

Determination of the EOQ is shown graphically in Figure 11-3. In this example the minimum annual total cost is attained with an EOQ of about 950. Note the flatness of the total cost curve near the optimum, suggesting that the order quantity can be adjusted over quite a range with little effect on total cost.

Problems with EOQ Analysis.   This formula has been used for many years, with very few American companies asking the critical question: Why does the setup cost have to be so high? Japanese companies, led by Toyota, developed techniques such as (1) designing dies and tooling so they could be switched quickly and cheaply, and (2) including simple cards (*kanban*) in each small lot that, when the succeeding process started using the lot, were sent to the preceding step to direct making another small lot with no further paperwork (see the "just-in-time" discussion later in the chapter). This permitted drastic reductions in overall costs by reducing both setup and holding costs.

As an example of the savings possible, consider a Japanese improvement where the time to change tooling was reduced from $2\frac{1}{2}$ hours (150 minutes) to 3 minutes, only a fiftieth (0.02) of the original value. The preceding EOQ equation calls for a lot size that is only $(0.02)^{0.5}$, or 0.14 of the previous value, so that the total cost of setup and storage, and the floor space required, was cut to about one-seventh of its original level. These savings represented such a significant part of the ability of the Japanese to produce cheaper products that many American firms have instituted similar systems.

Break-Even Charts

Break-even analysis divides costs into their fixed and variable components to estimate the production levels needed for profitable operation. **Fixed costs** are those assumed to be independent of production level, at least in the range of production volume of interest. They include lease payments, insurance costs, executive salaries, plant heating and lighting, and the like. **Variable costs** are those assumed to vary directly with the level of production, such as direct labor, direct materials, and power for production equipment. Some **semivariable costs** may be divisible into fixed and variable components. Selling costs, for example, may consist of both salary (fixed) and commissions (variable). To determine your break-even point, use the equation below:

$$\text{Break-even point} = \text{fixed costs}/(\text{unit selling price} - \text{variable costs})$$

$$R = U \times S = TC_1 = F_1 + U \times V \qquad \qquad \textbf{(11-3)}$$

$$BE_1 = U = F_1/S - V_1 \qquad \qquad \textbf{(11-4)}$$

Example

Consider the example in Figure 11-4, where a plant may produce and sell U units of product up to a plant capacity of 2,000 units. Fixed costs F_1 of \$100,000 must be paid regardless of the level of production. The selling price is assumed a constant $S = \$250$, regardless of volume, so that total revenue $R = U \times S$. The unit variable cost V_1 is assumed to be a constant \$150. Each unit sold therefore makes a *contribution* C_1 of

$$C_1 = S - V_1 = \$250 - \$150 = \$100$$

toward paying the fixed costs and providing a profit. The *break-even point* BE_1 is the production level U where total costs TC equal total revenue R:

$$R = U \times S = TC_1 = F_1 + U \times V$$

$$BE_1 = U = \frac{F_1}{S - V_1} = \frac{\$100,000}{\$250 - \$150} = 1,000 \text{ units}$$

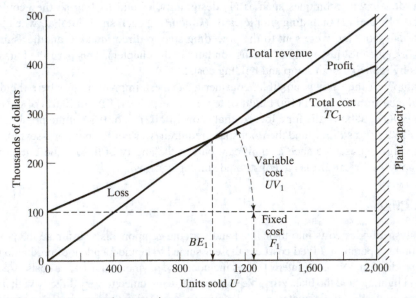

Figure 11-4   Break-even chart.

Automation normally involves increasing the (fixed) investment in production equipment in order to make production more efficient (i.e., to reduce the variable cost).

Figure 11-5 shows the effect in our example of increasing fixed cost by $80,000 to a total $F_2 = \$180,000$ in order to reduce the variable cost by $50.00 to $V_2 = \$100$. Our new contribution C_2 is ($250 – $100), or $150 per unit, and our new break-even BE_2 is (180,000/150), or 1,200 units. This is not the point at which automation is justified, since at 1,200 units the plant would make a profit ($20,000) without the added fixed cost of automation. The increased investment will be justified only at the point where the two total costs are equal; in the figure you should be indifferent between fixed cost (automation) levels F_1 and F_2 at a production level I of 1,600 units, because the profit will be the same ($60,000) in either case.

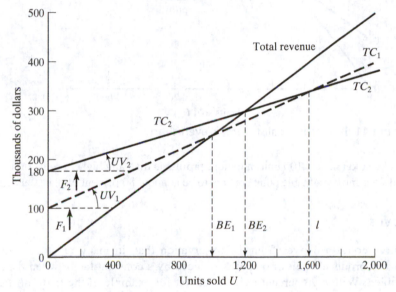

Figure 11-5   Break-even chart showing the effect of automation.

Automation will usually increase the break-even point, increasing the vulnerability to low sales levels, but beyond a certain point of production it will also increase profitability. Owners of efficient, highly automated plants will therefore strive to keep their plants busy and may negotiate a lower price (still above their variable cost) for discount chains or foreign shipment if they think this added volume can be achieved without affecting their current sales.

The break-even charts in the figures are idealized; Figure 11-6 represents a more realistic situation. Typically, the revenue line is really curved, since price may have to be reduced to increase the volume of sales. Also, up to some point the incremental cost to produce additional units may decrease because of economies of scale. At some other point, costs may increase as a step function as some additional fixed cost is added (such as the supervision and other overhead for a second shift). As you near plant capacity, incremental costs may increase as less-efficient backup equipment and less-trained workers are pressed into service. In the situation represented by Figure 11-6, you would not seek to add a second shift until you were confident that almost

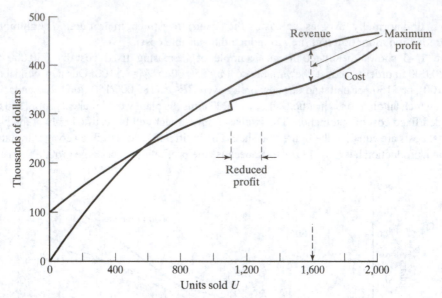

Figure 11-6   A more realistic break-even chart.

1,300 units could be sold (since 1,100 units are more profitable than any higher production quantity less than about 1,300), and your most profitable point appears to be at about 80 percent of plant capacity (1,600 units).

Learning Curves

The learning curve concept derives from the observation that, in many repetitious human activities, the time required to produce a unit of output is reduced by a constant factor when the number of units produced is doubled. With a 90 percent learning curve, for example, if the first unit takes 1,000 labor hours to produce, the second will take 900 hours, the fourth 810, the eighth 729, and so on, as shown in Figure 11-7. If it takes Y_1 time periods to make the first unit, the time Y_n to produce the nth unit can be found from

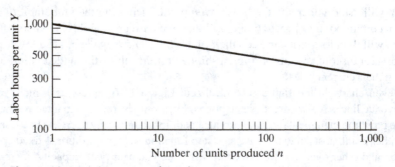

Figure 11-7   A 90 percent learning curve.

$$Y_n = Y_1 n^{-b}$$
$$\ln Y_n = \ln Y_1 - b \ln n \qquad \text{(11-5)}$$

The exponent b can be found for any learning curve rate k by setting $n = 2$:

$$\frac{Y_2}{Y_1} = k = 2^{-b} \qquad \text{(11-6)}$$

so that for the 90 percent learning curve, for which $k = 0.9$,

$$\ln(0.9) = -b \ln(2); \quad b = \frac{-\ln(0.9)}{\ln(2)} = 0.152$$

This relationship was developed in the aircraft industry, and its most common use has been there. Other applications are in the automobile industry, electronics assembly, and repetitive construction. Improvements are from a combination of factors, including increased worker skill, better work methods, better tooling and equipment, and organizational improvements. Tasks with greater manual and less mechanical content, however, tend to show a faster reduction in time required (lower percent learning curve). Table 11-1 gives a sample of the percent learning curves found in various industries in a study done in 1964. While today's learning curves would probably be different for these products (thanks to automation), these curves are similar to more recent studies of more complex products that showed integrated circuits to have a 72 percent learning curve and heart transplants to be 79 percent. Between competitors in mass production of similar products it appears that the competitor with the largest market share would always enjoy a production cost farther down the learning curve, and therefore it would be able to sell at a lower price or enjoy a higher profit. Fortunately, in many cases the learning curve may end in a plateau, permitting competition on other grounds.

Table 11-1 Sample of the Percent Learning Curves Found in Various Industries

Industry	Percent Learning Rate
Volkswagen, 1945–1949	60
Volkswagen, 1950–1954	80
Twenty light-alloy products	80
Home construction	73–86
Welding of thin steel	70
Airplane production	70–75
Shipbuilding	74–90
Vehicle bodies	70–80
German armament industry	65–82
Railway carriages	75–93

Source: Based on J. R. DeJong, "Increasing Skills and Reduction of Work Time—Concluded," *Time and Motion Study*, October 1964, pp. 20–33.

Note that the learning curve applies only for a continuous sequence of activity; if production stops at the end of one batch or lot and resumes later, the time to produce the first unit of the new batch will be greater than that for the last unit of the previous batch, and the learning curve will begin again at that point. This would seem to encourage large production batches, which carry with them large inventory levels. In the factory of the future, however, more and more production will be automated and accomplished by numerically controlled (CNC) machinery that does not "forget" how to produce a part, so that there will not be the loss of efficiency present when a human worker returns to a job done before, but now partially forgotten.

PRODUCTION PLANNING AND CONTROL

Introduction

Any activity whose success is dependent on the coordination and cooperation of many people will benefit from careful planning and control, and the manufacturing environment requires the interaction of many people and machines. Often, there is an exact sequence of operations that must be performed, and any deviation from this sequence will result in a scrapped part. The manager cannot focus solely on number of parts produced or even the cost per part, because quality, due-date performance, and efficiency are also scrutinized by upper management. Machines break down, parts are scrapped, raw material arrives late, and salespeople insist on the delivery of unscheduled rush orders.

In manufacturing it is essential to strike a strategic balance between idle resources and idle inventory. If inventories are very low, a worker may be *starved* for parts whenever the preceding workstation slows down, breaks down, sets up for a new product, or switches to a different product that does not require processing by the worker in question. The shorter these disruptions are, the shorter the idle period will be. If we wish to prevent idle periods, we must hold enough inventory between stations to keep the worker busy whenever the feed is unexpectedly disrupted.

The costs of idle resources are widely recognized. Management often believes that the wages of an idle worker have been wasted. There is the fear that more workers or more machines will be needed if one sits idle for a while. Worse, there is a belief that if someone has no work to do, it is time to have a lay off, even if that capacity is clearly needed to fill future demand.

The costs of idle inventory are just beginning to be recognized by many people. In the past, *inventory holding costs* were considered to consist primarily of the interest on working capital and the rental of warehouse space. This neglects the increased delay that long production runs of each product tend to produce in beginning another product run, increasing lead time (the time it takes from order placement to shipment). Long lead times may cause impatient customers to take their business elsewhere, a well-hidden cost. In rapidly advancing technologies, product may be obsolete even before it is shipped. Since new quality problems often remain hidden until discovered by a customer, a pipeline full of defective product could be a major liability.

Steps in Production Planning

The first step in planning of any type is to identify the goals you wish to achieve. The trade-off between idle machines and idle inventory will exist whether or not management cares to acknowledge it, and the schedule will enforce a given trade-off level, whether or not it is appropriate to the particular industry.

One company serving a seasonal market may decide to level their resource load by carrying more inventory. Another may decide to provide better service by carrying less inventory and more resources. Ideally, this should be a conscious decision rather than a random one. Once an inventory strategy is selected, the company should establish a procedure for quoting delivery dates that are in fact achievable. If customers know that the product cannot be delivered on the desired date, they may adjust their own schedules, or plan to order earlier next time. If they find out at the last minute, they may get upset and go elsewhere next time. If demand exceeds capacity, it may be necessary to consider other orders already promised when calculating a reasonable promise date.

The next several steps break down the production process into the required tasks and figure out when each one has to start. Sometimes, one task can be accomplished with any of several different resources (equipment and/or workers), in which case the assignment of tasks to resources can happen at planning time or at execution time. If several tasks need the same resource during the same period, the plan should provide workers with a means to determine priority.

These several steps have a traditional set of names. **Process planning** (routing) determines the sequence of operations needed to produce the product. **Loading** sets aside the necessary time on each machine or workstation to process the desired quantity. **Scheduling** establishes when each step of the work will be performed. **Dispatching** is the official authorization to do the work. In **flexible manufacturing systems** (discussed later) these conventional steps may occur automatically under computer control. Finally, *production control* is the system whereby deviations from the planned schedule are reported to the production planning and control office so that schedule adjustments can be made.

There is no such thing as a perfect plan, simply because the data on which the plan is based may change before the plan is executed. The time to process a given part is a statistical quantity that may vary unpredictably from time to time. Also, machines break down and parts get scrapped. Orders may be canceled or top-priority rush orders may be added. Sometimes, the database itself is in error. On the positive side, there may be a *learning curve* (discussed earlier in this chapter) such that the worker learns how to do a task faster than the database indicates, and the scheduling system should take advantage of such efficiencies.

Production Planning and Control Systems

Materials Requirements Planning (MRP). Materials requirements planning refers to a set of time-phased order-point techniques to support manufacturing schedules. MRP began development in the 1950s as the cost of computer calculation began to decrease. At its simplest, it provides a schedule for ordering raw material and parts and performing production operations to provide the products of production (end items) on time. MRP begins with a **master production schedule (MPS)** that identifies when end items must be available to meet customer or other commitments.

For example, assume that a customer has been promised that one unit of product A will be shipped six weeks from now, a need that becomes part of the MPS. The next document needed is the **bill of materials (BOM)** for product A, which shows that it is produced from one unit of material M, one component C, and two parts P. In addition, we need to know the supplier lead times for each item we have to buy, and the sequence and duration of such production activities as machining, assembly, and testing that will take place in our plant. The relationships are illustrated in Figure 11-8.

MRP converts this information into instructions for purchasing to place orders now for one unit of component C, one week from now for one unit of material M, and two weeks from now for two units of part P.

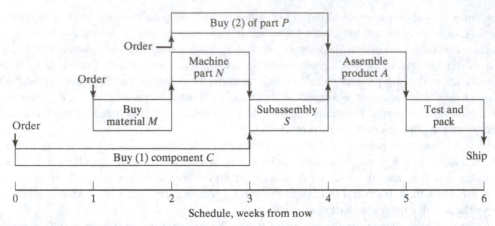

Figure 11-8   Schedule of production lead times for product *A*.

This will provide the suppliers the normal time required (three, one, and two weeks, respectively) to get them to us when we need them. Further, production planning is advised of the need to schedule the start of machining in two weeks, subassembly in three, assembly in four, and testing in five weeks. Similar requirements for all of the end items shown on the MPS are combined into comprehensive purchasing and production schedules, after the **inventory file** has been checked to identify material or subassemblies currently on hand or on order.

Manufacturing Resource Planning (MRP II).   The original MRP acted as if each order could be scheduled independently of the others, assuming that enough capacity existed to assign simultaneous orders to different resources. However, the capacity of both equipment and skilled workers is usually limited. The past decade has seen the shift from the simple MRP to one incorporating machine capacity and personnel planning, and a trend toward total integrated manufacturing control systems. As this took place, the terminology and the acronym MRP II have gradually replaced the earlier MRP. The *Tool and Manufacturing Engineers Handbook* lists the following information as being provided in a modern MRP II computer system:

- Customer demand activity
- Production plans
- Production schedules and their execution
- Purchasing management
- Inventory management
- Product cost reporting
- Support of and financial applications of accounts receivable, accounts payable, general ledger, and payroll

Enterprise Resource Planning (ERP)

ERP is a company-wide computer software system used to manage and coordinate all the resources, information, and functions of a business from shared databases. MRP and MRP II are predecessors of ERP. The development of these manufacturing, coordination, and integration methods and tools made today's

ERP systems possible. Both MRP and MRP II are still widely used, independently and as modules of more comprehensive ERP systems, but the original vision of integrated information systems as we know them today began with the development of MRP and MRP II in manufacturing. Today's ERP systems can cover a wide range of functions and integrate them into one unified database, and are used in almost any type of organization—large or small. For instance, functions such as human resources, supply chain management, customer relations management, financials, manufacturing, and warehouse management were at one time stand-alone software applications, usually housed within their own database and network. Today the ERP system allows all these applications to fit under one umbrella.

ERP asks people to change how they do their jobs. That is why the value of ERP is so hard to pin down. The software is less important than the changes companies make in the ways they do business. If you simply install the software without changing the ways people do their jobs, you may not see any value at all—indeed, the new software could slow you down by simply replacing the old software that everyone knew with new software that no one does.

Synchronized Manufacturing. In 1979, Eliyahu Goldratt developed OPT, a proprietary capacity-sensitive scheduling software that was supposed to correct the deficiencies of MRP and MRP II. In the process of implementing that software, he discovered that many of the scheduling problems found in industry were the result of not properly recognizing the relationships between inventory and capacity. The concepts that he developed came to be known by the same name as his software (OPT) and are best described in his book, *The Goal.* The same concepts are now advocated by many practitioners under the generic name *synchronized manufacturing* or *drum-buffer-rope,* and they can be used quite effectively in combination with MRP software packages, as described in *Regaining Control* by Burgess and Srikanth.

Just-in-Time (JIT). JIT is a method involving very small raw material or in-process inventory quantities, small manufacturing lots, and frequent deliveries, such that a small batch of each component or subassembly is produced and delivered "just in time" to be used in the next production step. It was initially developed by the Toyota Motor Company and later was adopted by other Japanese companies. The Toyota system uses a series of cards called **kanban** (pronounced kahn-bahn), the Japanese for a *visible record or plate,* to direct production. Following is a description by Reda:

> This card (kanban) is primarily used to signal the need to either deliver (withdrawal kanban) or produce (production-ordering kanban) more parts. A withdrawal kanban specifies the quantity required at succeeding processes (which are to be withdrawn from preceding processes), while a production-ordering kanban orders preceding processes to produce replacement parts.

Figure 11-9 provides examples of kanban cards, and Figure 11-10 illustrates the mechanics of their use. The quantity ordered on each kanban is typically very small, so defects and production problems in one location can cause the entire production line to shut down. Toyota reduced the incidence of such disruptions with a number of related innovations:

- Smooth production schedules for final assembly of the end item, with little month-to-month variation
- An incessant effort to eliminate the causes of defects
- Plant layout in FMC such as those already described under *group technology*
- Workers able and willing to work at different processes as demand requires

Withdrawal Kanban

Store Shelf No. 5E215 Item Back No. A2-15	Preceding process
Item No. 35670S07	*FORGING* *B-2*
Item Name *DRIVE PINION*	
Car Type *SX50BC*	Subsequent process
	MACHINING *M-6*

Box Capacity	Box Type	Issued No.
20	B	4/8

Production-ordering Kanban

Store Shelf No. F26-18 Item Back No. A5-34	Process
Item No. 56790-321	*MACHINING* *SB-8*
Item Name *CRANK SHAFT*	
Car Type *SX50BC*-150	

Figure 11-9   Examples of kanban cards. (From Hussein M. Reda, "A Review of 'Kanban'—the Japanese 'Just-in-Time' Production System," *Engineering Management International*, Volume 4, Issue 2, April 1987, Pages 143–150. Used with permission)

- Worker involvement in identifying and correcting problems as they occur, through mechanisms such as the *quality circles* described in Chapter 12
- Reduction in the number of suppliers, offset by great emphasis on the quality and delivery schedules of those that remain

Parts coming from other plants are typically delivered one truckload at a time in the JIT system, and the production line often is fed directly out of the truck that just arrived. This is particularly feasible in Japan, where manufacturers tend to be located in the Tokyo area and to have long-term relationships with trusted suppliers also located in the same area. In the United States, with suppliers often thousands of miles away, some adaptations have to be made. To make JIT work better, General Motors has encouraged its major suppliers to build parts plants near GM assembly plants. To keep assembly plants constantly informed of the location of supplies en route to them, some American trucking firms now carry radios capable of relaying their status to the plant at any time via the *Geostar* satellite. At the new (1992) General Motors Opel plant in Eisenach, Germany, final assembly schedules are transmitted electronically to the nearby Lear seating plant; car seats are built four hours before they are needed and trucked to the Opel plant.

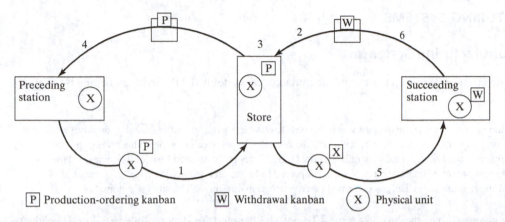

| P | Production-ordering kanban | W | Withdrawal kanban | X | Physical unit |

Figure 11-10   Mechanics of a simple kanban cycle. (1) Part produced at preceding station and P-kanban attached to it is sent to the store. (2) When the part is needed at a succeeding station, a W-kanban is sent to the store to withdraw the part. (3) At the store, the P-kanban is removed from the part and the W-kanban attached to it. The P-kanban is then collected in a "production-ordering" box. (4) At short time intervals, the P-kanban is then sent to the preceding station, constituting a production order. (5) The part with the W-kanban goes to the succeeding station to meet the demand. (6) The W-kanban is detached from the part and collected in a "withdrawal" box. (Source: Adapted from Hussein M. Reda, "A Review of 'Kanban'—the Japanese 'Just-in-Time' Production System," *Engineering Management International*, 4(2), 1987, pp. 143–150. Used with permission).

Model Changeover.   Production planning is especially vital in mass production where major changes are necessary between annual (or other periodic) models. In an example reported by Treece in *Business Week,* General Motors Corporation shut down the production lines at their Oshawa, Ontario, plant on November 19, 1993, to install new welding robots and other machinery to build the 1995 Chevy Lumina, and it was 87 days until the first production line started up on February 19, 1994—slowly. By April 1, it had built a total of only 288 cars and did not expect to reach its full line speed of 60 cars an hour until August. The Toyota plant in Georgetown, Kentucky, on the other hand, made the changeover to the 1992 Camry in 18 days, and the Honda plant in Marysville, Ohio, changed over to the 1994 Accord in just three days, and reached full speed in just six weeks. Treece explains:

> Toyota and other Japanese auto companies make new-car designs as compatible as possible with existing equipment. To make that easier, they tend to use more flexible automation—such as welding robots that can be reprogrammed. And their factories have extra space beside the assembly line, so that new gear can be tested well before it is needed.

Weeks when plants are shut down or run below design speed are weeks when profits are not made, and American car companies have played catch-up. They have gradually installed more flexible equipment and enabled faster change overs.

MANUFACTURING SYSTEMS

Flexible Manufacturing Systems

A 1985 U.S. government report on the flexible manufacturing systems (FMS) industry begins by explaining the following:

> Automation in manufacturing in the past was only considered where large quantities (mass production) were required, such as in the automotive industry and in the household appliance industries. This level of production automation could be accomplished only by transfer machines and dedicated lines of machines and then only for production of a limited variety of different parts. However, on a worldwide basis, 75 percent of all metalworking manufacture takes place in small batch production, limiting the benefits of automation.

A lot has changed over the last 30+ years. Today, the demand for differentiated (tailored) products rather than mass-produced ones, and worldwide competition in almost every industry, join to create a compelling need for more efficient means of producing small batches of high-quality products. At the same time, the computer revolution in both hardware and software makes possible computer control of machining and other manufacturing operations to reduce the cost of setup for small batches. Robots and other computer-controlled devices for handling and transferring work between machines, automatic guided vehicles (AGVs) for movement of work and tools, automated storage systems, and the development of computer-integrated manufacturing techniques and software combine to make FMS possible.

Some Definitions

The U.S. Department of Commerce recognizes four basic categories of flexible manufacturing technology: stand-alone machines, the flexible manufacturing cell, the flexible manufacturing system, and the fully automated factory. Each is described next:

> The **stand-alone machine** is typically a machining center or turning center with some method of automatic material handling, such as multiple pallets or chuck [tool] changing arrangements. These provisions permit the machine to operate unattended for extended periods (often a full eight-hour shift), changing tools and work pieces under direction of the machine control. This computer control can also initiate and control features such as probing, inspection, tool monitoring, and adaptive control.
>
> The **flexible manufacturing cell (FMC)** normally incorporates more than one machine tool, together with pallet changing equipment such as an industrial robot, to move work into the cell, between machines in the cell, and out of the cell.
>
> The **flexible manufacturing system (FMS)** includes at least three elements: a number of workstations, an automated material handling system, and system supervisory computer control....Automatic tool changing, in-process inspection, parts washing, automated storage and retrieval systems (AS/RS), and other computer-aided manufacturing (CAM) technologies are often included in the FMS. Central computer control over real-time routing, load balancing, and production scheduling distinguishes FMS from FMC.

The **fully automated factory (FAF)** or "factory of the future" represents the full development of all aspects of computer integrated manufacturing (CIM). In the FAF, all functions of the factory will be computer controlled, integrated, and, to varying degrees, self-optimizing.

Advantages of an FMS

Hartley provides a good description of the FMS installed by Yamakazi, a manufacturer of machining centers (themselves FMS components) at its main plant in Oguchi, Japan:

- It consists of 18 machining centers instead of the 36 needed previously.
- Previous employment was 106 direct and 80 indirect workers. The new FMS "is manned by one person in the computer room, one person in the tool room, and four people at the loading/unloading station. These people are needed on two shifts only, the third shift being unmanned. Therefore the number of workers required is down from 190 to 12."
- "The components for a complete machine [used to take] three *months* to pass through the machine shops, four weeks in assembly, and one week for inspection and adjustment. Now, the time spent in the machine shop has been reduced to only four days—for an average process time of 24 hours— while assembly now takes two weeks [because workpieces are available when needed], and inspection one week. Thus, a machining center can now be produced in under four weeks, whereas it previously took over four months."
- Estimated capital costs (in U.S. dollars) compared as follows: land and building was reduced from $2.43 MM to $1.28MM; machinery and equipment increased from $5.12 MM to $9.60 MM; work in process [working capital] decreased from $3.20 MM to $0.13 MM; total investment therefore only increased from $10.75 MM to $11.01 MM.
- Although the total capital costs above were comparable, labor costs dropped from over $2.56 MM to about $0.17 MM, so that average annual profit (assuming three-shift operation) increased from $1.02 MM to $2.43 MM

Ranky states that FMS provides the following benefits if designed and used successfully:

- Productivity increases, which means there is a greater output and a lower unit cost, on 45 to 85 percent smaller floor space.
- Quality is improved because the product is more uniform and consistent.
- The intelligent, self-correcting systems (machines equipped with sensory feedback systems) increase the overall reliability of production.
- Parts can be randomly produced in batches of one or in reasonably high numbers, and the lead time can be reduced by 50 to 75 percent.
- FMS is the only available manufacturing environment to date where the time spent on the machine tool can be as high as 90 percent and the time spent cutting can again be over 90 percent. Compare this with stand-alone NC machines, where the part, from stock to finished item, spends only 5 percent of its time on the machine tool, and where the actual productive work takes only 30 percent of this 5 percent.

Lean Manufacturing

The principles and practices of **Lean Manufacturing** are simple and have evolved over the past century, beginning with Taylor and Gilbreth (from Chapter 2). While they have been developed by trial and error over many decades, and many prominent men and women have contributed to their development, the principles and practices are often not easy to implement. Implementation requires organizational commitment and support by management, and participation of all the personnel within an organization, to be successful. The critical starting point for lean thinking is value, and value can be defined only by the customer.

Henry Ford was one of the first people to develop the ideas behind Lean Manufacturing. He used the idea of "continuous flow" on the assembly line for his Model T automobile, where he kept production standards extremely tight, so each stage of the process fitted together with each other stage, perfectly. This resulted in little waste. But Ford's process was not flexible. His assembly lines produced the same thing, again and again, and the process did not easily allow for any modifications or changes to the end product—a Model T assembly line produced only the Model T. It was also a *push* process, where Ford set the level of production, instead of a *pull* process driven by consumer demand. This led to large inventories of unsold automobiles, ultimately resulting in lots of wasted money.

Other manufacturers began to use Ford's ideas, but many realized that the inflexibility of his system was a problem. Taiichi Ohno of Toyota then developed the Toyota Production System (TPS), which used JIT manufacturing methods to increase efficiency. Toyota was able to greatly reduce cost and inventory. As Womack reported in his book, Toyota used this process successfully and, as a result, eventually emerged as one the most profitable manufacturing companies in the world, eventually becoming the largest automotive company in the world. Due to the success of this production philosophy, many of these methods have been copied by other manufacturing companies. TPS is known more generically as Lean Manufacturing.

This system was developed between 1948 and about 1975 in Japan as Toyota was returning to the production process following World War II. As the system spread, the name changed several times from TPS to (now) Lean Manufacturing. The main goal of Lean is to eliminate waste (*muda*). There are seven kinds of waste targeted in Lean: defects, overproduction, transportation, waiting, inventory, motion, and overprocessing.

According to Womack in *Lean Thinking*, "lean thinking must start with a conscious attempt to precisely define value in terms of specific products with specific capabilities offered at specific prices through a dialogue with specific customers." Specifying value accurately is the critical first step in lean thinking and avoiding *muda*. Lean is basically all about getting the right things to the right place at the right time in the right quantity, minimizing waste while being flexible and open to change.

The key Lean Manufacturing principles are as follows:

- *Perfect first-time quality.* Quest for zero defects, revealing and solving problems at the source
- *Waste minimization.* Eliminating all activities that do not add value and safety nets, and maximizing use of scarce resources (capital, people, and land)
- *Continuous improvement.* Reducing costs, improving quality, increasing productivity, and sharing information
- *Pull processing.* Products are pulled from the customer end, not pushed from the production end

- *Flexibility.* Producing different mixes or greater diversity of products quickly, without sacrificing efficiency at lower volumes of production
- Building and maintaining a long-term relationship with suppliers through collaborative risk-sharing, cost-sharing, and information-sharing arrangements

---

Lean is So Much More than Manufacturing

Like many quality improvement programs (e.g., 6σ, Total Quality Management, etc.), after starting as a manufacturing approach, Lean has expanded to most sectors of the global economy. Starting over 10 years ago, lean became a common approach to process improvement in the healthcare sector. Since Lean is focused on driving out waste, Lean projects in healthcare vary widely. Examples include anything from streamlining patient check-in to development and deployment of checklist forms for transferring patients between departments and reducing use of continuous observation to improving rates of timely discharge. Lean is a particularly powerful approach in healthcare as it recognizes the complexity of the healthcare system while aligning staff to focus on the importance of customer needs—in this case patients.

But healthcare is not the only sector to embrace Lean approaches. Many areas of the service economy have benefited from Lean implementations. One reason Lean is so beneficial here is that service companies generally perform worse than manufacturing companies on key metrics, such as percentage of product defective. For example, think about the percentage of manufactured products that you purchase that do not perform as expected due to a manufacturing defect. Chances are that is a pretty low percentage, probably far below 1 percent. Now think about the number of times you have a service experience that is defective (e.g., server brings the incorrect meal, flight is delayed or cancelled). Typically the defect rates of service experiences are an order of magnitude or more than those in manufacturing. For this reason, Lean's approach to quickly fixing obvious problems through techniques such as Kaizen events, provides an organization quick wins that build support and momentum. This is especially important in service companies, which are often larger than manufacturing companies, and see even greater benefits in the "bottom-up" approach common to Lean.

Sources

Institute for Healthcare Improvement, *Going Lean in Healthcare* (Cambridge, MA, 2005).
Lawal, Adegboyega K, Thomas Rotter, Leigh Kinsman, Nazmi Sari, Liz Harrison, Cathy Jeffery, Mareike Kutz, Mohammad F Khan, and Rachel Flynn, "Lean management in health care: definition, concepts, methodology and effects reported (systematic review protocol)", *Systematic reviews*, 3, 103. doi:10.1186/2046-4053-3-103, 2014.
Liker, Jeffrey K, and Karyn Ross. *The Toyota Way to Service Excellence: Lean Transformation in Service Organizations* (New York: McGraw-Hill Education, 2017).

Supply Chain Management

Supply chain management (SCM) is a series of processes that go into improving the way a company finds the raw components it needs to make a product or service and delivers it to customers. The goals are to lessen the time to market, reduce the cost to distribute, and supply the right products at the right time. The concept of supply chain management is based on two core ideas. The first is that practically every product that reaches an end user represents the cumulative effort of multiple organizations. These organizations are referred to collectively as the supply chain.

The second idea is that while supply chains have existed for a long time, most organizations have only paid attention to what was happening within their organization. Few businesses understood, much less managed, the entire chain of activities that ultimately delivered products to the final customer. The result was disjointed and often ineffective supply chains.

Supply chain management, then, is the active management of supply chain activities to maximize customer value and achieve a sustainable competitive advantage. The following are five basic components of SCM:

- *Plan.* Define a strategy for managing all the resources of the supply chain
- *Source.* Choose suppliers to deliver the goods and services required for production
- *Make.* Manufacturing step
- *Deliver.* Logistics of getting products to customers
- *Return.* Network for receiving defective and excess products from customers

Each of the five major supply chain steps is comprised of many tasks, and many have their own software. The different manufacturing systems discussed above might also use SCM. Since the wide adoption of Internet technologies, all businesses can take advantage of Web-based software and Internet communications. Instant communication between vendors and customers allows for timely updates of information, which is key in the management of the supply chain and optimizes costs and opportunities for everyone.

DISCUSSION QUESTIONS

11-1. Why is a robust manufacturing capability so important to the United States in the "postindustrial society"?

11-2. What are some of the positions that engineers fill in a large manufacturing plant? How have these roles changed over the last decade?

11-3. Research how manufacturing firms are using additive manufacturing (3D printing). How does this change the needed skills of a manufacturing engineer?

11-4. Discuss some of the factors that would be most important in selecting a site for (**a**) a cement plant, (**b**) a research "think tank," (**c**) a software company, and (**d**) a furniture manufacturer.

11-5. Distinguish between (**a**) product layout, (**b**) process layout, and (**c**) group technology. When is each the best design?

11-6. List some of the characteristics that make the Dunn and Savastano Orthodontics business compete effectively with other practices.

11-7. Distinguish between MRP and ERP.

11-8. Describe how the kanban is used in the just-in-time production system.

11-9. Outline how lean manufacturing has been applied to other industries. What other approaches discussed in this chapter have moved beyond manufacturing? Why have others not made the transition?

11-10. What is the relationship between FMS and the group technology concept introduced earlier in the chapter?

11-11. Discuss how the management functions of planning, organizing, leading, and controlling relate to the production planning process.

PROBLEMS

11-1. (a) If it costs $2.00 per unit to store an item for one year and $40.00 setup cost every time you produce a lot, and you use 1,000 units per year, how many lots of what size should be manufactured each year? (b) How would your answer change if the setup cost can be reduced to $10.00?

11-2. Setup for a stamping operation required a time-consuming fixture installation and testing that took 4 hours each time a different part was to be produced; typically, 12 hours' production was made for inventory of a given stamping before the machine was stopped to permit setup for a new part. After careful process analysis, fixtures and transfer methods were revised to permit setup in 15 minutes. Discuss the implications for this change on (a) optimum batch size, (b) order frequency, and (c) machine and labor productivity.

11-3. A production plant with fixed costs of $300,000 produces a product with variable costs of $40.00 per unit and sells them at $100 each. What is the break-even quantity and cost? Illustrate with a break-even chart.

11-4. A machine tool salesperson offers the plant of Problem 11-3 equipment that would increase their fixed cost by $180,000, but reduce their variable cost from $40.00 to $25.00. Should the plant accept this suggestion if they can sell their entire plant capacity of 10,000 units per year at $100 each? Illustrate by modifying the break-even chart of Problem 11-3.

11-5. A plant is beginning production of a light-alloy product and finds that it takes 400 hours to produce the first item. How many hours should it take to produce each of the following: (a) the second item; (b) the eighth item; (c) the 37th item? (*Hint:* Refer to Table 11-1.)

11-6. The first two units of a product cost a total of $9,000 to produce. If you believe an 80 percent learning curve applies, how much would you expect the *fourth* unit to cost?

SOURCES

Amrine, Harold T., Ritchey, John A., Moodie, Colin L., and Kmec, Joseph F., *Manufacturing Organization and Management,* 6th ed. (Englewood Cliffs, NJ: Prentice Hall, Inc., 1993).

Burgess, Susan and Srikanth, Mokshagundam L., *Regaining Control: Get Me to the Shipping Dock on Time* (New Haven, CT: Spectrum Publishing, 1989).

Cohen, Stephen S. and Zysman, John, "Manufacturing Innovation and American Industrial Competitiveness," *Science,* 39, March 4, 1988, p. 1114.

Cunningham, James A., "Using the Learning Curve as a Management Tool," *IEEE Spectrum* (June 1980): 45.

"GM's German Lessons," *BusinessWeek,* December 20, 1993, p. 67.

Goldratt, Eliyahu M. and Cox, Jeff, *The Goal: A Process of Ongoing Improvement,* rev. ed. (Croton-on-Hudson, NY: North River Press, Inc., 1986).

Grove, Andrew presentation, President, Intel Corporation, to the plenary session "A New Era in Manufacturing" at the American Society for Engineering Education annual conference June 21, 1988, in Portland, OR.

Hartley, John, *FMS at Work* (Kempston, Bedford, UK: IFS [Publications] Ltd., 1984), pp. 157–160.

Jaikumar, Ramchandran, "Postindustrial Manufacturing," *Harvard Business Review,* 64:6, November–December 1986, p. 70.

Kearney, A. T., Inc., *Countdown to the Future: The Manufacturing Engineer of the 21st Century* (Dearborn, MI: Society of Manufacturing Engineers, 1988).

Muther, Richard, informal presentation to the Department of Engineering Management, University of Missouri–Rolla, October 10, 1988.

National Research Council, Manufacturing Studies Board, *Toward a New Era in U.S. Manufacturing: Need for a National Vision* (Washington, DC: National Academy Press, 1986).

Ranky, Paul, *The Design and Operation of FMS* (Kempston, Bedford, UK: IFS (Publications) Ltd., 1983), p. 4.

Reda, Hussein M., "A Review of 'Kanban'—the Japanese 'Just-in-Time' Production System," *Engineering Management International* 4, 1987, pp. 145–146.

Smith, David B. and Jan L. Larsson, "The Impact of Learning on Cost: The Case of Heart Transplantation," *Hospital and Health Services Administration* (Spring 1989): 85–97.

Treece, James B., "Motown's Struggle to Shift on the Fly," *BusinessWeek*, July 11, 1994, pp. 111–112.

University of Missouri–St. Louis, *Metalworking* and *Plastics-working* Machine Operators, http://www.umsl.edu/services/govdocs/ooh9899/141.htm

Veilleaux, Raymond F. and Petro, Louis W., eds., *Tool and Manufacturing Engineers Handbook, Volume 5: Manufacturing Management,* 4th ed. (Dearborn, MI: Society of Manufacturing Engineers, 1988), pp. 2–17.

Wailgum, Thomas, *Supply Chain Management Definition and Solutions. CIO Nov. 20, 2008.* http://www.cio.com/article/40940/Supply_Chain_Management_Definition_and_Solutions.

Welch Jr., John F. "Competitiveness: The Real Stuff of American Engineering," *Gateway Engineer,* February 1990, p. 9.

Womack, James P. and Jones, Daniel, *Lean Thinking: Banish Waste and Create Wealth in Your Corporation* (New York: Simon & Schuster, 1996), p. 19.

Yoder, Brian L., *Engineering by the Numbers* (Washington, DC: American Society for Engineering Education, 2017).

12

Managing Quality and Production Operations

PREVIEW

Quality and Production operations are a set of interrelated activities that are involved in manufacturing products. Quality management focuses on delivering products and services that meet (or exceed) customer expectations, including work to understand what those expectations include. Production management focuses on carefully managing the production operations to produce and distribute products and services. Production is the creation of goods and services. A major focus of production operations is on efficiency and effectiveness of processes. The use of quality concepts with production operations includes substantial measurement and analysis of internal processes.

The operations manager's objective is to build a production system that identifies and satisfies the customers' needs. This chapter begins by defining product quality and introducing the categories of quality costs. The statistics of the measurements of variables and attributes are introduced and applied to control charts and inspection sampling. This is followed by a discussion of the contributions of various quality

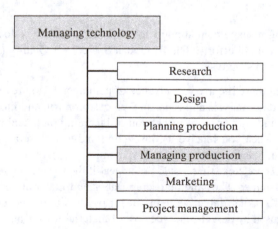

approaches, including total quality management, Taguchi, Deming, and others. The chapter then explores detailed concepts of production management including productivity measures and time standards, maintenance, and other functions in manufacturing.

LEARNING OBJECTIVES

When you have finished studying this chapter, you should be able to do the following:

- Define quality.
- Describe the quality revolution.
- Discuss some of the tools of quality.
- Recognize the methods of work measurement.

ASSURING PRODUCT QUALITY

Some Definitions

Quality has been described as *fitness for use* or *customer satisfaction*. It may be divided into two categories. ***Quality of design*** measures the extent to which *customer satisfaction* is incorporated into the product design through the specification of proper materials, tolerances, and other precautions. Quality of design will vary to some extent with the intended customer: One would not expect the same features in a stripped-down Ford Focus and a fully equipped Lincoln Navigator. ***Quality of conformance*** (or *quality of production*) measures how well the quality specified in the design is realized in manufacture and delivered to the customer. The customer may be an internal customer, as the next process on the production line, or an external customer. Some consider a third aspect of quality, measuring how the product is applied or employed, and what that does to its properties.

Quality Costs

An important step in getting management support for improving quality is documenting the total cost of poor quality and of quality control efforts. The American Society for Quality (ASQ) has established four categories of costs to help in this analysis:

1. *Prevention costs* are those incurred in advance of manufacture to prevent failures, such as quality planning, training, data analysis and reporting, process control, and motivation programs.
2. *Appraisal costs* include the costs of inspection of incoming parts and materials (whether by your supplier or by you when you receive it), inspection and test of your product in process and as a finished product, and maintenance of test equipment.
3. *Internal failure costs* are those that would not appear if there were no defects in the product before shipment to the customer. They include scrap (labor and material spent on unrepairable items), rework (the cost of making defective items fit for use, including necessary retesting), downtime and yield losses caused by defects, and the cost of material review and disposition of defectives.

4. *External failure costs* are those caused by defects found after the customer receives the product. These include the costs of investigating and adjusting complaints, the costs of replacing defective product returned by the customer, price reductions ("allowances") offered to compensate for substandard products, and warranty charges. The total costs to your customer in downtime and other damages may be much higher, and these may drive your customer to seek a more trustworthy supplier.

When these quality costs are added up, they usually total far more than management realized—often of the same magnitude as total company profit. Typically, the prevention costs are found to be a very small percentage of the total. When a concerted effort is instituted to develop a comprehensive quality program, to find the primary reason for failures, and to modify design, processes, and employee training and motivation to minimize failures, savings in failure costs are commonly many times the cost invested in prevention. Even appraisal costs are reduced, since top-quality product does not require the same intensive level of inspection. Figure 12-1 shows the relationship of these components of quality cost with quality level. This classic figure suggests that there is some optimum economic quality short of 100 percent conformance that should be striven for. Merino points out that this may be a valid conclusion where quality is "inspected in" through intensive inspection and test, but that the modern approach of continuous improvement of product design and of the processes used in manufacture makes it possible

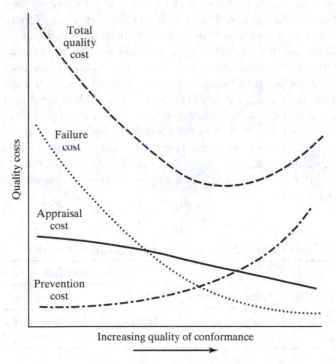

Figure 12-1   Effect of quality improvement on quality costs.

to approach very close to 100 percent conformance without excessive prevention and appraisal costs. In fact, most current thinking about quality highlights that it is truly impossible to inspect in quality and processes must be designed for zero defects.

Statistics of Quality

Statistics consists of gathering, organizing, analysis, and use of data. The methods of statistical quality control were developed in the United States in the 1930s and 1940s (largely at Bell Laboratories). However, they received their most intensive application in Japan after World War II, as a result of visits of the American statistical quality control experts Deming and Juran at the invitation of General Douglas MacArthur's occupation forces. Only when the Japanese brought their quality and reliability to a level that threatened the U.S. economy did domestic industry begin to pay attention. Unfortunately, many U.S. engineers and business leaders are poorly prepared to effectively manage quality improvement efforts. This lack of preparation was highlighted over 30 years ago by the then ABET (formerly the Accrediting Board for Engineering and Technology) president Gordon Geiger when he noted the limited preparation in statistics of U.S. engineers and citizens, typically less than one course in the subject. Since quality improvement efforts always rest on a foundation of understanding the statistical nature of processes, this is a problematic limitation for American business.

Lester Thurow, former dean of MIT's Sloan School of Business, highlighted this gap in statistical understanding in discussing the problem faced by a Japanese firm when they built a plant in North Carolina. Although they were accustomed to using high school graduates for statistical quality control in Japan, they could not find a high school or college graduate able to do the job and thus had to hire someone with a graduate degree. Thurow concludes by asking, "How can you win in a technical era with mathematical illiterates?" Statistical methods are used to evaluate some quality characteristic, such as the diameter of a hole, the weight of a package, or the tensile strength of a metal strip. Two types of statistical methods are used in quality control (Table 12-1). Variables methods involve measuring the quality characteristic (such as the hole diameter) on a sample of the item being controlled, then using a continuous probability distribution such as the normal distribution for analysis. Attributes methods involve counting as defective those items that do not fall within a stated specification, then using the fraction defective in a sample in discrete probability distributions such as the binomial or Poisson for analysis. Each probability distribution is described by a *measure of central tendency* (average) and a *measure of dispersion* (spread).

Table 12-1   Some Statistical Methods Used in Quality Control

Class of Statistics	Action Involved	Probability Distributions	
		Type	Examples
Variables	Measuring	Continuous	Normal, Exponential
Attributes	Counting	Discrete	Binomial, Poisson

Example

Consider a hole with a specified diameter x of 1.250 $\pm$ 0.010 inches (i.e., holes from 1.240 to 1.260 inches will meet specifications). Assume that this diameter was actually measured as 1.235, 1.245, 1.250, 1.256, and 1.259 inches in a sample of five items. Using the variables approach, the *mean* value (termed "x-bar") would be found as the sum of these n values (6.245 inches) divided by the number of measurements ($n = 5$), or 1.249 inches:

$$\bar{x} = \frac{1}{n}\sum_{j=1}^{n}\bar{x}_j \tag{12-1}$$

Using the attributes approach in this same example, one would simply determine (perhaps using a "go/no go gauge") that only one of these five holes was defective (fell outside the specified range of 1.240 to 1.260). The sample would then be assigned a fraction defective p of 1/5, or 0.20; the mean value "p-bar" would then be found by averaging the p values for a large number of samples.

The most common measure of dispersion from the mean value is the *sample standard deviation s*, which is the square root of the *sample variance V. V* is an unbiased estimator of the population variance and can be found as the sum of the squares of the deviations from the sample mean, divided by one less than the sample size:

$$V = \frac{1}{n-1}\sum_{j=1}^{n}(x_j - \bar{x})^2 \tag{12-2}$$

Example

$$V = \frac{(1.235 - 1.249)^2 + (1.245 - 1.249)^2 + \cdots + (1.259 - 1.249)^2}{5 - 1}$$

In the previous example $= \dfrac{0.000196 + 0.000016 + 0.000001 + 0.000049 + 0.000100}{4}$

$$= 0.0000905$$

The sample standard deviation s (an estimate of the population standard deviation σ) is the square root of V, or 0.0095.

In the variables method, the *range R*, which is the difference between the highest and lowest values in a sample, is also a common measure of dispersion. In the previous example, the sample range would be 1.235 to 1.259, or 0.024 inch. Ranges are easier to calculate, but provide less information per sample than does the standard deviation.

Process Control Charts

In production operations it is important to ensure that a process is "in control," which means that it continues to produce items with unchanged quality characteristics. **Process control charts** (Figure 12-2) are used to point to potential problems that need attention. Control charts consist of three parallel lines: a central line representing the mean value of a quality characteristic; an *upper control limit* (UCL), normally three standard deviations (3σ) above this mean value; and a *lower control limit* (LCL), normally three standard deviations below the mean. If the process stays in control, 99.73 percent of all future observations should fall between the UCL and the LCL, symmetrically dispersed about the mean. A great deal can be learned about such things as raw material or operator changes, tool wear, or changes in machine settings by observing measurements that fall out of the control limits, bunch on one side of the central line, or follow some other nonrandom pattern.

For example, if you wish to control some measurable quality characteristic x using statistics, you may wish to maintain a control chart based on the mean (average) value $\bar{x}$ of measurements of random samples of five items. To do this, you would take samples of five items at regular periods until you had about 25 samples; use this information to calculate values for the central line (mean of these mean values, or $\bar{\bar{x}}$) and control limits, and construct an $\bar{x}$ chart similar to Figure 12-2. As you continue making the product, you continue to take samples of five items and enter their mean value on the control chart to assure that nothing has changed in your process or materials.

To control the variation (dispersion) about this mean value, you would maintain one of two types of charts: an "*R*-chart," measuring the difference between the highest and lowest value within each sample of five, or a "sigma-chart" (δ-chart), measuring the standard (root mean square) deviation of measurements from the mean value.

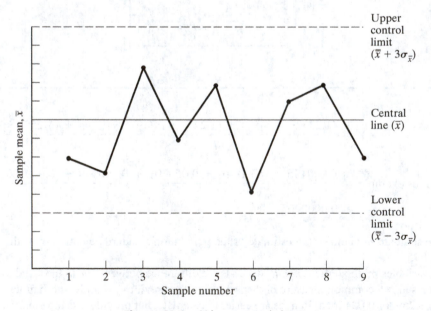

Figure 12-2 *X*-bar control chart.

If you are using the attributes approach (counting, but not measuring defects), you can control the level of these defects by using either a *p*-chart (which measures sample *fraction defective* and is based on the binomial probability distribution) or a *c*-chart (which measures *defects per sample* and is based on the Poisson probability distribution). (A defective item is one that contains one or more defects.) In the attributes approach, separate charts to monitor variation (dispersion) are not needed.

Inspection and Sampling

Examining a product to determine if it meets the specifications set for it, or *inspection*, is certainly the original method of quality control, and it is still the most common. Inspection may be performed on the raw materials and parts you receive from suppliers (*acceptance sampling*), on your finished product, or on your goods-in-process (before you invest the cost of the next production step in them). Examination of every item (100 percent inspection) may seem desirable, but it is often expensive unless it can be done automatically, and is often no more effective than sampling since human inspectors cannot be expected to be continually vigilant and catch all defects. Most inspection is therefore done by sampling lots (batches) of product and accepting or rejecting the lot, depending on the number of defectives in the sample. Sampling rules can be developed statistically for each situation, but it is more common to consult an established sampling table.

For example, suppose that you were using the common military standard MIL-STD-105E for acceptance sampling by attributes (counting, not measuring) of a lot of 2,000 items, and you considered 1.0 percent defective an *acceptable quality level* (AQL). If you looked up the normal "general inspection level II" for this lot size and AQL in MIL-STD-105D you would be directed to follow these steps:

- Take a random sample of 125 items from the lot.
- Accept the lot if it contained no more than three defective items.
- Reject the lot if it contained four or more defective items. Then, either 100 percent inspect it (sort out the defectives) or return it to the producer as unsatisfactory.

Or, you might choose the *double-sampling* alternative:

- Take a sample of 80 items.
- Accept the lot with no more than one defective.
- Reject the lot with four or more defectives.
- With two or three defectives, take a second sample of 80 items and accept the lot if the total defectives in the two samples were four or less; otherwise, reject it.

In *multiple sampling*, from 1 to 7 sequential small (32-item) samples would be used in this same situation, with a more complex decision rule.

Sampling from a larger lot or to assure a higher quality (smaller AQL) would require a larger sample and a different decision rule; the reverse would permit a smaller sample. Sampling plans are also available on a variables basis, where you can take a smaller sample, but use the actual measured value of your quality characteristic for lot acceptance; other plans are used for continuous rather than batch production.

TOTAL QUALITY MANAGEMENT

The quality revolution has been sweeping industry for over 40 years and has also made inroads in government, health, and higher education. The emphasis on quality is key to managing production operations and to achieving excellence in today's global economy. The focus is on customer-driven standards. Over this time, the quality movement has matured. The new quality systems have evolved beyond the foundations of the early users and now incorporate the ideas of Deming, Taguchi, Juran, Philip Crosby, Armand Feigenbaum, and many others, blending these techniques and tools with other tools. Some of these include the following:

- Affinity diagrams
- Brainstorming
- Pareto charts
- Check sheets
- Flowcharts
- Theory of Constraints
- Quality function deployment
- Statistical methods (as explained earlier)
- Run charts

ISO 9000

The ISO 9000 and ISO 14,000 families are among ISO's most widely known standards. ISO 9000 and ISO 14,000 standards are implemented by over one million organizations in over 170 countries. ISO 9000 has become an international reference for quality management requirements in business-to-business dealings, and ISO 14,000 is well on the way to achieving as much, if not more, in enabling organizations to meet their environmental challenges.

The ISO 9000 family addresses various aspects of quality management and contains some of ISO's best-known standards. The standards provide guidance and tools for companies and organizations who want to ensure that their products and services consistently meet customer's requirements, and that quality is consistently improved.

The ISO 14,000 family is primarily concerned with "environmental management." This means what the organization does to accomplish the following:

- Minimize harmful effects on the environment caused by its activities, and to
- Achieve continual improvement of its environmental performance.

The vast majority of ISO standards are highly specific to a particular product, material, or process. However, the standards that have earned the ISO 9000 and ISO 14,000 families a worldwide reputation are known as "generic management system standards."

Source: http://www.iso.org., January 2019.

Six Sigma

In the mid-1980s, with former chairman Bob Galvin at the helm, Motorola engineers decided that the traditional quality levels—measuring defects in thousands of opportunities—did not provide enough *granularity*. Instead, they wanted to measure the defects per million opportunities. Motorola developed this new standard and created the methodology and needed cultural change associated with it. Six Sigma helped Motorola realize powerful bottom-line results in their organization—in fact, they documented more than $16 billion in savings as a result of their Six Sigma efforts. The goal of Six Sigma is to increase profits by eliminating variability, defects, and waste that undermine customer loyalty. Following Motorola, Six Sigma was widely adopted by companies across all segments of the economy.

The benefits of Six Sigma are accomplished through the use of two sub-methodologies: DMAIC and DMADV. The Six Sigma DMAIC Process (define, measure, analyze, improve, control) is an improvement system for existing processes falling below specification and looking for incremental improvement. The Six Sigma DMADV process (define, measure, analyze, design, verify) is an improvement system used to develop new processes or products at Six Sigma quality levels. Six Sigma efforts are led by "belts" with traditional project managers trained in Six Sigma tools becoming certified as "Black Belts" and senior program managers seeking certification as "Master Black Belts."

The exact implementation of quality varies with each industry and, indeed, with each organization pursuing it, but it always is customer-driven.

In the early and mid-1980s, many industry and government leaders saw that a renewed emphasis on quality was no longer an *option* for American companies, but a *necessity* for doing business in a more competitive global market. The **Malcolm Baldrige National Quality Award** was established in 1987 by Congress to recognize U.S. organizations for their achievements in quality and business performance and to raise awareness about the importance of quality and performance excellence as a competitive edge. The criteria for the Baldrige Award have played a major role in achieving the goals established by Congress. They are now widely accepted around the world as the standard for performance excellence. The criteria are designed to help organizations use an integrated approach to organizational performance management. They are a set of questions that focus on critical aspects of management:

- Leadership
- Strategic planning
- Customer focus
- Measurement, analysis, and knowledge management
- Workforce focus
- Operations focus
- Results

Each year Baldrige awards are given to at most three organizations in five different categories—manufacturing, service, small business, education, and health care. Some of the past winners have included Motorola, Inc., Milliken & Company, Ritz Carlton, Eastman Chemical, Concordia Publishing House, Henry Ford Health System, Xerox Corporation-Business Products & Systems, IBM Rochester,

Pal's

In 2001 Pal's Sudden Service, a quick-service restaurant in the East Tennessee region, won both the Malcolm Baldrige Quality Award and the Tennessee Quality Award in the small business category. As the company went through the award processes, they realized that first and foremost, they were a manufacturing concern. They take raw product, process it, and create a totally unique product. Today they continue to use the Baldrige criteria to improve by learning from the best current management techniques. The process helped Pal's to identify their seven key business drivers:

- Quality of products and services
- Hospitality
- Cleanliness and sanitation
- Training and development of all employees
- Value creation for internal and external stakeholders
- Speed
- Accuracy

Pal's metrics are nearly unheard of in the foodservice world: a car served at the drive-thru every 18 seconds, one mistake every 3,500 orders, and customer satisfaction at nearly 98 percent. These numbers have become a benchmark for foodservice operators across the country.

Sources: Adapted from Pal's, May 2004; http://www.baldrige.nist.gov, December 2005; and http://www.palsweb.com, December 2012.

Solar Turbines Incorporated, and Pal's Sudden Service. The criteria provide a framework that assists in planning and measuring performance for all organizations. The criteria also assist with other continuous improvement processes, such as ISO, Lean, and Six Sigma.

The various quality approaches, including ISO, Six Sigma, and Lean (Chapter 11), all share the concept of continuous improvement. The Baldrige Criteria help you identify areas within your organization that are most ripe for improvement, and continuous improvement is an integral part of the cyclical steps of Lean and Six Sigma. For example, in a Lean environment, you work continuously to identify and eliminate waste-generating processes. In the control stage of a Six Sigma project, you generate and monitor data continuously to identify needs for further improvement.

One difference is that the Baldrige Criteria serve as a comprehensive framework for performance excellence. They focus on business results as well as organizational improvement and innovation *systems*. Lean and Six Sigma methodologies drive waste and inefficiencies out of *processes,* whereas ISO 9000 is a series of *standards* for an efficient quality conformance system. Overall, ISO 9000 registration covers less than 10 percent of the Baldrige Criteria.

In Europe there is a similar organization, EFQM (formerly the European Foundation for Quality Management), which gives the EFQM Excellence Award. In Japan the Deming Prize, established in December 1950 in honor of W. Edwards Deming of the United States, was originally designed to reward Japanese companies for major advances in quality improvement. Over the years it has grown to where it is now also available to non-Japanese companies, albeit usually operating in Japan, and also to individuals recognized as having made major contributions to the advancement of quality.

Taguchi Methods

Specifications have traditionally been treated on an all-or-nothing basis—a measurement is either "in specification" and completely acceptable or "out of spec" and completely unacceptable. The first three (normal) distributions of a quality measurement shown in Figure 12-3:—(a) narrow spread, centered on the specification midpoint m; (b) narrow, but off center; and (c) wider spread but centered—would therefore be almost equally acceptable. Even the "uniform distribution" in Figure 12-3d would be equivalent.

Genichi Taguchi believes instead that there is some "loss to society" whenever a quality characteristic deviates from its ideal value. In one common model used by Taguchi, that loss is assumed to be proportional to the square of the deviation from some target value T. In Figure 12-4, T is taken as the specification midpoint, and the dollar loss L varies with the actual value y as

$$L = k(y - T)^2, \qquad\qquad \textbf{(12-3)}$$

where k is a cost coefficient.

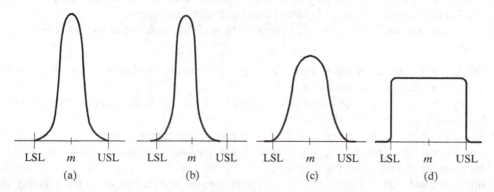

Figure 12-3 Four distributions of a quality characteristic in terms of the upper and lower specification limits (USL and LSL) and specification midpoint m.

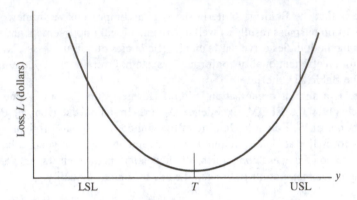

Figure 12-4   Illustration of the Taguchi loss factor.

Taguchi emphasizes the need for a continuous quality improvement program to reduce the variation of product performance characteristics about their target values. His methods include extensive experimentation in which product and process parameters are varied in a statistical matrix of tests. Results are then evaluated using ANOVA (analysis of variance) methods to identify the values that lead to least variation. These tests also show the parameters that cause most of the variation, and adjusting these parameters then leads to the most cost-effective design modifications and process improvements. Taguchi methods have been widely adopted in a variety of industries globally.

Deming's 14 Points

W. Edwards Deming, a statistician with the Bell System, was invited to Japan in 1950 to help their industrialists improve their reputation for poor-quality goods. He convinced them that they could make their goods the highest quality in the world, and they proceeded to do so by using his methods. Today they award the coveted Deming Prize each year for greatest improvement of quality. Returning to the United States, Deming found American industry to be slower to respond, until Japanese competition became a real threat to corporate survival when his ideas gained increasing acceptance.

Deming is best known for the *14 points* developed in his 1982 book *Out of the Crisis*, which are summarized here:

1. Create "constancy of purpose" that encourages everyone to cooperate in continually improving quality and meeting customer needs.
2. "Adopt the new philosophy" of defect prevention instead of the concept of "acceptable quality level" monitored by defect detection.
3. End dependence on mass inspection by building in quality from the start.
4. End the practice of purchasing solely on price; develop long-term relationships with single suppliers based on product quality and trust.
5. Improve constantly and forever the system of production and service, to improve quality and productivity, and thus constantly decrease costs.

6. Institute modern methods of training in the organization's philosophy and goals as well as job performance.

7. Institute leadership, so that supervisors become coaches to help workers do a better job.

8. Eliminate fear, which impedes employee performance.

9. Break down barriers between departments.

10. Eliminate numerical goals and slogans—they don't work.

11. Eliminate work standards and numerical quotas, which emphasize quantity rather than quality.

12. Remove barriers that rob employees of their pride of workmanship.

13. Institute a vigorous program of education and training.

14. Structure the organization for quality.

Deming adopted the PDCA (Plan, Do, Check, Act) Cycle, which was originally conceived by Walter Shewhart in the 1930s. This model provides a framework for continuous improvement of a process or system. The PDCA Cycle is similar to the steps of the engineering problem-solving technique introduced in Chapter 5.

Lean Six Sigma: The Best of Both Worlds?

While various approaches to quality improvement may come and go, one truth has held throughout the years: statistics provides the most solid foundation on which to build quality. This is the reason that Six Sigma approaches (described previously) have endured for over two generations. However, one common complaint about the approach when the author was a practicing Six Sigma Black Belt and Master Black Belt in industry was that it was "too slow." A common reason for this complaint is the amount of time needed to collect reliable data on problems before performing analysis to determine root causes and develop solutions.

To combat this complaint, a new Six Sigma approach was developed that incorporates tools and approaches from lean manufacturing (discussed in Chapter 10). This approach is known as Lean Six Sigma (LSS). The American Society for Quality defines LSS as a "fact-based, data-driven philosophy of improvement that values defect prevention over defect detection".[1] This definition would also fit the standard DMAIC process defined previously, so what makes LSS different? First is the focus on *eliminating waste* commonly used in lean approaches, rather than just reducing variation. By identifying each of the eight kinds of waste evident in a process, LSS is often able to more quickly make a difference for the organization. Second, and more importantly, it is the incorporation of lean tools into the Six Sigma toolkit. These tools (e.g., kaizen) are generally less technical than the statistical approaches of Six Sigma and can more quickly implement changes. These changes often make a meaningful improvement to the operation while additional data is collected, and final solutions are developed and implemented using traditional Six Sigma tools (e.g., root cause analysis and designed experiments).

Source
1. American Society for Quality. (No Date). Six Sigma Definition—What is Lean Six Sigma? I ASQ. Retrieved from https://asq.org/quality-resources/six-sigma

Quality Teams

Production workers are the final determinants of quality, and their willing and informed involvement in the quality effort is essential. One approach to achieving this is through the institution of *quality improvement teams*. In this technique, workers are gathered into small groups, which meet together, perhaps one hour a week, over an extended period. These quality teams (other names are sometimes used) are taught some basic methods of statistics and problem-solving tools, as mentioned previously, and then they proceed to identify problems within their work area, develop alternatives, and formally propose a solution to management.

The Japanese call these teams quality circles or *kaizen* teams, and they had more than five million such circles in operation at any one time, involving an average 10 people each. These have produced tens of millions of suggestions for improving products and production methods. Much of the success of kaizen came about because the system encouraged many small-scale suggestions. Many American firms adopted the Japanese model of quality circles in the 1970s and 1980s, but this model has almost died out in the United States being replaced by the quality teams or just teams.

PRODUCTIVITY

Productivity Defined

Productivity can be defined as output produced per unit of resources applied. **Productivity** is a measure of the efficiency with which an organization performs its activities. **Efficiency** is achieved by using the fewest inputs to generate a given output. The **effectiveness** of these operations is achieved when the organization pursues the appropriate goals. A simple measure in productivity might be units of production per labor hour, or per-labor dollar:

$$\text{Productivity} = \text{Output/Input}$$

For example, a primitive farmer in a developing country may barely be able to feed their family by the sweat of their brow, while a large industrial farm can feed a community; the industrial farm is therefore considered much more productive. Among the reasons for this farm productivity is the application of resources other than human labor—much more land, much more equipment and fuel for it, the finest seed and ample water and fertilizer. The industrial farm is therefore capital intensive, whereas the primitive farm is labor intensive.

Manufacturing Productivity—International Comparisons

For a quarter of a century after World War II, the United States was the industrial giant of the world; it enjoyed the highest productivity among the major industrial nations. As a result, American workers enjoyed a standard of living that was the envy of the world. As of 1960, our major international competitors had lower manufacturing productivity than the United States, but they achieved annual rates of productivity growth from 1960 to 1973 much higher than that of the United States, and significantly higher for several countries through the 1980s.

According to *The Competitiveness and Innovative Capacity of the United States* report sent to Congress in January 2012, after decades of losing manufacturing jobs, the manufacturing sector began adding jobs. In the 25 months leading up to the report manufacturing has added nearly a half million new jobs and 120,000 of those came in the first three months of 2011. Importantly, these tend to be high-paying jobs with good benefits. Even with these improvements in the manufacturing sector, there is much more work to do to ensure America remains competitive. At this writing an updated version of the report is not available, but other economic data points to continued movement of certain manufacturing sectors back into the United States, although generally these operations are far more automated and employ relatively few workers due to the high levels of productivity.

Globalization and technology are just two of the many factors that have changed the face of manufacturing. Energy, sustainability, resources, and talent are also becoming priorities for manufacturers as these companies look to future growth. Governments around the world are recognizing the need to develop and support an over-arching strategy in support of manufacturing for growth, economic stability, and global competitiveness. Policy changes, public-private collaboration, and talent development must be viewed as critical success factors for manufacturing. Today's engineering student will be playing a major role in determining whether America can continue to meet this challenge as the twenty-first century continues.

Work Measurement

Work measurement is the art and science of determining reasonable and fair times for performing various work tasks. These are called time standards. Work measurement is rooted in the concept of a fair day's work.

A time standard is the time required for a qualified employee working at a normal pace under capable supervision experiencing normal fatigue and delay to do a defined amount of work of specific quality when following the prescribed method. It is a measure of how long the task should take. Time standards are usually set on a per-piece basis. Tasks that do not vary according to the number of parts made such as setup and tear-down are measured separately and are often prorated over an average batch size.

Companies find many uses for time standards including estimating costs, estimating equipment needs, scheduling, line balancing, capacity analysis, evaluating automation costs, planning staffing levels, methods comparisons, pricing, revealing production problems, evaluating employees, setting piece rates, and compliance with contractual requirements.

A variety of methods are used to develop time standards. They vary as to cost, speed, convenience, accuracy, and precision. Methods for determining time standards are often divided into informal time standards and engineered time standards. Informal time standards are less precise, often less accurate, and more subject to bias than engineered time standards. However, informal time standards are usually less expensive, quicker, and easier to set than engineered time standards. In some situations, such as with a proposed product, the informal time standards are the only ones feasible.

Informal Time Standards

Methods for developing informal time standards include educated guesses, use of historical data, timing one cycle of a task, and work sampling.

Estimates and educated guesses from engineers, supervisors, operators, or others may be the quickest, cheapest, and most obvious method to develop time standards. Such estimates are subject to bias. For example, operators may consciously or unconsciously pad their estimates so they can easily meet the standard. Estimates are often somewhat imprecise. There can be large variations among the estimates and sincere disagreement about what is the correct time standard.

Historical data can be used to develop time standards. If 40 minutes were charged to a task the last time it was done, it makes sense to believe that 40 minutes is a reasonable estimate of the time needed. If there are several times recorded, one might use the average, the longest, or the shortest time. However, historical data often hides such things as five minutes taken up by a chat with a coworker, ten minutes used looking for tools, or an operator working unusually fast to meet a deadline (then slowing down on the next task to recover). Data may also be poorly documented; operators may not be careful about recording when they start and finish each task.

Timing one whole cycle of a task, start to finish, gives an estimate of the time the task should take. It does not take into account the pace of the operator and may miss variations in the method. It may include inappropriate times (such as interruptions) and may miss work that is not done every cycle (such as replacing a blade or brushing away scraps).

Work sampling can be used to determine what percentage of time operators spend on different tasks. When one knows the percentage of time applied to the task and the number of times the task was accomplished in a given number of hours, one can figure the average time to do the task. This method is often useful in less structured work environments such as service and professional work.

Engineered Time Standards

Engineered time standards are precise and accurate. Each portion of the work measured is observed closely to eliminate unnecessary motion, to improve work place arrangement, and to institute better work methods where possible. The prescribed method is documented and the time is measured. Engineered time standards can always substitute for informal time standards but not vice versa.

Allowances are included in engineered time standards for personal time, recovery from fatigue, and unavoidable delays (usually referred to as Personal, Fatigue, and Delay or PF&D). Personal time needs may include biological needs, coughing fits, work-related wash up, and putting on and removing protective clothing. Recovery from fatigue includes scheduled breaks, stretching, and extra time allowed due to job difficulty. Unavoidable delay may include miscellaneous shift-related operations such as start-up and cleanup, some breakdowns, fire alarms, team meetings, and other communication with supervisors and others.

Engineered time standards are developed by stopwatch time study, predetermined time systems, and standard data developed using the previous two techniques.

Stopwatch time study refers to a specific method where work is carefully observed, documented, broken into segments called elements, and several cycles are timed with a stopwatch or similar timer.

The time study analyst rates the operator's pace in comparison to a normal pace and factors that into the measurement. The number of cycles timed is dependent on the desired accuracy and the time variation from cycle to cycle.

Predetermined time systems break tasks into motions and subtasks for which times have been determined. The analyst determines which motions and subtasks apply and the times are looked up and totaled (often by a computer program). There are over 50 predetermined time systems available. Many are evolved from Methods-Time Measurement, which is also known as MTM-1. Some are quite precise, measuring times to the millionth of an hour. These often take a long time to apply (50 or more times the actual time of the task). Others are less precise but are applied more quickly and provide accurate times on longer tasks. Some predetermined time systems are developed for specific types of work such as clerical work, machining, or assembly under binocular microscopes.

Standard data uses previously developed time standards to determine time standards for similar work. It can save a great deal of cost and effort when many similar tasks are to be analyzed. For example, suppose there are 70 circuit cards to be analyzed. They all go through the same steps but have different numbers of resistors, capacitors, connectors, and integrated circuits in varying locations. Analysts could develop a base time per circuit card and additional times per resistor, capacitor, connector, and integrated circuit. Then to obtain the time for a specific card one would count the components, multiply them by their additional times, and add the base time. This is much quicker than studying each of the 70 cards separately.

Improper, sloppy, or haphazard application of standard data can lead to imprecise and biased standards. It is important to understand what is included in the previously determined standards and what variations may be prorated in. For example, does the standard "Put Tray in Oven" include opening and closing the oven, prorated times for various sized trays, recording the time and temperature, and removing the tray from the oven? Or does "Put Tray in Oven" mean moving the tray 20 inches to a conveyor belt that goes through the oven (which would result in a short time standard)? Does the worker need to walk through two doors to another room to put the tray in the oven and is that included?

Operators Often Resist Time Standards

There are a number of reasons for which operators resist and reject time standards. These include a resistance to change, past abuses of time standards, poor practitioners, fairness issues, misunderstandings, dislike of being measured, and not knowing why companies need time standards. People fear that time standards will make them look bad.

Operators may fear that higher efficiency and productivity will lead to less work, loss of jobs, and loss of overtime pay—and sometimes they do in the short run. In the long run, higher efficiency and productivity are required to keep the company in business.

The amount of resistance will vary from company to company and person to person depending on circumstances. It will be higher where poor standards have had an effect on pay and promotion. It is important that a manager be aware of this resistance and work to reduce it. It helps to explain why your company needs time standards and how they will be used.

Source: Dr. Jean Babcock.

MAINTENANCE AND FACILITIES (PLANT) ENGINEERING

A wide variety of functions and activities required for the effective functioning of a manufacturing plant, but not concerned directly with production of the product, fall under an organization headed by the plant or facilities engineer or the maintenance manager or superintendent (see Figure 11-1). In the first two subsections following, we discuss several types of maintenance and some aspects of maintenance management; in the third subsection, we outline briefly some of the other activities that fall under the umbrella of facilities or plant engineering.

Scope of Maintenance

The *Maintenance Engineering Handbook* identifies the following primary functions of the maintenance (engineering) activity:

1. Maintenance of existing plant equipment
2. Maintenance of existing plant buildings and grounds
3. Equipment inspection and lubrication
4. Utilities generation and distribution
5. Alterations to existing equipment and buildings
6. New installations of equipment and buildings

Some of these functions, such as major alterations or additions to buildings and equipment, occur so irregularly that it is not economical to staff for such activity, and these are contracted out; even so, a plant engineer is normally appointed as project engineer to monitor progress of contractor activity to assure that changes will meet the needs of the plant. Maintenance of some items (elevators, computers, office equipment, rewinding of burned-out motors) is so specialized that it is normally contracted out. Some custodial activities, such as washing windows, care of grounds, and office janitorial service, may be contracted out if it is found to be more cost-effective. Contract services of these types may provide better methods and supervision of these ancillary activities than the plant affords, and they often have lower labor costs than those in the plant.

To give an idea of the variety of maintenance concerns in a plant of any size, the following topics are listed, each of which is the subject of a separate chapter in the *Facilities and Plant Engineering Handbook*.

- Roofing
- Flooring
- Refrigeration
- Air conditioning, heating, and ventilation
- Special-purpose rooms and their environment
- Electric-circuit protection
- Utilities
- Transportation equipment
- Materials handling systems
- Elevators
- Painting

- Corrosion protection
- Applied biology (insect, animal, and other pest control)
- Lubrication of machine tools

Types of Maintenance

The mainstream activities of maintaining plant equipment can be divided into corrective, preventive, and predictive maintenance; each is considered next. *Corrective maintenance* is simply repair work, made necessary when something breaks down or is found to be out of order. This is the activity that most of us think of when maintenance is mentioned. When equipment breaks down, especially machinery on which an integrated production line depends, the costs of lost production mount and the pressure is on the maintenance team to get the equipment fixed and back into operation. Effective maintenance engineering requires thinking through the most likely types of breakdowns, assuring an adequate inventory of the most commonly needed or critical replacement parts, and providing spare capacity where breakdowns cannot reasonably be avoided.

Many mechanical systems *wear out*. Their failure rates increase with time and the quality of performance falls off because bearings become loose, gears wear, O-rings and belts deteriorate, and grease hardens. These types of problems are reduced by periodic inspection, lubrication, and identification and replacement of worn parts. Efficient *preventive maintenance* requires documentation of all equipment to be included in the program and establishment of the most cost-effective schedule for inspection. Inspection checklists need to be established for each type of equipment, and inspectors must be trained to make simple repairs when problems are observed. Computers are useful to print out lists of inspection tasks that are due and maintain data on the time and material costs of inspection to support periodic analysis and revision of the preventive maintenance plan. To some extent, preventive maintenance can be deferred or "scheduled around" more urgent corrective maintenance, but deferring it too long invites breakdowns and higher costs.

Predictive maintenance is a preventive type of maintenance that involves the use of sensitive instruments (e.g., vibration analyzers, amplitude meters, audio gages, optical tooling, and pressure, temperature, and resistance gages) to predict trouble. Critical equipment conditions can be measured periodically or on a continuous basis. This approach enables maintenance personnel to establish the imminence of need for overhaul. Where diagnostic systems are built into equipment, production workers can observe warning signs during operation, catching incipient failures long before maintenance workers would see them.

Some Maintenance Management Considerations

Size of Maintenance Staff.   Production supervisors naturally would like maintenance specialists of all types available immediately when a breakdown occurs, since the cost of idle maintenance time does not come out of their budget. When a plant is in full production and profits are high, it is easy to build the maintenance staff to a comfortable level, but when demand slows and costs are being trimmed, short-sighted managers will find maintenance an easy target for drastic cuts. Good management balances the cost of additional maintenance personnel against the probable costs of production loss and equipment damage if adequate maintenance is not provided.

Work Orders.   To keep control over maintenance costs, work is not ordinarily performed without a supporting work order signed by a supervisor. The work order states the problem, estimates the cost of repair, and provides space for workers to document the time they spent on the problem and any materials or parts they used in solving it. Completed work orders provide data to analyze maintenance costs of each type of equipment, so that cost-saving decisions such as redesign or replacement can be made.

Work Scheduling.   In larger plants, maintenance scheduling is the responsibility of a separate unit of the maintenance organization; in smaller plants, this is done directly by the supervisor. Schedules are only estimates and may have to be changed if a breakdown emergency takes place.

Repair Parts Inventory.   As industry becomes more automated, it has more complex operating equipment with more parts that can fail and require replacement. Like any other inventory, this can tie up large sums of money that might be put to more productive use, and good judgment and periodic review are required. Where equipment vendors will provide prompt repair service at an acceptable price, this eliminates the need for parts inventories as well as for specialized training, so that after-sale service is a real consideration in purchasing equipment.

Total Productive Maintenance (TPM)

Total productive maintenance (TPM), a concept originated by the Japanese, is an integrated, top-down, system-oriented, life cycle approach to maintenance, with the objective of maximizing productivity. Directed primarily to the commercial manufacturing environment, TPM does the following:

1. Promotes the overall effectiveness and efficiency of equipment in the factory
2. Establishes a complete preventive maintenance program for factory equipment based on life cycle criteria
3. Is implemented on a "team" basis involving various departments to include engineering, production operations, and maintenance
4. Involves every employee in the company, from the top management to the workers on the shop floor. Even equipment operators are responsible for maintenance of the equipment they operate.
5. Is based on the promotion of preventive maintenance through "motivational management" (the establishment of autonomous small-group activities for the maintenance and support of equipment)

The objective of TPM is to eliminate equipment breakdowns, speed losses, minor stoppages, and so on. It promotes defect-free production, just-in-time (JIT) production, and automation. TPM includes continuous improvement in maintenance.

Other Facilities and Plant Engineering Functions

Some of the other activities that are often included in the responsibilities of the plant engineer, facilities engineer, or maintenance superintendent (often for lack of a better place to locate them) are as follows:

1. Plant security (guards, fences, locks, theft control, emergency planning)
2. Fire protection (fuel and chemical storage, fire detection and extinguishment, loss prevention, and risk management)

3. Insurance administration
4. Salvage and waste disposal
5. Pollution and noise abatement
6. Property accounting

OTHER MANUFACTURING FUNCTIONS

Human Resources (Personnel) Management

The many concerns for and about employees are centered in the personnel or industrial relations or (more recently) human resources department. A typical personnel department in a single-plant company employing several thousand persons might include the following sections:

1. Recruiting and employment (human resource planning, recruiting, interviewing, testing, transfers, and layoffs)
2. Equal employment opportunity (affirmative action, minority records and reports, complaint investigation)
3. Industrial relations (contract negotiations, contract administration, grievances, and arbitration)
4. Compensation (job analysis and evaluation, wage surveys, incentives and performance standards, managerial and professional compensation)
5. Education and training (orientation, skills training, management training, career planning, tuition assistance, organizational development)
6. Health and safety (industrial hygiene, safety engineering, first aid and medical, workers' compensation)
7. Employee benefits (insurance, pensions, profit sharing, food service, dependent day care, social programs)

Of special interest to the engineer is the safety engineering activity (under the health and safety section). As an example of the hazards of concern to the safety engineer and industrial hygienist, consider the following chapter titles from Hammer's *Occupational Safety Management and Engineering*:

Acceleration, falls, falling objects, and other impacts
Mechanical injuries
Heat and temperature
Pressure vessels
Electrical hazards
Fires and fire suppression
Explosions and explosives
Hazards of toxic materials
Radiation
Vibration and noise

Safety personnel are involved in (1) identifying and analyzing hazards, (2) recommending protective devices and warnings, (3) providing safety training, (4) interpreting the Occupational Safety and Health Act (OSHA) and other codes and standards to management and other personnel, and (5) workers' compensation insurance activity. In some of these areas, safety personnel share functions with fire prevention and other security personnel. They are also closely involved with plant insurance activities, since future plant fire, workers' compensation, medical, and liability insurance premiums will depend on the success of occupational safety and health programs.

Purchasing and Materials Management

Importance of Purchasing.   Purchasing is a vital contributor to producing a quality product at a profit. Half the value of the typical industrial product consists of materials and components purchased from other organizations. If a firm is making a 10 percent profit on its product, a dollar saved in more efficient purchasing has the same effect on profit as $10.00 in added sales. Moreover, a quality defect in a supplier's component incorporated into the product has the same impact on your reputation as a mistake made internally. Zenz lists the following steps as being performed by purchasing in the large majority of organizations:

1. Recognition of need
2. Description of requirement
3. Selection of possible sources of supply
4. Determination of price and availability
5. Placement of the order
6. Follow-up and expediting of the order
7. Verification of the invoice
8. Processing of discrepancies and rejections
9. Closing of completed orders
10. Maintenance of records and files

The Engineer in Purchasing.   A survey of 12 purchasing manager associations showed about half of the respondents were college graduates; of these, 58 percent majored in business administration, and 17 percent in engineering. Certainly, an engineering education is of great value in the purchasing of highly technical components. Interestingly, one of the early articles written on purchasing is entitled "The Engineer as a Purchasing Agent" and appeared in a 1908 engineering publication, *Materials Management*. This is a more comprehensive organizational viewpoint in which all activities involved in bringing materials into and through the plant are combined under a materials manager. These activities commonly would include purchasing, inventory control, traffic and transportation, and receiving; they may include warehousing/stores and even production control. Purchasing is often responsible for make-or-buy analysis, value engineering/analysis (see Chapter 10), incoming inspection, and reclamation and salvage.

Even where these activities do not fall under a single manager, they must be performed in concert with each other. Engineering and purchasing personnel in particular must work closely together. Design engineers must be careful not to make specifications for purchased materials and components so restrictive that suppliers with less expensive, but satisfactory products are ruled out; purchasing, on the other hand, must not make its decisions solely on price when a slightly more expensive choice may bring quality, reliability, delivery, or customer acceptance worth much more than the price differential.

DISCUSSION QUESTIONS

12-1. Distinguish between quality of design and quality of conformance.

12-2. Research to find out why 6σ allow for a 1.5σ shift in the mean when determining process performance? Does this make the organization to appear to perform better than the customer is experiencing?

12-3. (a) In your School of Engineering, determine which engineering curricula require a course in statistics, and estimate the proportion of engineers that graduate as literate in statistics, *or* (b) survey engineers and other employees in your company to estimate how many were trained in statistics in school or on the job.

12-4. Distinguish between the statistics of *attributes* and the statistics of *variables*, and comment on how they are applied in process control charts and in sampling.

12-5. The involvement of production workers in quality circles seems to conflict with the concept of Frederick W. Taylor, founder of scientific management, that managers should define how work is to be done, and workers should simply perform as they are instructed and trained. Which management thinkers discussed in earlier chapters expressed ideas consistent with Quality Circles?

12-6. State the essence of Taguchi's teaching.

12-7. Examine Deming's 14 points, and select three or four you consider most effective in assuring product quality. Explain your reasoning.

12-8. What do you see as some of the organizational changes that have occurred as quality has matured into the twenty-first century?

12-9. For most of the middle of the twentieth century, U.S. factories were the most productive in the world. What has happened to change this?

12-10. Distinguish between the methods used in (a) time study of existing tasks, (b) time study of proposed new tasks, and (c) work sampling. Why are they different?

12-11. Identify three types of maintenance, and distinguish the role of each one in promoting quality.

12-12. Discuss how total productive maintenance (TPM) relates to other concepts developed in this chapter.

12-13. Where might engineering knowledge and skills be valuable in each of the following functions: (a) human resources management, (b) purchasing, and (c) other materials management activities?

12-14. Discuss how the management functions of planning, organizing, leading, and controlling relate to the production process.

12-15. What is the importance of quality in different industries? Why does it vary?

PROBLEMS

12-1. Tomatoes are packaged in a can designed to hold a nominal 28 ounces of product. Five cans sampled randomly were found to contain 28.3, 27.3, 29.1, 28.5, and 27.8 ounces of product. (a) Calculate the mean value and range of sample data, and (b) estimate the variance and standard deviation of the sample.

12-2. An "*x*-bar" control chart is developed for recording the mean value of a quality characteristic by use of a sample size of three. The control chart has control limits (LCL and UCL) of 1.000 and 1.020 pounds, respectively. If a new sample of three items has weights of 1.023, 0.999, and 1.025 pounds, what can we say about the lot (batch) it came from?

12-3. The specification limits for the length of a meter stick are 1000 ± 0.5 mm. To understand if the manufacturing process is in control, once every hour the lengths of five meter sticks are measured. Samples from the last 16 hours of production are provided in the table below.

   a. Construct an appropriate control chart. Does the process appear to be in control?

   b. Does the process appear to be meeting customer specifications?

   c. What is (are) your recommended action(s) for this process?

Hour	Average Length	Range
1	999.70	0.40
2	999.46	0.40
3	999.38	0.20
4	999.60	0.30
5	999.86	0.50
6	999.52	0.20
7	999.60	0.50
8	999.56	1.00
9	999.68	1.90
10	1000.08	0.60
11	999.88	0.40
12	1000.12	0.40
13	999.70	1.40
14	999.80	0.40
15	999.68	0.70
16	999.46	0.50

12-4. Under normal use conditions, the Mean Time to Failure of a heating coil can be modeled using the normal distribution with a mean of 15 months and a standard deviation of 2 months. Determine the probability that a heating coil will wear out after 12 months.

SOURCES

Amrine, Harold T., Ritchey, J. A., Moodie, C. L., and Kmec, Joseph F., *Manufacturing Organization and Management*, 6th ed. (Englewood Cliffs, NJ: Prentice Hall, Inc., 1993), p. 170.

Blanchard, Benjamin S., Verma, Dinesh, and Peterson, Elmer L., *Maintainability: A Key to Effective Serviceability and Maintenance Management* (New York: John Wiley & Sons, Inc., 1995), p. 17.

"Competitiveness and Innovative Capacity of the United States." U.S. Department of Commerce in consultation with the National Economic Council, January 2012.

Cremer, James M., "The Engineer as a Purchasing Agent," *Cassier's Magazine*, August 1908, pp. 322–332.

Deming, W. Edwards, *Out of the Crisis* (Cambridge, MA: The MIT Press, 1982).

Eder, W. Ernst, "Total Quality Management—Defining Customers and Quality," *American Society for Engineering Education 1993 Annual Conference Proceedings*, p. 1388.

Geiger, Gordon H., "Reinventing Grinter's Wheel," unpublished presidential address to the Accrediting Board for Engineering and Technology annual meeting, November 12, 1987.

Goldratt, Eliyahu M. and Cox, Jeff, *The Goal: A Process of Ongoing Improvement*, 2nd ed. (Great Barrington, MA: North River Press, 1992).

Hammer, Willie, *Occupational Safety Management and Engineering*, 3rd ed. (Englewood Cliffs, NJ: Prentice-Hall, Inc., 1985).

Juran, J. M. and Gryna, F. M., *Quality Planning and Analysis*, 3rd ed. (New York: McGraw-Hill Book Company, 1993), p. 3.

Kaushal, A., Mayor, T., and Riedl, P, "Manufacturing's Wake-Up Call," www.booz.com. October 2012.

Lewis, Bernard T. and Marron, J. P., eds., *Facilities and Plant Engineering Handbook* (New York: McGraw-Hill Book Company, 1974).

Merino, Donald N., "Optimizing the Cost of Quality Using Quality Economic Models," *American Society for Engineering Education 1991 Annual Conference Proceedings*, p. 94.

Morrow, L. C., ed., *Maintenance Engineering Handbook* (New York: McGraw-Hill Book Company, 1957), pp. 1–4.

Quality Costs—What and How (Milwaukee, WI: American Society for Quality Control, 1971).

Taguchi, Gen'ichi, Elsayed, E.A., and Hsiang, Thomas C., *Quality Engineering in Production Systems* (New York: McGraw-Hill Book Company, 1989). For a very readable commentary, see Raghu N. Kackar, "Taguchi's Quality Philosophy: Analysis and Commentary," *Quality Progress*, December 1986, pp. 21–29.

Thurow, Lester, unpublished remarks on *60 Minutes* television program, February 7, 1988.

U.S. Dept. of Defense, *MIL-STD-105E, Military Standard: Sampling Procedures and Tables for Inspection by Attributes*, May 10, 1989.

Zenz, Gary J., *Purchasing and the Management of Materials*, 5th ed. (New York: John Wiley & Sons, Inc., 1981), p. 100.

13

Engineers in Marketing and Service Activities

PREVIEW

Production, as discussed in the preceding chapters, does not end the engineer's involvement in the product life cycle. The more technical the product is, the more engineers are involved in listening to the customer, marketing the product, and supporting its use in the field. In this chapter, the nature of engineering involvement in the function of marketing is analyzed, and engineering involvement in after-sales service is discussed.

The second major topic of the chapter explores the roles of engineers with organizations in the service industry. Today, many engineers work in industries producing a service rather than a physical product and this sector of the economy continues to grow in scale and importance.

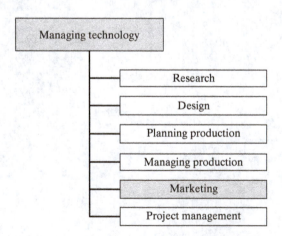

MARKETING AND THE ENGINEER

It might be easy to assume that marketing is well outside the realm of anything an engineering manager, much less an engineer, might need to work on. This assumption might be further reinforced by countless Dilbert cartoons depicting engineering as where the hard work gets done and marketing as "liquor and guessing." However, perhaps no common engineering assumption could be further from the truth. Given the complex nature of products and services in our economy, it is growing increasingly common for engineers to play a role in marketing and even hold positions in the marketing group. In addition, in a hyper-competitive market, successful marketing can be what separates successful companies from those who struggle. In this sense, marketing is vital to any firm since it is the activity that produces revenues that sustain the enterprise. For this reason, engineering managers should understand the basics of how marketing functions within any organization, how it has changed over the past 10–20 years, and how it might continue to change in the future.

What Is Marketing?

Marketing, like engineering, comes in many different variations. These variations change based on the product being marketed, the customer being sold to, the way customers acquire the product and the types of information available to the marketer. At its most fundamental, Merriam-Webster defines marketing as, "the process or technique of promoting, selling, and distributing a product or service." Armstrong and Kotler define marketing as "engaging customers and managing profitable customer relationships." For the student thinking about how engineers participate, the second component of the second definition should provide insight. Engineers are frequently called upon to support the management of customer relationships. Table 13-1 provides some examples of the activities this might include for industrial products. We will discuss potential roles in consumer products later in this chapter.

THE PROCESS OF MARKETING

Fundamentally, the goal of any marketing activity is to position an organization's products or services ("offerings") in the marketplace and gain sales. To achieve this goal, marketing typically seeks to differentiate an organization's offerings from others in the market. The process used to accomplish this should be focused on the customer. It can be summarized in four steps, depicted in Figure 13-1.

Table 13-1 Engineering Activities in Marketing Industrial Products

Type of Product	Description	Engineering Activities
Installations	Large, durable custom constructions	Selling and performance of design service; cost estimation and construction supervision
Accessories	Shorter-lived capital goods (equipment)	Seller's engineers design for general customer
Raw materials	Extractive and agricultural products	Assessment of quality
Process materials	Goods that change form in production	Buyer's engineers establish specifications
Component parts	Catalog items that do not lose identity in production	Supplier's engineers design for general customer and introduce to user's engineers
Fabricated items	Custom-made items	Buyer's engineers design and specify; seller bids on manufacture
Maintenance, repair, and operating items	Consumed in process of production or use	Repair parts and methods specified by maker's engineers; users have little engineering involvement
Services	Involve only incidental product use	For engineering services, engineers sell as well as perform

The first step in the process is traditionally known as *market research*. In this step, members of the organization engage with customers and potential customers to understand their needs and how those needs are currently being met by the offerings of the market. This is a technique that works well in established categories such as automobiles or hotel amenities. However, in markets that have not yet been defined, it has limitations. Steve Jobs, the deceased former CEO of Apple, famously often spoke of their philosophy in product design. One of his key points was usually that customers don't know what they want until you hand it to them. In other words, by creating something new that people didn't even know they wanted, one can create entirely new markets. Along these lines, a quote often attributed to Henry Ford is that if he had asked his customers what they wanted, they would have told him to build a faster horse, not the Model T car.

Apple's approach led to some of the most ubiquitous products of our time, including the iPhone, and for Apple to be the most valuable company in the world (as measured by market capitalization) for most of 2018. A more complete discussion of these longer-range implications of marketing is provided later in this chapter.

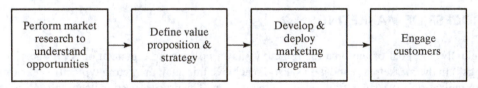

Figure 13-1 Overview of the marketing process.

The next three steps in the process depicted in Figure 13-1 utilize the results of this market research to capture sales from customers, first by building a strategy to approach potential customers and demonstrate the value of the organization's offerings. This is often referred to as *the value proposition*, a succinct way to define why customers would buy the organization's offerings. In the next step, we operationalize this strategy in the form of a marketing program that captures customers' attention. In today's markets, this type of integrated marketing program may include print and other forms of traditional media (i.e., television and radio), direct marketing through mail, and almost certainly includes electronic marketing through websites, social media, and email campaigns. The result of this program is the final step in the figure, the engagement of customers that results in profitable relationships.

THE 4PS OF THE MARKETING MIX

Engineers may assume that the development of the marketing program is simple or highly unstructured. In successful organizations, this is simply untrue. Typically, the structure that is utilized follows the foundational model of marketing mix, or 4Ps, first proposed by McCarthy in 1960 and still widely utilized today. These 4Ps are product, price, place, and promotion. They are summarized as follows:

- Product—What is the offering that the organization will provide to customers? Decisions in this area will typically include how the offering will be designed, packaged, desired quality level, features, etc.
- Price—What the customer is expected to pay for the offering? The goal is that the price is in line with the customer's perceived value of the offering.
- Place—How the customer will access the offering? Decisions regarding place include how the offerings will be distributed, where they will be marketed, how inventory will be maintained, etc. Place is also commonly referred to as channel. Examples of channels include retail and direct-to-consumer and because of internet offerings, channels are changing rapidly. For example, only recently have mattresses been made available through a direct-to-consumer channel by companies such as Casper and Leesa (as of late 2018); prior to that, consumers went to a retail location to select their mattresses.
- Promotion—How will the organization reach the customer? Decisions here include the development of the message, media strategy and message frequency, and price-driven promotions.

MARKETING AND ENGINEERS—PARTNERSHIPS, R&D, AND TECHNICAL SALES

To be successful at higher levels of an organization, engineering managers will find it necessary to partner with marketing functions. One key reason for this is that marketing (and market research) often defines what engineers should be building in the future to keep the company profitable. This ties directly to the R&D function discussed in Chapters 9 and 10. For this reason, one function performed by engineers for their customers and clients is the continuous updating of their technical competency. A remarkable feature of the last decades, and likely to be a feature of the next few, has been the rapid discovery, development, and commercialization of new technologies. To support these new technologies, thousands of new products, processes, and materials have been developed and offered to the marketplace. Often, the buyers of

Electronic Marketing

The power of the web and social media platforms has fundamentally changed how marketing is done in the Western world. One major change, as noted by the U.S. Small Business Administration is that electronic marketing is a significant business leveler, allowing small- and medium-sized companies to compete with the giants on the same global playing field. This leveling is enabled by electronic tools such as email marketing, search engine optimization (SEO), paid search and social media placement that enable organizations of all sizes to precisely target specific customers.

Simply by being a consumer, we are all familiar with email marketing. Similarly, the controversy surrounding the 2016 U.S. Presidential election has raised the profile of social media-based ad campaigns through platforms such as Facebook, YouTube, and Snapchat. The ideas of SEO and AdWords are closely inter-related and less commonly understood. For that reason, we discuss them here more completely. SEO is the process that marketing organizations go through to:

1. Understand how customers search for their offerings—notably what words they search for in a web or smart phone-based search engine.
2. Redesign their website to ensure that customers find the organization highly ranked in the search results.
3. Gain free traffic to their website to gain more sales.

It is the third step that critically differentiates SEO from platforms such as Google's AdWords where organizations pay for high ranking placement in searches by buying certain words. This form of paid search marketing is a highly analytical version of marketing suitable to an engineering background (see "An Engineer Running Marketing" section). It also requires high levels of computing expertise and is the reason that Gartner predicting Chief Marketing Officers would soon spend more money on information technology than Chief Information Officers.

these advancements are full partners in their development, having described their needs for such improvements to their suppliers, aided in the design or application of new developments, and/or cooperated in their testing. But far more frequently, it has been the seller of new technological advances who has been left with the task of teaching its scientific bases so that designers can incorporate them with understanding and confidence. In this context, the technical salesperson, typically an engineer, functions as a teacher and consultant.

Technical salespeople serve a second function as a sensor of market needs. As a part of their everyday activities, they are constantly assessing the technological sophistication of their customers, and they are exposed to the constraints and frustrations that retard their customers' progress in competing in their marketplaces. Technical salespeople may note the need for higher speeds, better discrimination, lower prices, better maintainability, or other directions in which their companies can improve the product from the standpoint of their customers. This sensor function acts as a guideline for determining the most important features to be developed in making their products more acceptable and preferred. In a sense, the technical salesperson becomes a voice for their customers to their own firm, and smooths these relationships for the future.

An Engineer Running Marketing?

Given the stereotypical image of marketing in most companies, why would an engineer be the head of marketing? The stereotype of an engineer doesn't fit the picture of the freewheeling creative type that most associate with marketing professionals. But that was the experience of the Schell while in his last industry position before joining the faculty. This experience happened at a smaller, high growth company (ranked on both the Inc 500 and Inc 5000) after what had already been a varied career spanning over a decade. It occurred when the company's head of marketing announced his departure with limited notice and Schell was given the responsibility for marketing operations in addition to his duties running the technology, project management, and organizational development teams. The choice was an interesting decision by the CEO given the importance of marketing to the company's continued growth and the multi-million-dollar annual budget.

What was the result? A noticeable growth in marketing effectiveness and subsequent sales during the six-month period reporting to an engineer while the company searched for an experienced marketing professional to serve as VP of Marketing. The reasons for this perhaps surprising result were three-fold. First, the marketing group had not been enjoying success in the period leading up to the departure of the group's leader. Due to these challenges, they were hungry for change and open to new ideas. Second, the core team had been with the company for a number of years, knew their products, and were professionals who wanted to make a difference. Finally, as discussed in the section on Electronic marketing, the skills required for running a marketing program have changed radically since the growth of e-commerce and social media. Marketing is now more often about data collection and analysis than it is concerned with development of the "right creative." While messaging still matters a lot, in the web-enabled marketing world, messages can be quickly deployed and tested with great numbers of potential customers. The results of this testing are then analyzed using statistical models and other traditional engineering tools to identify which creative is most effective for which customer segment.

This type of data-driven decision making is now the norm in marketing, rather than the image of marketing professionals displayed in shows like Mad Men. This move to data is the reason Schell's experience is becoming less common, as more technical professionals transition to marketing roles.

ENGINEERS IN THE SERVICE ECONOMY

Importance of Service-Producing Industries

According to the World Bank, as of 2015, the service sector accounts for nearly 79 percent of the value of the U.S. economy and over 86 percent of all employment. For engineers, that rate is substantially lower, near 50 percent, but critically important and growing. In fact, the author spent most of his 15-year industry career in the service sector, always in an engineering or engineering management role. These engineers are in sectors such as government services, colleges and universities, healthcare, and financial services. Many of the engineers in the service sector work in information technology and computer-related occupations. In the computer applications division, computer engineers use software systems, network design, and consulting skills to provide services. Throughout the government, engineers contribute services that include water purification, waste management, law enforcement, transportation, national defense, and demographical statistics. College educators and researchers are essential to the growth of engineering by providing a service to all future engineers.

The United States is said to have become a "postindustrial society." Although, as has been argued, the United States cannot afford to permit its manufacturing industry to be less than the best, it is true that manufacturing no longer employs mostly Americans. According to the U.S. Bureau of Labor Statistics, employment growth for the immediate future is projected to be concentrated in the service-providing sector of the economy. Within this sector, two industry groups are expected to account for half of all wage and salary employment growth in the economy: (1) education and health services; and (2) professional and business services. In the goods-producing sector, employment is expected to grow in construction; and employment is expected to decline in natural resources, mining, and manufacturing.

Engineers have traditionally been concentrated in slow-growing manufacturing industries, in which they will continue to be needed to design, build, test, and improve manufactured products. However, increasing employment of engineers in faster-growing service industries should generate most of the employment growth. Overall job opportunities in engineering are expected to be favorable because the number of engineering graduates should be in rough balance with the number of job openings over this period.

Characteristics of the Service Sector

The wide variety of service-producing industries makes identifying common characteristics hard, but a few traits are normally present. Most services are intangible, whereas manufactured goods are not. They are usually performed in real time, often in the presence of the customer. Services can seldom be inventoried; they must be performed on a schedule that fits the needs of the customer—a challenge in staffing that anyone who has worked at a fast-food restaurant can understand.

Most professional or consulting services are customized, personalized, and labor intensive. Others, such as airline transportation or telephone or electrical service, are standardized and very capital intensive. Some have both aspects: a stockbroker in the "front office" needs to have a customized approach to the needs and interests of each client; in the "back room," the clerical functions of record keeping and stock transfer need to be carried out efficiently and accurately—more in line with the mass production philosophy of manufacturing.

Like manufacturing, service-producing industries prosper by providing value for the customer, but often in a more immediate and personalized manner. Peters and Waterman identify Delta Airlines, Marriott (hotels), McDonald's (restaurants), Disney Productions, Wal Mart (stores), and both Bechtel and Fluor (in project management) as service-producing industries that have achieved excellence in performance.

Some Specific Service Industry Examples

Computer Applications.   The rapid spread of computers and information technology has generated a need for highly trained workers to design and develop new hardware and software systems and to incorporate new technologies. These workers—computer systems analysts, engineers, and scientists—include a wide range of computer-related occupations. Job tasks and occupational titles used to describe this broad category of workers evolve rapidly, reflecting new areas of specialization or changes in technology as well as the preferences and practices of employers.

Although they are increasingly employed in every sector of the economy, the greatest concentration of these workers is in the computer and data processing services industry. Firms in this industry provide nearly every service related to commercial computer use on a contract basis. Services include customized computer programming services and applications and systems software design; the design, development, and production of prepackaged computer software; systems integration, networking, and reengineering services; data processing and preparation services; information retrieval services, including online databases and Internet services; on-site computer facilities management; the development and management of databases; and a variety of specialized consulting services. Many work in other areas, such as for government agencies, manufacturers of computer and related electronic equipment, insurance companies, financial institutions, and universities.

Computer hardware engineers usually need a bachelor's degree in computer engineering or electrical engineering, whereas software engineers are more likely to hold a degree in computer science or software engineering. Computer engineering programs emphasize hardware and may be offered as a degree option or in conjunction with electrical and electronics engineering. As a result, graduates of a computer engineering program from a school or college of engineering often find jobs designing and developing computer hardware or related equipment, even though they also have the skills required for developing systems or software.

Government Service.   The largest share of the budgets of local governments (except perhaps for education) and the biggest share of the 13.8 million people employed by state and local governments fall in the area of public works. Running this domain in each local government is a city engineer or public works director, who is usually, except in smaller towns, a registered professional civil engineer; in larger cities, a number of their department heads and professional staff will also be engineers. The responsibility of the public works director will include all or part of the following:

- Streets, highways, and bridges: their specification, maintenance, lighting, traffic control systems, and snow removal. (Detailed design and construction of new roads and bridges are more likely to be subcontracted to engineering consulting firms, who will be led by engineers, and construction firms, who may be.)
- Water purification and distribution; sewage retrieval and pickup; solid waste disposal
- Parks, playgrounds, airport, and/or cemetery operation
- Zoning, building inspection, and code enforcement
- Vehicle maintenance

At the state level, the largest employment of engineers is in the state highway (or transportation) department, but others are employed in energy, environmental, and various regulatory functions.

Engineering and Sustainability—Green Engineering.   According to the U.S Environmental Protection Agency (EPA) green engineering can be defined as environmentally conscious attitudes, values, and principles, combined with science, technology, and engineering practice, all directed toward improving local and global environmental quality. Green engineering embraces the concept that decisions to protect human health and the environment can have the greatest impact and cost-effectiveness when applied early to the design and development phase of a process or product. Green engineering encompasses all of the engineering disciplines, and is consistent and compatible with sound engineering design principles.

Sustainable Energy and Environments

The United States and other developed nations have a tremendous reliance on energy, water, and natural resources to manage, maintain, and grow business, industry, and the economy. Over recent history the supply and availability of our natural resources (i.e., groundwater, oil, coal, and natural gas) has led to an escalation in pricing, an increase in international dependence, and in many cases disputes involving the rights, or environmental impacts that may be associated with the recovery and use of these resources. Engineers have a key role in developing new renewable energy producing products and services, and in determining strategic methods to conserve, harvest, and redirect beneficial uses of all of our natural resources. The movement toward sustainability has evolved in the United States, and this movement requires new engineered products and services. As an example, in 2011, the U.S. Department of Energy reported that renewable sources of energy accounted for about 9.3 percent of total U.S. energy consumption and 12.7 percent of electricity generation (U.S. Department of Energy, EIA) and this number continues to grow. Some of the new products being developed involve new methods for fuel generation (e.g., biofuels or blended ethanol fuels), photovoltaics, sophisticated energy monitoring devices, and more efficient home products.

Source: Mark J. Flint, PE President, Watermark Engineering Group, Inc.

College Teaching and Research.   Although these two careers are not necessarily related, they require the same basic preparation: graduate education. New assistant professors in engineering schools overwhelmingly are expected to have a "terminal degree" (usually Ph.D. or D.Sc.), and if they are teaching graduate subjects, they will normally be expected to do research and publish literature in their field—that is, to be a scholar. Engineers working in research or advanced design soon find they need more technological depth than what a B.S. degree provides, and most of them will seek an M.S. degree; senior researchers, even in industry, may find a doctorate desirable.

The position of professor of engineering can be a very satisfying one, with a great deal of personal freedom in the way time is spent and the subject areas pursued in teaching and research. Any engineer with an above-average academic capability can find opportunity in a faculty career.

Biomedical Engineering and the Healthcare.   Employment in healthcare is projected to continue to increase rapidly as the U.S. population ages. But the reason more people are reaching such an age is the rapid expansion of medical knowledge and capability, and the technology that underlies it. The medical explosion has increased the need for medical specialists of all types: chemists, biochemists, and pharmacists. The increasing technical complexity has required the service of a new breed of engineer—the biomedical engineer. We will look at this profession in some detail to provide one example of the increasing specialization of modern technology and of its practitioners; other, more numerous classes of engineer can be subdivided even further into specialties. Attinger provides a definition and identifies the growing subdivisions of the field:

> Biomedical engineering can be broadly defined as the application of engineering concepts, methods, and techniques to biology and medicine. Because of the breadth of the field, several subspecialties have been emerging: Bioengineering is concerned with the quantitative analysis, both theoretical and experimental, of the structural and functional properties of the components of biological systems....

Medical engineering, or biomedical technology, deals with the design, development, application, and evaluation of the instrumentation, computers, materials, diagnostic and therapeutic devices, artificial organs and prostheses, and medical information systems for use in medical research and practice....

Clinical engineering uses engineering concepts and technology to improve health-care delivery in hospitals and clinics....

Health care systems engineering deals with problems in the analysis of health-care concepts and health-care systems, such as [the] socioeconomic and psychosocial determinants of health. It is also concerned with the design and implementation of more efficient and less costly modes of health-care delivery....

Biochemical engineering, agrobioengineering, and genetic engineering are now emerging as new subspecialties in the rapidly developing field of biotechnology....

Work done by biomedical engineers may include a wide range of activities such as the following:

- Artificial organs (hearing aids, cardiac pacemakers, artificial kidneys and hearts, blood oxygenators, synthetic blood vessels, joints, arms, and legs)
- Automated patient monitoring (during surgery or in intensive care; healthy persons in unusual environments, such as astronauts in space or underwater divers at great depth)
- Blood chemistry sensors (potassium, sodium, O, CO, and pH)
- Advanced therapeutic and surgical devices (laser system for eye surgery, automated delivery of insulin, etc.)
- Application of expert systems and artificial intelligence to clinical decision making (computer-based systems for diagnosing diseases)
- Design of optimal clinical laboratories (computerized analyzer for blood samples, cardiac catheterization laboratory, etc.)
- Medical imaging systems (ultrasound, computer-assisted tomography, magnetic resonance imaging, positron emission tomography, etc.)
- Computer modeling of physiologic systems (blood pressure control, renal function, visual and auditory nervous circuits, etc.)
- Biomaterials design (mechanical, transport, and biocompatibility properties of implantable artificial materials)
- Biomechanics of injury and wound healing (gait analysis, application of growth factors, etc.)
- Sports medicine (rehabilitation, external support devices, etc.)

DISCUSSION QUESTIONS

13-1. Outline the detailed tasks that should be happening in the third step of the marketing process shown in Figure 13-1.

13-2. How has the wide adoption of social media changed the final step in Figure 13-1? Provide some examples of how you have been engaged as a consumer.

13-3. What is the role of the engineer in the 4Ps of marketing? How does this change for the engineering manager?

13-4. Using the example in this chapter, why is it a good idea for someone with an engineering background to be involved running marketing? Why is it not a good idea?

13-5. How should electronic marketing be used for commercial products (i.e., business to business sales)? How does this differ for consumer marketing?

13-6. In what ways are goods-producing and service-producing industries generally different?

13-7. Do some research to explain how big data and mobile devices are changing the ways that service companies do business.

13-8. Find an application of engineering concepts and techniques to the health services in the literature, and write a summary describing it.

13-9. Discuss how the management functions of planning, organizing, leading, and controlling relate to the marketing process.

SOURCES

Armstong, G. and Kotler, P., *Marketing: An Introduction, 13th ed.* (Pearson, 2016).

Attinger, E. O., "Biomedical Engineering" field definition in *Peterson's Graduate Programs in Engineering and Applied Sciences 1988* (Princeton, NJ: Peterson's Guides, Inc., 1987), p. 285.

McCarthy, E J. *Basic Marketing: A Managerial Approach* (Homewood, Ill: Irwin, 1960).

Marketing [Def. 2]. (n.d.). In Merriam Webster Online, Retrieved December 24, 2018, from https://www.merriam-webster.com/dictionary/marketing.

Opsata, Margeret, "Working for the State and Uncle Sam," *Graduating Engineer*, March 1991, pp. 48–52.

Peters, T. J. and Waterman, R. H., Jr., *In Search of Excellence: Lessons from America's Best-Run Companies* (New York: Harper & Row Publishers, Inc., 1982).

Tosh, John J., presentation at the 1988 Human Resources Conference of the American Management Association.

Wessel, David, "U.S. Workers Excel in Productivity Poll," *Wall Street Journal*, October 13, 1992, p. 2.

World Bank national accounts data, OECD national accounts data files, 2017, https://data.worldbank.org/indicator/.

Part IV

Managing Projects

14

Project Planning and Acquisition

PREVIEW

Project management is the last of the management functions this text examines. This chapter first considers what makes a project a project. Then the process of proposal anticipation and preparation that is essential to the life of the project-driven organization is discussed. The main body of the chapter is a description of some of the management tools used in project management, how it is like engineering management and how it is different. The key processes and deliverables of project management include: defining scope; developing a schedule, including milestones; creating a work breakdown structure; utilizing network scheduling systems (program evaluation review technique [PERT], critical path method [CPM], and other variations); and working with a budget. Finally, key tools for monitoring and controlling the project schedule and cost are introduced.

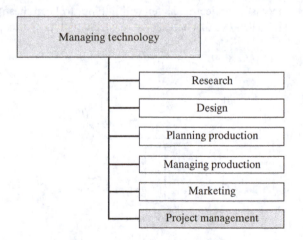

CHARACTERISTICS OF A PROJECT

A **project** represents a collection of tasks aimed toward a single set of objectives, culminating in a definable end point and having a finite life span and budget. A project is a one-of-a-kind activity, aimed at producing some product or outcome that has never existed before. (Of course, there have been earlier aircraft or buildings, but none of them were exactly like the one being created by *this* project.) Responsibility for a project is normally assigned to a single individual, assisted by a close-knit project team. The term "program" is sometimes used interchangeably with "project," but more often a program is larger and consists of a number of projects.

Formal project management methods received their greatest impetus in U.S. aerospace programs and complex construction projects of the 1960s, and the methods have spread to many other complex, dynamic activities. Project management methods should be considered (1) where close interaction of a variety of technologies, divisions, or separate organizations is required; (2) when completion within a tight schedule and budget is necessary; and/or (3) for activities involving significant technical and/or economic risk to the organization.

The three essential considerations in project management are (1) time (project schedule), (2) cost (in dollars and other resources), and (3) performance or quality (the extent to which objectives defined in the scope are achieved). The successful project manager will attempt to keep these three in balance like the legs on a stool (Figure 14-1). To achieve this balance, all projects will have a scope, schedule, and budget.

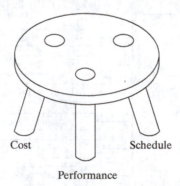

Cost   Schedule

Performance

Figure 14-1 The "three-legged stool" of successful project management.

Since achieving maximum performance is often possible only at the expense of cost and schedule, difficult trade-off decisions involving compromises are often necessary.

Programs and Projects

To achieve their strategic objectives, organizations will define and implement new programs. A program will consist of a series of related projects that collectively make up the program and seek to achieve a larger strategic objective. Projects differ from an organization's normal operations in that operations are continuous, while projects have a fixed scope. When the scope of work is completed, the project ends and those working on it move on to other work. In addition to a defined scope, all projects will also have a schedule and a budget.

For example, General Motors might have a program to design new cars that utilize new technology, meet new environmental and safety standards, and appeal to the customer. A specific project within this program might be for Chevrolet to develop a new mid-sized, four-door sedan with a target sale price under $25,000. Programs and projects may also be service oriented. For example, a Fortune 500 firm might have a strategic objective and related program to energize and train its workforce. One project within this program might be to train all engineers in the firm's approach to project management.

Source: Charles "Chick" Keller, retired University of Kansas professor, and Black & Veatch senior executive. 2012.

THE PROJECT PROPOSAL PROCESS

Every type of project should be preceded by a detailed description of what is to be accomplished, together with a proposal or estimate of the time and cost required. This process has been carried farthest by aerospace and other R&D organizations that depend on a continuing sequence of external project awards for their livelihood, and we will discuss the proposal process in that context.

Preproposal Effort

First, successful organizations of this type begin work long before a request for proposal (RFP) is received from a potential customer. The successful project-driven organization is continuously identifying new business opportunities—areas of technology or types of activity where attractive projects are likely to be funded. The firm estimates the resources and capabilities that will be required to meet expected future needs of potential customers, compares them with the resources it has on hand, then proceeds to develop the necessary technical skills and acquire other needed resources (or at least identify sources for them) in advance.

Successful firms also maintain new business (marketing) groups that seek to identify specific customer needs as early as possible, well in advance of their issuing an RFP, so that the firm can make a preliminary bid decision (a decision whether to invest the resources it will take to prepare a proposal on a major project). Creating winning proposals is an expensive, time-consuming process that should be begun only on potential projects that the firm believes fit its needs and for which it feels it has a reasonable possibility of capture. Other projects should be declined to permit the firm to concentrate on the opportunities the firm is most likely to win compared to its competition.

If the bid decision is favorable, successful firms will try to get an early start on developing a response to their best estimate of what the RFP will ask for. In many aerospace project opportunities, the firm that waits until it has an RFP in hand before evaluating the potential project and starting on a proposal has little chance to capture the award.

Proposal Preparation

By the time the RFP arrives, management often has appointed a proposal manager, who has prepared a budget for the proposal process and a letter ready for release calling on functional managers to provide members of the proposal team. The RFP is quickly examined to be sure it holds no surprises, and the tentative decision to prepare a bid is reconfirmed.

An RFP from the U.S. government typically includes a cover letter, a **statement of work** (which specifies the work to be performed), the required schedule, specification of the length and content desired in the proposal, and a stack of standard clauses (sometimes called "boilerplate") covering legal aspects of doing business with the federal government that may be several times as thick as the rest of the RFP. Then the RFP will "call out" specifications whose provisions then become a legally binding part of the contract. These specifications, in turn, call out other, "second tier" specifications that must be complied with.

A well-prepared "kickoff meeting" for the proposal team launches the proposal process. A representative of senior management may give a short pep talk on the importance of the project to the company and introduce the proposal manager, who will do much or all of the following:

- Give an overview of what the RFP asks for.
- Provide the best estimate from company intelligence as to what the customer *really* wants and the factors the customer will use in determining the contract winner.
- Identify the organization, schedule, and labor-hour allocations for the proposal effort.
- Provide handouts giving—in as much detail as preparation time has permitted—management's concept of how the project might be carried out, and instructions to the project proposal team.

Proposal personnel are usually experienced people, and so they can work rapidly with minimum guidance. They interact and prepare drafts of their parts of the proposal, which are reviewed by engineers and management. Proposal preparation usually must undergo a sequence of reviews and revisions on a tight time schedule.

Proposal Contents

The RFP will often specify separate management, technical, and cost proposals and their expected contents:

- The management proposal typically discusses the company, its organization, its relevant experience, and its management methods and control systems, and describes the personnel proposed to lead the project.
- The technical proposal outlines the design concept proposed to meet the client's needs, with special emphasis on the approach planned to resolve the most difficult technical challenges posed by the project.

- The cost proposal not only includes a detailed price breakdown, but often also discusses aspects of inflation, contingencies, and contract change procedures.

The proposal package is critically reviewed by company senior management not involved in the creation of the proposal, then revised, printed, and delivered to the customer.

Internally Driven Projects

The prior sections have discussed how companies that execute projects for external organizations, like consultants and construction firms, develop their project pipeline. For most organizations, project development is based on organizational strategy (Chapter 4) and responding to new market opportunities (Chapter 13). Because most organizations will have several strategic priorities, there will typically be a variety of projects. In fact, there are often multiple projects to support a single corporate strategy. For instance, if a firm wants to increase sales through new products, they will likely try to develop multiple new products for simultaneous release, as Apple is famous for with their "big events."

When a firm is focused on internal projects, rather than trying to predict RFPs, they are constantly looking for new project ideas that will support their strategies and take advantage of new market opportunities. Effective organizations will develop methods to capture these new project ideas, prioritize them, and then execute on the selected projects. How projects are executed is discussed in the following sections and Chapter 15.

PROJECT PLANNING TOOLS

The Project Management Institute (PMI) identifies five phases in project management:

- **Initiating** the project includes the steps previously described and the preliminary scope.
- **Planning** includes refining the scope and scheduling, which are described in the next section.
- **Executing** has the project manager as the leader of the project team, as described in Chapter 15.
- **Monitoring and controlling** the progress is done on a continuous process with a reporting process and a change process.
- **Closing** the project includes obtaining the customer acceptance, final documentation, and a final report.

Each organization may have different interpretations of project management, but they will all follow a similar structure. As shown in Figure 14-2, this structure is often iterative in nature. Members of the project team will learn new things in project execution that cause them to loop back and adjust plans, just as

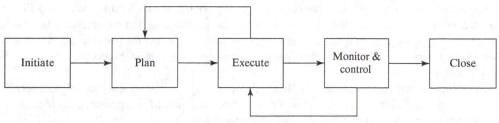

Figure 14-2 Phases of managing a project.

new information will be discovered in project monitoring and control that cause adjustments to how the project is being executed.

Since a project is a set of activities that has never been done before, planning is extensive and critical. There are three essential elements on every project plan: scope, schedule, and budget. As shown in Figure 14-1, these elements are interdependent, and a change in one may require a change in another to restore the balance of the project. All of these elements are tied together by quality:

- Quality of the deliverables of the project—do these deliverables meet the expectations of the customer?
- Quality of the project process—how well does the project management process work, and how should it be improved?

Scope

Scope is a statement that defines the boundaries of the project. It says what is going to be accomplished and what is not going to be done. The **Scope Statement** is an essential element of any project. In the engineering world, this is often called the **statement of work**. The scope includes the problem or opportunity, goal, objectives, success criteria, assumptions, risks, and obstacles. It is usually one to two pages and may include appendices stating the risk and financial analysis. It is important to include all appropriate stakeholders in the development of the scope. The customer, whether an internal manager or an external customer, needs to approve the scope. A standard documentation method is typically used to simplify the decision-making process.

The **problem** or **opportunity statement** is what the project addresses. It is a statement of fact. This sets the priority of the project that management addresses.

The **goal** is what you intend to do to address the problem or opportunity. A project generally has one overarching goal. If there is more than one goal, there might be more than one project. The goal gives purpose and direction to the project and defines what is to be done so that everyone understands what is to be accomplished. It is short and to the point.

The **objectives** further define the goal. They clarify the goal with more exact boundaries for the project. Each objective statement should contain four parts:

- An outcome—what is to be accomplished
- A time frame—the expected completion date
- A measure—metrics that will quantify success
- An action—how the objective will be met

The definition of these objectives is similar to the objectives in Chapter 4. It is important to realize that a project is dynamic and objectives may change; however, the boundaries set by this scope are still there.

The **success criteria** answer the question of what the project is going to accomplish. The success criteria will say when the project is done. They must be quantifiable and measurable and accepted by the customer, either internal or external. Quite often, a success criterion is how the project affects the bottom line of the company.

Assumptions, **risks**, and **obstacles** are often unintentionally ignored by the project planners. The listing of these factors will alert the project team or senior management to any potential problems. At times these items may be hard to define, but they need to be discussed.

    Scope creep is a term that refers to the incremental expansion of the scope of a project, which may include and introduce more requirements that may not have been a part of the initial planning of the project, while nevertheless failing to adjust schedule and budget. Scope creep may happen in small projects, as well as large projects. Scope creep may be introduced by technologists adding features not originally contemplated. Scope creep may also occur when the customer has a difficult time making a decision.

Schedule

A **work breakdown structure (WBS)** is a product-oriented "family tree" of work effort that provides a level-by-level subdivision of the work to be performed in a contract. There is no one correct WBS for each project. The WBS provides a common framework or outline that can be used to accomplish the following tasks:

- Describe the total program/project effort
- Plan and schedule effort
- Estimate costs and budgets
- Support network schedule construction
- Assign responsibilities, and authorize work
- Track time, cost, and performance

Example

Figure 14-3 illustrates a simplified work breakdown structure for developing a jet engine. The top level is the entire project or program, which can be given a unique number or code, such as "XYZ," to distinguish it from other projects. The first item at the second level (XYZ-1) is traditionally the "end item" to be delivered, in this case the jet engine itself. Other items at the second level might be the training of user maintenance and repair people (XYZ-2), creating the necessary ground support equipment (GSE) for starting and maintaining the engine (−3), system testing (−4), and the project management (−5) needed to integrate and manage all these activities. Each of these second-level items is divided further. For example, the engine is divided into the major subsystems of fan (XYZ-1.1), compressor (−1.2), and turbine (−1.3); the fan further into the fan assembly (XYZ-1.1.1), full-scale fan rig (−1.1.2), and so forth. Similarly, project management (XYZ-5) might be divided into project management per se (−5.1), configuration management (−5.2), and reliability engineering (−5.3).

    The second dimension, shown on the lower left of the figure, is the functional organization, and this also can be coded. For example, manufacturing may be coded 1000 and engineering 2000; the latter might be further divided into 2100 for plant engineering, 2200 for design engineering, 2210 for mechanical design, 2220 for analytical design, and so forth. These two dimensions meet in a cost account such as XYZ–1.1.1–2210, which represents all the mechanical design (2210) performed on the fan assembly (−1.1.1). The cost account consists of one or more *work packages*, which form the ultimate unit by which work is controlled. The work package incorporates a set of tasks to be performed, a schedule, and a budget in labor-hours and other costs. The work package is the responsibility of one person (the "work package manager") from the organizational unit in which most of the work in the package will be done, and where possible it is structured to have a short duration and defined end point.

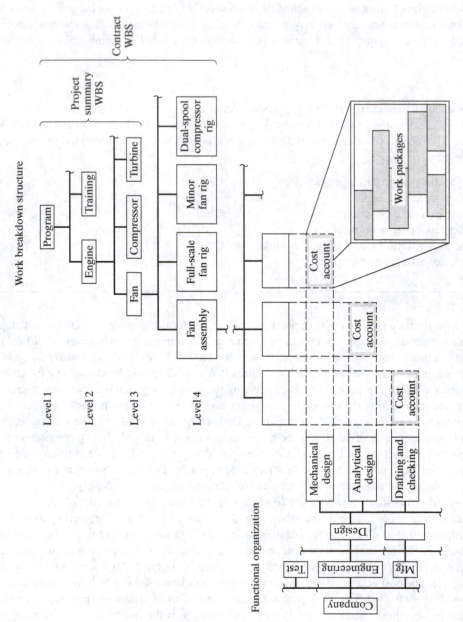

Figure 14-3 Integration of work breakdown structure.

296

Every project activity that consumes resources is included in some work package, permitting progress on a particular end item of the work breakdown structure to be evaluated. Beginning with the work package database, a computer-based management information system can easily tailor reports in either the project or the organization dimension, and it can summarize information by adding work package data together to provide reports in less detail for any level of management or for the customer.

Milestones are the key dates for major project phases or activities. Examples from a typical aerospace project are project "go-ahead" (start), design reviews ending each design phase, 90 percent drawing release, start of each major test phase, delivery of first prototype and first production item, and the customer's required operational capability date (see Figure 14-4 for an example).

Such a schedule is essential for detailed planning, since reaching a major milestone point typically requires the coordinated efforts of a great many people. For example, a major design review may require completion to a specified level of component or subsystem design by dozens of design groups (from subcontractor organizations as well as your own departments), analyses of reliability, maintainability, producibility, safety, and other aspects of the design, and plans for testing, training operators, production tooling, and logistic support. In turn, accomplishing all of these analyses and reports will require "backing up" in time from each major milestone to hundreds of earlier supporting schedule points, at which specified information must be transferred between project entities to enable the receiving group to do their design or analysis, all in order to meet the major milestone.

There are two ways to build a project schedule: bar charts (or Gantt charts) and network diagram. **Bar charts** are simpler to use and have been used longer. Henry L. Gantt, one of the pioneers of the scientific management movement, is generally credited with initiating the concept of a class of charts in which the progress of some set or sequence of activities or resources in the vertical dimension is plotted against time in the horizontal dimension. The first reported application was in 1915, when Gantt was keeping track of the time between ordering and delivery of each lot of ammunition produced in the United States for Allied forces in World War I.

Milestone	\multicolumn{12}{c}{20XX}											\multicolumn{12}{c}{20XX}										
	1	2	3	4	5	6	7	8	9	10	11	12	1	2	3	4	5	6	7	8	9	10	11	12
Project go-ahead			↑																					
Complete project plan				↑																				
Preliminary design review								↑																
90% design release															↑									
Prototype complete																↑								
System test complete																		↑						
Final design review																					↑			
Production release																								↑

Figure 14-4 Typical milestone schedule.

In construction, a milestone is often at the end of the stage, such as when the last beam is placed at the top of a building. This is marked with either an American flag or a tree. Here they have both. This is referred to as topping off.

Source: Morse, Fall 2012.

Bar charts have found many other applications. In the job-shop or batch-production environment, bar charts schedule the use of production machines, and elsewhere they plan and control work crews. In project management, it is tasks or activities (project performance) that must be charted against time (project schedule). Three things must be established in the project planning process before bar charts can be created:

1. The tasks (activities) needed to complete the project
2. The precedence relationships of the tasks (which tasks must be complete before other specified tasks can begin)
3. The expected duration of each task

Example

Table 14-1 illustrates these three items (plus a "resources required" for later use) for a simple project: building a single-story residence on a concrete slab by first prefabricating wall panels (with electrical wiring and plumbing inside) and roof trusses. Time durations have been given in weeks, assuming a five-day work week, so that an eight-hour day is 0.2 week. Given these durations and precedence relationships, a simple bar chart can be drawn in which each task is represented by a solid bar (Figure 14-5).

Table 14-1 Information for Planning House Project

Task	Follows Task(s)	Duration (Weeks)	Task Description	Resources Required
A	Start	1.0	Clear site	3
B	Start	0.6	Obtain lumber and other basic materials	1
C	Start	2.0	Obtain other materials and components	1
D	B	2.0	Prefabricate wall panels	4
E	B	0.9	Prefabricate roof trusses	3
F	A, B	1.0	Form and pour footings and floor slab	3
G	D, F	0.3	Erect wall panels	4
H	E, G	0.2	Erect roof trusses	4
J	C, H	0.5	Complete roof	3
K	J	2.0	Finish interior	4
L	J	1.0	Finish exterior	2
M	L	0.4	Clean up site	1
N	K, M	0.2	Final inspection and approval	1

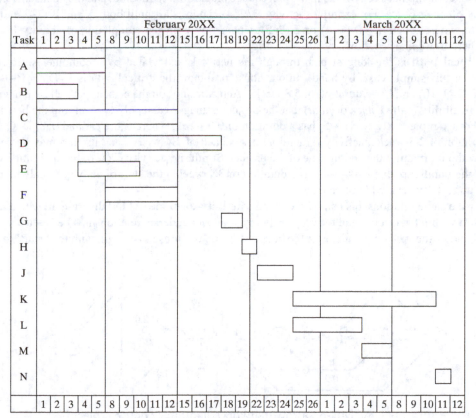

Figure 14-5 Bar chart of house project.

Bar charts are easy to understand and use, and they provide a good tool for managing small projects without an excessive number of tasks.

About 1958, two similar systems for **network-based project scheduling** were devised: the **program evaluation review technique** (PERT) was created by Booz, Allen, and Hamilton (management consultants) and Lockheed Aircraft Corporation for use in development of the Polaris ballistic missile, and the **critical path method** (CPM) was developed by the DuPont Company for chemical plant construction. In the intervening years, the features of each have been added to the other.

A network can be portrayed by either of two graphical techniques: the **activity-on-arrow** (AOA) or the **activity-on-node (AON)** diagram. Due to the power of project management software, AON has become the dominate technique. Figure 14-6 is the AOA diagram for the house project, based on the same data (Table 14-1) used for the bar chart in Figure 14-5. The arrows represent *activities* or *tasks*, which have time durations and consume resources (dollar cost and use of people and equipment); the circles represent *events*, which indicate the start and/or end of one or more activities. An activity may be given its own symbol (such as A in Figure 14-6), or it can be designated by its predecessor and successor events (activity 1, 2 or 1–2 instead of A). No activity may begin until all activities ending in its predecessor event have been completed. *Dummy activities*, shown by dashed arrows, such as 3–2 in Figure 14-6, simply show a precedence relationship between events in the precedence diagramming method, and they consume neither time nor any other resource. For example, activity F (2–4) cannot start until both A and B (1–2 and 1–3) are complete, whereas activity D (3–4) or E (3–5) depends only on the completion of activity B (1–3). Durations (here in weeks) are shown below each arrow.

The **critical path** is the longest path through the network, calculated by a computer software algorithm (or, in this simple case, by hand). In our house problem, the critical path is B–D–G–H–J–K–N (1–3–4–5–6–7–9–10) and has a duration of 5.8 weeks. Activities not on the critical path allow a degree of scheduling flexibility (called *slack* or *float*) that the project manager can apply to obtaining the best use of resources. For example, activity E (3–5) has a duration of 0.9 week, whereas the parallel path D–G (3–4–5) has a duration of 2.3 weeks; activity E therefore has a slack of 1.4 weeks, and its start may be delayed that much without affecting the ending date of the project. Similarly, activity C (1–6) has a duration of 2.0 weeks, and the parallel path 1–3–4–5–6 has a duration of 3.1 weeks; the start of activity C may therefore be delayed up to 1.1 weeks without penalty.

Exactly the same relationships can be shown on a **activity-on-node (AON)** diagram, in which activities are shown within the circles, and the arrows simply show dependency relationships between activities. Figure 14-7 shows the AON equivalent for the house project. All paths must begin with the "start" symbol

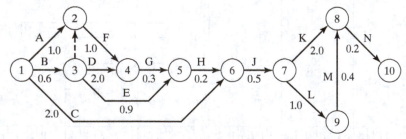

Figure 14-6   Activity-on-arrow (network) diagram of house project.

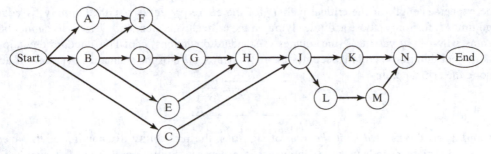

Figure 14-7 Activity-on-node diagram of house project.

and terminate at the "end." No equivalent to the dummy activity of the arrow diagram is required. The AON approach is the method used by most project management software packages.

A special feature developed with **PERT** is the treatment of activity durations (and therefore total project duration) as random variables rather than constants. To use this feature, estimators are asked to provide three estimates of the duration of any activity that might vary:

1. An *optimistic time (a)* that would be improved upon only once in 100 attempts
2. A *most likely time (m)* that would occur most often if the activity were repeated many times (statistically, the *mode*)
3. A *pessimistic time (b)* that would be exceeded only once in 100 attempts

The developers of PERT assumed that the probability distribution of possible durations of an activity fits a *beta distribution*, which need not be symmetrical (*m* need not be equidistant between *a* and *b*). The expected time (or mean value) t_e in the beta distribution can be approximated by

$$t_e = \frac{a + 4m + b}{6}.$$

(14-1)

Example

If an activity were estimated to have an optimistic time of 10 weeks, a most likely time of 13 weeks, and a pessimistic time of 19 weeks, one would predict an expected (mean) time of

$$t_e = \frac{10 + 4(13) + 19}{6} = \frac{81}{6} = 13.5 \text{ weeks.}$$

Assuming that the optimistic (*a*) and pessimistic (*b*) estimates for duration of an activity are three standard deviations on either side of the mean t_e, the standard deviation σ for the activity becomes

$$\sigma = \frac{b - a}{6} = \frac{19 - 10}{6} = 1.5 \text{ weeks.}$$

The expected length of the critical path T_e for the entire project is obtained simply by adding the expected times t_e for (only) those activities lying on the critical path. Standard deviations cannot be added in the same way—only variances (the squares of standard deviations) can. The standard deviation σ_T of the total project duration therefore becomes the *root mean square* of the standard deviations of activities lying along the critical path:

$$\sigma_T = \sqrt{\sum(\sigma^2)}$$

(14-2)

According to the *central limit theorem* of statistics, the probability distribution of the average or sum of a set of variables tends toward (approaches) the symmetrical *normal distribution*, even though the original variables fit other distributions. Knowing the expected time (mean) and standard deviation for the critical path permits us to draw a normal distribution fitting those two criteria.

Example

If the mean duration of the critical path T_e were calculated as 58.0 weeks and its standard deviation as 3.0 weeks, the critical path length would have the probability distribution shown in Figure 14-8. Then, if you had a contract to complete the project in 61.0 weeks (which is $(61.0 - 58.0)/3.0$, or 1.0, standard deviation longer than the mean of 58.0 weeks), you would estimate an 84 percent probability (50 percent + 34 percent) of completing the project within that time. The probability of completion within a 52-week year, 2.0 standard deviations (6.0 weeks) less than the mean T_e, on the other hand, would only be $(50.0 - 34.1 - 13.6) = 2.3$ percent.

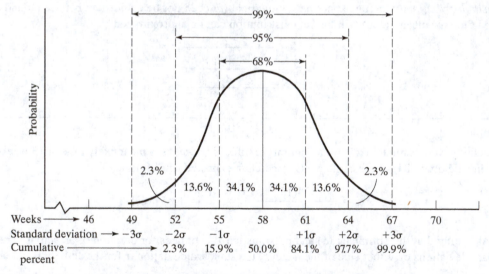

Figure 14-8   Normal probability distribution of project critical path.

PERT calculations normally consider only the longest (critical) path. If there is a second near-critical path with a duration close to the critical one, ignoring it may lead to an overly optimistic estimate of the probability of completion. This error can be eliminated by using the Monte Carlo (simulated sampling) method and averaging many simulated trials of the project, as discussed under "Simulation" in Chapter 5. The major errors in estimating project duration, however, lie in the accuracy with which the three estimates (or even one) can be provided for each activity, and in the assumption that variations in individual activities are independent of one another. As a result, most projects that employ PERT use only one estimate of each task and avoid this calculation.

Resource allocation is also a part of the scheduling process. The bar chart schedule for the house project (Figure 14-5) was prepared by beginning each activity at the earliest possible time, but this may not lead to the best use of resources. In our data for the house project (Table 14-1), we included resources required in the right-hand column. If we assume that all workers are skilled in all areas and therefore interchangeable, the personnel required in our "earliest possible" schedule will vary with time, as shown in Figure 14-9. Eleven workers are needed, the last three for less than a week's work. They may not be available or may be expensive to import and train for that brief period.

Figure 14-10 provides a modified schedule that requires only eight workers, obtained by delaying the start of activity E by its slack (float) of 1.4 weeks. Project management software with resource allocation provisions can help in scheduling tasks that employ a variety of resources that may be in limited supply: qualified welders or other craftsmen, large cranes or other expensive equipment, or fixed delivery rates for common-use materials. The software proceeds through time by selecting among alternative activities that could begin at the same point, using a specified heuristic scheduling rule (such as first scheduling tasks with the least slack or with the earliest values of late finish time). The schedules that result are not always the shortest possible, but they are much better than would be obtained by a random selection among tasks.

We cannot complete our sample "house project" within the minimum 5.8-week time with fewer than eight workers. However, we may prefer a profile that does not reduce workers to three for a short period (February 22–24). This might be possible by reexamining our initial logic and perhaps concluding that some of the finishing work of activities K and L could begin before the roof (activity J) is complete.

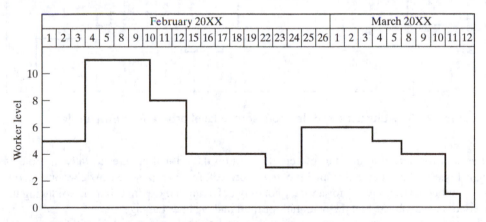

Figure 14-9   Resource-level profile for house project.

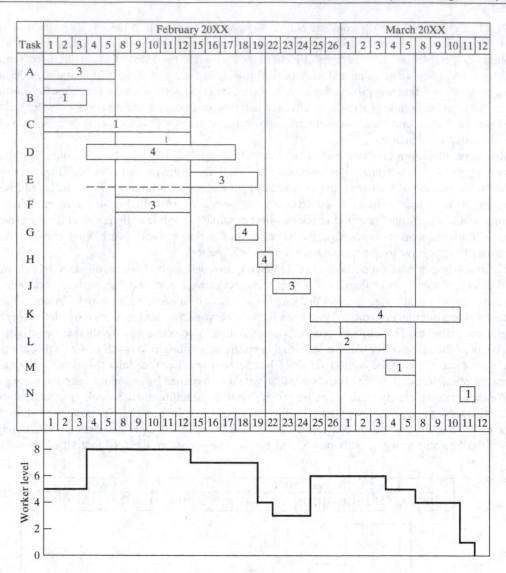

Figure 14-10   Adjusted schedule and resource-level profile for house project.

Another reason for deferring the start of activities not on the critical path, especially in projects of long duration, is to defer the expense they involve. In our project, for example, we can delay buying other materials and components (activity C) for a week, not only deferring the expense, but also reducing the storage space needed on the job site and the potential for theft and weather damage.

Today there are a number of project management software packages that simplify these processes. One of the most common project management software tools is Microsoft Project.

MONITORING AND CONTROLLING

In Chapter 8, the essential management function of control was discussed and has been defined simply as "compelling events to conform to plans." The first step in control of projects is to establish objectives; this is done in the planning process. Earlier in this chapter essential project planning methods were introduced, including the scope, work breakdown structure, bar (Gantt) charts, and network scheduling systems such as CPM and PERT.

Projects have a **budget** and limited resources. Resources may be people, money, material, or machines, and they are dedicated to the project. For the project manager, these are fixed resources, but management may adjust them up or down. Keeping within the scope of the project is important for the impact on the cost.

Reducing Project Duration

Project managers often find during initial planning that the predicted project duration, found by summing up activity durations along the critical path, is too long to meet the required project completion date. Also, slippages along the critical path early in the project may predict a delay in project completion.

The first approach to reducing project duration to meet a desired completion date is to reexamine the logic used in sequencing activities. The project schedule may have been created by assuming that some activity X (grading a roadbed, for example) must be completed before activity Y (paving the road) can begin, and this sequence is later found to be on the critical path. The project manager may conclude that paving can begin as soon as half the roadbed is graded. Activities X and Y might be divided into smaller activities X_1, X_2, Y_1, and Y_2, such that Y_1 could begin as soon as X_1 was complete, and Y_2 as soon as both X_2 and Y_1 were complete, as shown in Figure 14-11a.

Precedence diagrams such as Figure 14-11b are an alternative tool of network planning, often used in construction, which can simplify the expression of a variety of precedence relations between activities such as the roadbed/grading example just described. Figure 14-13b, for example, shows a start-to-start delay of three weeks and a finish-to-finish delay of two weeks between grading and paving.

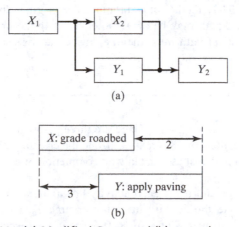

(a)

(b)

Figure 14-11   (a) Modified Gantt and (b) precedence diagrams.

In another variation, known as **fast tracking**, initial phases of manufacture or construction are begun before the design is complete for the remaining phases. The danger in this is that early work may have to be redone to accommodate unforeseen changes and unexpected results. Fast tracking should be attempted only when the advantages of time saving are compelling, and then only by experienced design and construction teams.

Crashing the Project

When we say that an activity will take a certain number of days or weeks, what we *really* mean is this activity *normally* takes this many days or weeks. It could be less time, but to do so would **cost more money**. Spending more money to get something done more quickly is called "**crashing**." There are various methods of project schedule crashing, and the decision to crash should only take place after all of the possible alternatives have been analyzed. The key is to attain maximum decrease in schedule time with minimum cost.

Another method of shortening a project is by reducing the duration of some activity along the critical path by applying more resources to it, or "crashing" it. The *normal time* T_n originally estimated for the activity is usually the time associated with the lowest cost (the *normal cost* C_n) to complete that activity. However, for many activities there will be some shorter duration (the *crash time* T_c) that can be achieved at some higher *crash cost* C_c by using overtime, larger crews, more expensive equipment, or subcontractors. Each such activity along the critical path therefore has a *slope* defined as

$$\frac{C_c - C_n}{T_n - T_c} \tag{14-3}$$

(the negative of the normal algebraic slope definition) in terms of dollars per unit time reduction. The prudent project manager will add resources to "crash" the activities along the critical path having the lowest slope, as long as the cost of reduction is less than the benefits realized by a shorter project duration. This must be done with care, since reducing the duration of one path through the network often results in a parallel critical path having the same duration. The original project schedule could have two critical paths or the new critical path may be longer after crashing the original critical path. When this occurs, an activity on each critical path must then be crashed at the same time to speed final project completion.

Example

Consider the project shown in Figure 14-12. There are three paths through the network: A–C–E and B–D–E, with normal time durations of 18 weeks each, and the critical path A–D–E, with a normal time duration of 19 weeks. The normal cost associated with completion of all activities in their normal time is $17,200.

In addition to normal cost and time, each activity has associated with it a higher cost C_c that would be required to complete the activity in a shorter time T_c. For example, activity B can be completed in four weeks for a normal cost of $3,000, or in two weeks for an additional $1,000.

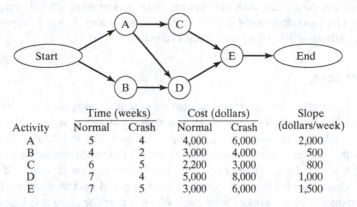

Activity	Time (weeks)		Cost (dollars)		Slope (dollars/week)
	Normal	Crash	Normal	Crash	
A	5	4	4,000	6,000	2,000
B	4	2	3,000	4,000	500
C	6	5	2,200	3,000	800
D	7	4	5,000	8,000	1,000
E	7	5	3,000	6,000	1,500

Figure 14-12 Network and data for "crashing" example.

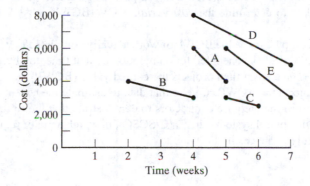

Figure 14-13 Illustration of slopes in crashing example.

Assuming that costs are linear between these extremes, each activity then has a slope in dollars per week of reduction, as tabulated in the right-hand column in Figure 14-12 and shown graphically in Figure 14-13. The project manager will wish to crash the activity with the smallest slope *among those on the critical path*.

In our example, activity D would be reduced by one week at a cost of $1,000, which is less than the cost of crashing A or E (and more than the $500 cost of crashing activity B or the $800 of C, but they would be of no help). This reduces critical path A–D–E to 18 weeks (and path B–D–E to 17 weeks). Note that path A–C–E remains at 18 weeks, and so it becomes a second critical path.

If we wish to shorten the project further, we must find a solution that shortens both paths A–C–E and A–D–E. Three solutions to save a second week are possible: to shorten activity A ($2,000), to shorten both C and D ($800 + $1,000, or $1,800), or to shorten E ($1,500). Naturally, we will choose the third solution, and we will complete the project in 17 weeks at a cost of $19,700 ($17,200 + $1,000 + $1,500).

Business is more complex today than ever before, so project managers must become more rational in their decision making by using the most effective tools and techniques. Crashing a project needs to be the alternative only after all possible options have been evaluated.

Earned Value System

In planning a project, activities are scheduled against time, and in budgeting the project, a cost for each activity is estimated. The **budgeted cost of work scheduled** (BCWS) can therefore be represented as a cumulative cost curve versus time. As the project progresses, labor, material, and other costs are carefully recorded. However, this *actual cost of work performed* (ACWP) cannot be compared effectively with the work scheduled to be accomplished by the same date, since some tasks will be behind schedule, and some ahead of schedule. What is needed is a third measure, the **budgeted cost of work performed** (BCWP).

This "**earned value**" system works as follows: When a task (or separable part of a task) has been completed, the project is considered to have "earned the value" (BCWP) originally estimated (budgeted) for that task or segment. It then becomes possible to compare the budgeted (estimated) and the actual costs of work performed to date to determine the *cost variance* CV (BCWP − ACWP) experienced thus far in the project.

Similarly, one can compare the value BCWP of work actually completed (performed) at some point in time with the value BCWS of work scheduled for completion by that time, to obtain the *schedule variance* SV (BCWP − BCWS). The analyst then projects the earned value (BCWP) curve to estimate the revised completion date, and projects the ACWP curve to that date to estimate the cost at completion (CAC). The U.S. Department of Defense may require contractors to demonstrate that their project control system will meet certain cost/schedule control system criteria (C/SCSC), in which a major requirement is the ability to calculate cost and schedule variances in this way.

Closing

The functions of project management are similar to engineering management, except for the last function of closing. Closing signals the completion of project work and the delivery of the results to the customer. Wysocki and others give six steps in closing:

- Get customer acceptance of deliverables.
- Ensure all deliverables are installed.
- Ensure documentation is in place.
- Get customer sign-off on final report.
- Conduct final evaluation.
- Celebrate success.

The closing phase is very important to the project management of this project and of future projects. Unfortunately it is the part most often ignored or omitted.

Project management provides a structured, yet flexible, framework suitable for all kinds of initiatives. Project management has over the last few decades emerged as a discipline that all types of companies cannot do without. If implemented correctly it becomes one of the key factors in corporate sustainability for success. It embraces not only the management and execution of individual projects but also of programs of projects. It has a control system that allows a company to monitor the results. Project management provides companies with a common language and methodology that facilitate the management of projects of all sizes.

Project Management Under Uncertainty: Agile Project Approaches[1]

This chapter has explained approaches of traditional project management and managing the "Golden Triangle" of scope, schedule, and cost. These methods work wonderfully when these constraints are known, but tend to struggle under times of greater uncertainty, such as software development or research projects. A key difference between software development and more traditional projects, like construction, is the nature of the project scope. In construction (or product launch), the scope of the project is given and fixed, and the project is managed to minimize "scope creep." In software development projects, scope is often fuzzy and subject to change as the project progresses. The issue of unknown scope is not limited to software, it occurs in many projects that are exploratory in nature. When project scope is undetermined and perhaps indeterminable, traditional project management approaches have significant problems. Agile project management was born from a desire to combat these problems and find more effective methods in highly uncertain environments. In 2001 a group of Information Technology (IT) professionals gathered to discuss problems with IT project management and created the Agile Software Development Alliance.[2]

The informal gathering of IT professionals discussed the need for greater flexibility in software development and codified it into 12 principles based upon a set of four core values:

- Focus on interactions with individuals rather than following a rigid set of processes and tools.
- Meet customers' requirements by delivering functioning software without the delays caused by providing extensive documentation.
- Cultivate a spirit of collaboration with the customer instead of hard bargaining in contract negotiations.
- Provide flexibility to adapt to customers' requests for changes as opposed to rigidly retaining the pre-defined scope. Be able to embrace scope creep as responding to a customer's need rather than fighting it.

Together the core values and principles are referred to as the "Agile Manifesto."[2]

There are three philosophical beliefs underpinning the agile principles.[3] These are, 1) Visibility, in that project outcomes must be visible to the people controlling the process; 2) Adaptation, such that scope changes must be recognized and responded to quickly in order to minimize disruptions; and 3) Inspection, in which the team members responsible for quality control must inspect and test functionality regularly and be able to detect issues that violate the specifications. The key tenet in Agile approaches is that flexibility is required in order to meet ever changing scope requirements. Since the project scope changes, the traditional project management concept that "better planning leads to better outcomes" is invalidated. Instead Agile utilizes an iterative approach to ensure that finite resources are always focused on delivering the next component that represents the highest value to the customer. The approach has moved beyond software development and been adopted for high uncertainty projects in many industries.

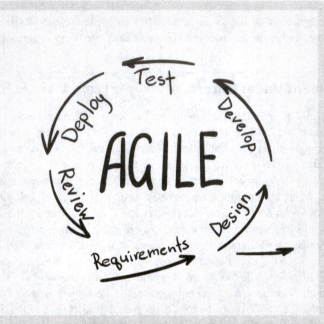

Source: Natata/Shutterstock

Sources

1. Eschenbach, T., N. Lewis, G. Nichols & W.J. Schell. *Using Agile Project Management to Maximize You and Your Coauthors' Productivity.* in *ASEE Annual Conference and Exhibition.* 2015. Seattle, WA.
2. Kwak, Y.H., *Brief History of Managing Projects*, in *The story of managing projects: An interdisciplinary approach*, E.G. Carayannis, Y.-H. Kwak, and F.T. Anbari, Editors. 2005, Praeger Publishers: Westport, CT.
3. Schwaber, K., *Agile project management with scrum.* 2015, [place of publication not identified]: Microsoft.

DISCUSSION QUESTIONS

14-1. Which of the following would be considered a project: (**a**) construction of a hydroelectric dam; (**b**) operation of a nuclear reactor; (**c**) development of an engine for the B-2 bomber through the first production prototype; (**d**) a production order for an additional 20 F-15 aircraft?

14-2. Often the most difficult part of any project is defining the project. Why is this the case?

14-3. In your words, what are the key differences between agile and traditional project management approaches?

14-4. Discuss how the management functions of planning, organizing, leading, and controlling relate to the project process. Which of these roles becomes the responsibility of the project manager?

PROBLEMS

14-1. Prepare (**i**) a milestone schedule and (**ii**) a work breakdown structure for one of the following: (**a**) construction of a steel and concrete highway bridge, or (**b**) movement of the printing department of a daily newspaper from one building to another (assuming two identical presses, with one press continuously available for use).

14-2. Establish tasks, times, and precedence relationships for the project selected in Question 14-1, and draw (**a**) a bar chart and (**b**) a network diagram schedule for it.

14-3. For the project outlined in the following table, prepare (**a**) a bar chart, (**b**) an AOA network diagram, and (**c**) a AON diagram. (**d**) What and how long is the critical path?

Task	Follows Task(s)	Duration (Weeks)
A	Start	3.0
B	Start	7.0
C	A	2.0
D	B	7.0
E	B, C	5.0
F	D, E	1.0

14-4. For the project outlined in the following table, prepare (**a**) a bar chart, (**b**) an AOA network diagram, and (**c**) AON network diagram. (**d**) What and how long is the critical path?

Task	Follows Task(s)	Weeks Duration	Manning Level
A	Start	6.0	3
B	Start	5.0	4
C	Start	5.0	3
D	A	3.0	2
E	A, B	6.0	5
F	D, E, C	1.0	2

14-5. For the project in Problem 14-4, (**a**) provide a worker-level profile, assuming that all tasks begin as early as possible; (**b**) repeat the profile, assuming that no more than nine people are available in any week and the manning level for a task cannot change; and (**c**) identify the project duration in each case.

14-6. Tasks X, Y, and Z must be completed in series to complete a project. The three time estimates (a, m, and b) for each task in days are X: 30, 45, and 66 days; Y: 24, 42, and 60 days; and Z: 26, 50, and 68 days. For each task, calculate (**a**) the expected time t_e and (**b**) the standard deviation σ. What is the (**c**) expected time T_e and (**d**) the standard deviation σ_T for the complete project?

14-7. If a project has an expected time of completion T_e of 45 weeks with a standard deviation σ_T of 7 weeks, what is the probability of completing it (**a**) within one year (52 weeks)? (**b**) within 38 weeks?

14-8. Tasks A, B, and C must be completed in series to complete a project. The three time estimates (a, m, and b) for each task in weeks are A: 8, 11, and 14 weeks; B: 7, 10, and 19 weeks; and C: 10, 19, and 22 weeks. For each task, calculate (**a**) the expected time t_e and (**b**) the standard deviation σ. What is the (**c**) expected time T_e and (**d**) the standard deviation σ_T for the complete project? What is the probability of completing the project in (**e**) 40 weeks? (**f**) 46 weeks?

14-9. The text shows how to reduce the duration of the project shown in Figure 14-12 from 19 to 18 to 17 weeks at the lowest cost. Continue this process, showing the most economical way to reduce project duration from 17 weeks, week by week, to the minimum possible duration. What is the minimum project total cost at this minimum duration?

14-10. For the project outlined on the following table, (**a**) draw a network diagram (AOA or AON, as you prefer), and (**b**) identify the critical path and duration. (**c**) Identify the task(s) you would crash and the incremental cost to reduce project duration by (**i**) one week; (**ii**) a second week; (**iii**) a third week.

Task	Follows Task(s)	Duration (Weeks)		Cost (Dollars)	
		Normal	Crash	Normal	Crash
A	(Start)	3	2	$500	$600
B	(Start)	2	1	400	450
C	A, B	5	3	600	750
D	B	5	4	550	640
E	C, D	4	3	400	550

14-11. The following project carries a penalty cost of $200/day ($1,400/week) for any delay in completion beyond 26 weeks. Any task can be accelerated by up to three weeks at a cost of $1,000 per week reduction. Draw an AOA diagram, identify the critical path and duration, and determine which task(s), if any, you will crash, and by how much, to minimize project cost.

Task	Follows Task	Duration (Weeks)
V	(Start)	11
W	V	8
X	W	10
Y	V	20
Z	(Start)	30

14-12. A project manager observes that they had a budgeted expenditure of $450,000 by a specific date and had only spent $425,000 by then. What should they conclude?

14.13. Consider the following network for a small project (all times are in days)

			Network		
Activity	Initial Node	Final Node	Optimistic Time	Pessimistic Time	Most Likely
A	1	2	1	3	2
B	1	3	4	6	5
C	2	4	1	3	2
D	3	5	7	15	8
E	4	6	4	6	5
F	4	5	6	14	10
G	5	7	2	4	3
H	6	7	6	14	10

a. Draw a AOA network diagram representing the project (3)
b. Use PERT to determine the critical path and associated time—label your answers (8)
c. The customer wants the project delivered in 20 days. What is the probability of meeting the customer's expectation? Use Figure 14-3 to develop your estimate.
d. If activity D had an *expected* completion time of 15 days, does this impact the expected project completion time? If so, how? (3)

SOURCES

Dougherty, Frank, Assistant for Contract and Quality Matters, U.S. Dept. of Defense, Directorate of Industrial Productivity/Quality, presentation on "Acquisition Streamlining" to the American Society for Quality Control Midwest Conference, St. Louis, MO, October 9, 1986.

Guide to the Project Management Book of Knowledge (PMBOK© Guide) 6th ed., Project Management Institute, Newtown Square, PA.

www.pmi.org Project Management Institute, Inc.

Wysocki, Robert K., *Effective Project Management: Traditional, Adaptive, Extreme,* 4th ed. (New York: Wiley Publishing, Inc., 2006).

15

Project Organization, Leadership, and Control

PREVIEW

In the first chapter on project management, the subject of the project itself was developed. The second chapter on project management begins by enumerating the elements needed in the project-driven organization and alternative organization designs. Special attention is given to the nature and functioning of the matrix organization. Next, the project manager's personal characteristics and career development, and the importance and content of a charter that defines responsibilities and authority, are examined. Then, the methods of motivating effective project performance through team building and conflict management are discussed. Next, the importance to the project performance of communicating with the customer at various levels of the organization is considered. Finally, the several types of fixed price and cost reimbursement contracts are defined and compared.

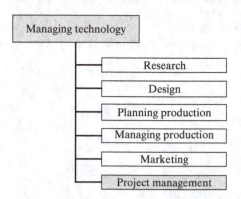

PROJECT ORGANIZATION

The Project-Driven Organization

Kerzner classifies organizations into two groups, depending on whether or not their dynamics are primarily project driven. He distinguishes them as follows:

> In a project driven organization, such as construction or aerospace, all work is characterized through projects, with each project as a separate cost center having its own profit and loss statement. The total profit of the corporation is simply the summation of the profits on all projects. In a project-driven organization, everything centers around the projects.

> In the non-project-driven organization, such as low-technology manufacturing, profit and loss are measured on vertical or functional lines. In this type of organization, projects exist merely to support the product lines or functional lines. Priority resources are assigned to the revenue-producing functional line activities rather than the projects.

The legal forms and the patterns of departmentation (subdivision) for traditional (non-project-driven) organizational styles were discussed in Chapter 6. In this section, we take up organizational alternatives for the project-driven organization or division.

Elements of the Project-Driven Organization

Every project-driven organization needs four different categories, or types of elements, and project organizations can be characterized by the number of these categories coming under the direct control of the individual project manager. These four categories are as follows:

1. *The project office.* Every project needs a "unifying agent" of some type that bears primary responsibility for the project. In a small project, the project manager may serve this function alone. A larger project with any substantial design or development responsibility will usually have a *project engineer*, responsible for the technical integrity of the project and the cost and schedule of engineering activities. Another member of the project office is usually (by whatever name) the

project administrator, responsible for project planning and control systems and documentation. Other functions frequently centered in the project office include design review and configuration/ change control.

2. *Key functional support.* Support from certain key functional areas is central to project success, and activities in these areas often must be tailored to the specific needs of the projects. These key support areas may include the following:
   - Systems analysis, systems engineering, and integration
   - Product design and analysis
   - Quality assurance and reliability
   - Production planning
   - Product installation and test
   - Training, logistics planning, and field support

3. *Manufacturing and routine administration.* These activities are less likely to be under direct project control—manufacturing because it is usually too expensive to replicate for each project, and the others because they are service activities done essentially the same way for all projects. Activities in this category commonly include the following:
   - Manufacturing
   - Accounting and finance
   - Purchasing and subcontracting, although project subcontract administration may fall under key functional support
   - Personnel and industrial relations
   - Plant facilities and maintenance

4. *Future business.* Activities such as non-project-specific R&D and marketing are necessary for the continuation of the project-driven organization. However, these activities are not properly part of specific projects, since projects by definition should have a definable end point.

Projectized Versus Functional Organizations

Conducting Projects Within the Functional Organization.   A functional organization (such as that of Figure 15-1) is subdivided at the top into functional areas. When a project involving several of these areas is to be conducted, a project manager is appointed to coordinate the activities of the various people working on the project. This person might be attached as staff to the general manager over all the functions involved, or may be the supervisor of the function most heavily involved in the project or a key subordinate of that supervisor, as shown in the figure. The designated project leader usually has no line authority over the bulk of project personnel—only over the immediate project office (category 1 of the four groups described in the preceding section) if one exists. He or she must therefore lead by persuasion and influence. Nonetheless, projects are often conducted from within the functional organization—especially when they are small, short in duration, and low in complexity, and when schedule is not critical.

The Projectized Organization.   In the fully projectized organization (Figure 15-2), the project manager is in direct control of all the elements needed to conduct that project (all of categories 1, 2, and 3). Such an organization is attractive for large, long-duration projects, especially those that are very complex,

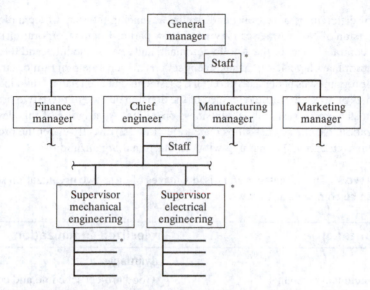

Figure 15-1 Typical functional organization (showing locations (*) from which a project manager or coordinator might lead a project).

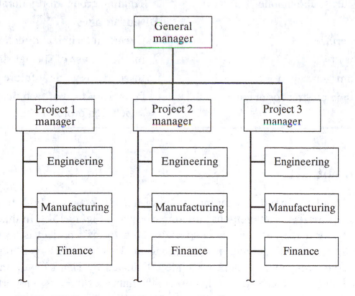

Figure 15-2 A projectized organization.

involve a number of different organizations, and require advancing the state of technology. For example, when the Space Division of North American Aviation (now part of Boeing Corporation) was awarded two multibillion-dollar contracts (one for the Apollo command and service modules and the other for the second stage of the Saturn launch vehicle), management set up two separate program organizations in different locations. Each program was under a division vice president, and each had its own manufacturing plant and staff of specialists of all kinds. Little remained of the division outside of these two programs except the division president's staff and a vestigial new business operation. A variation of this design, the *partially projectized* organization, has the key activities (categories 1 and 2) directly under the project manager and the supplemental ones (category 3) remaining with the functional organization.

Comparing the Two.   The advantages and disadvantages of these two organization structures for project management can be compared as follows:

Functional Organization	Projectized Organization
Advantages	**Advantages**
Efficient use of technical personnel	Good project schedule and cost control
Career continuity and growth for technical personnel	Single point for customer contact
	Rapid reaction time possible
Good technology transfer between projects	Simpler project communication
Good stability, security, and morale	Training ground for general management
Disadvantages	**Disadvantages**
Weak customer interface	Uncertain technical direction
Weak project authority	Inefficient use of specialists
Poor horizontal communication	Insecurity regarding future job assignments
Discipline rather than program oriented	Poor cross-feed of technical information between projects
Slower work flow	

Matrix Management

A composite organization structure that combines many of the advantages of both functional and projectized management is the **matrix management** structure, shown in Figure 15-3. In this system, the person assigned responsibility for a specific functional specialty on a specific project is accountable in two dimensions, reporting to both functional and project managers. This "two manager" reporting relationship defies the "unity of command" management principle articulated by Henri Fayol and, indeed, by most early management scholars. This leads to conflict unless the nature of these two reporting relationships is clearly understood.

Figure 15-4, which depicts a matrix structure simplified to show only one of many functional managers and one of several projects, helps explain the nature of these relationships and the benefits of the matrix organization. In the horizontal relationships, the project manager has control over the three key

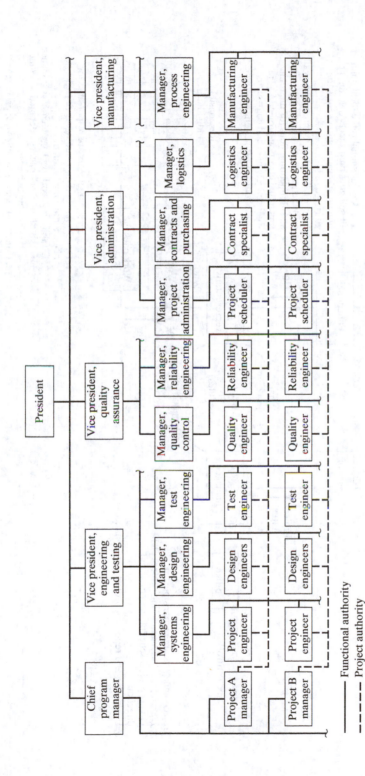

Figure 15-3 Typical matrix organization.

——— Functional authority
- - - - Project authority

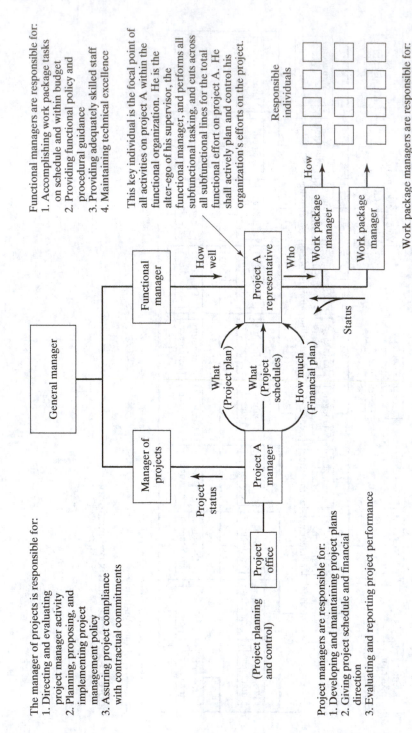

The manager of projects is responsible for:
1. Directing and evaluating project manager activity
2. Planning, proposing, and implementing project management policy
3. Assuring project compliance with contractual commitments

Project managers are responsible for:
1. Developing and maintaining project plans
2. Giving project schedule and financial direction
3. Evaluating and reporting project performance

Functional managers are responsible for:
1. Accomplishing work package tasks on schedule and within budget
2. Providing functional policy and procedural guidance
3. Providing adequately skilled staff
4. Maintaining technical excellence

This key individual is the focal point of all activities on project A within the functional organization. He is the alter-ego of his supervisor, the functional manager, and performs all subfunctional tasking, and cuts across all subfunctional lines for the total functional effort on project A. He shall actively plan and control his organization's efforts on the project.

Work package managers are responsible for:
1. Developing and maintaining work package plans for accomplishment
2. Establishing work package technical guidance
3. Establishing work package detailed schedule and operating budgets
4. Controlling and reporting work package performance

Figure 15-4 Matrix relationships. (From David I. Cleland and William R. King, *Systems Analysis and Project Management*, 3rd ed., McGraw-Hill Book Company, New York, 1983, p. 353.)

factors in project control: *what* has to be done (project tasks and performance), *when* it must be done (project schedule), and *how much* in the way of resources will be allocated (project budget). The project manager also provides a central focal point for customer contact, decisions on project changes, and project communication.

The functional manager, on the other hand, has primary responsibility for assignment of functional specialists, and he or she can therefore try to optimize the distribution of key specialists among the several project and nonproject assignments competing for them. The functional manager also is responsible for assuring the quality of work done on projects or elsewhere in that specialty, and for selecting, evaluating, and rewarding work done in that specialty. The specialists, in turn, benefit from communication with their peers and from having a congenial "technical home" to return to when a project assignment is complete.

The functional specialists assigned to the matrix organization (or assigned temporarily to a projectized organization) develop an understanding of other functions that enhances their personal growth in the organization. They also develop contacts outside their narrow function that persist for some time, and research has shown a correlation between communication outside the specialized working group and the effectiveness of a technologist.

Young distinguishes between shifting and fixed matrix organizations. In the "shifting matrix," common in the aerospace industry, personnel are shifted between projects according to the workload and project cycle, and the project team is disbanded at the end of the project. In the less common "fixed matrix" organization, the project manager is responsible for a successive series of projects, and functional personnel "are always assigned to the same project managers whatever the project."

The matrix organization has many applications aside from project management. Newman et al. generalize these applications:

> Matrix organization basically gives an operating manager two managers. One manager deals with mobilizing resources, techniques of production, and other aspects of creating the product (or service). The other manager is concerned with creating a product that pleases the customer—the right performance characteristics, quality, delivery time, and so on. To avoid too much attention to either the input side or the output side, the two managers negotiate doable instructions for the operating managers.

A classic application is a consumer products firm such as a soap manufacturer, where a *product manager* is assigned for each product line (e.g., home dishwasher detergents or industrial floor cleaners); the second dimension consists of the functional resources of R&D, production, sales, and advertising needed by all the product managers. Another example is an advertising firm with one dimension of *account executives* responsible for satisfying a specific client, and the other for the common resources such as "market research, copyrighting, artwork, television production, media selection, and other functions."

Organization Structure and Project Success

Larson and Gobeli describe an extensive study sponsored by the Project Management Institute (PMI) in which the effectiveness of five project management structures were compared:

1. *Functional organization.* The project is divided into segments and assigned to relevant functional areas and/or groups within functional areas. The project is coordinated by functional and upper levels of management.

2. *Functional matrix.* A person is formally designated to oversee the project across different functional areas. This person has limited authority over functional people involved and serves primarily to plan and coordinate the project. The functional managers retain primary responsibility for their specific segments of the project.

3. *Balanced matrix.* A person is assigned to oversee the project and interacts on an equal basis with functional managers. This person and the functional managers jointly direct workflow segments and approve technical and operational decisions.

4. *Project matrix.* A manager is assigned to oversee the project and is responsible for the completion of the project. The functional manager's involvement is limited to assigning personnel as needed and providing advisory expertise.

5. *Project team.* A manager is put in charge of a project team composed of a core group of personnel from several functional areas and/or groups, assigned on a full-time basis. The functional managers have no formal involvement.

(Their two functional structures are versions of the "functional organization" described previously, the balanced and project matrix structures are versions of the matrix structure, and the "project team" is a highly projectized organization.)

Larson and Gobeli mailed a questionnaire to PMI members, and they used data from 547 respondents who answered "a series of questions concerning a recently completed development project they were familiar with.... Respondents were asked to simply evaluate their project according to (a) meeting schedule, (b) controlling cost, (c) technical performance, and (d) overall performance with a response format of 'successful,' 'marginal,' and 'unsuccessful.'" While the functional matrix (2) was perceived as more successful than the functional organization (1), the other three (3, 4, 5) were perceived as more successful than either of these in all four measures (a, b, c, d). Of the latter three (3, 4, 5), the project matrix (4) was judged best in controlling cost and, along with the project team (5), in meeting schedule.

THE PROJECT MANAGER

Characteristics of Effective Project Managers

Project managers need enthusiasm, stamina, and an appetite for hard work to withstand the special pressures of project management. Where possible, project managers should have seniority and position in the organization commensurate with that of the functional managers with whom they must negotiate. Whether they are project coordinators within a functional structure or project managers in a matrix structure, they will often find their formal authority incomplete, and will need a blend of technical, administrative, and interpersonal skills to provide effective leadership within the project and secure sufficient resources and prioritization. Like engineers, project managers also have their own professional society, the Project Management Institute (pmi.org)

Technical Skills.   Many projects depend on effective application of certain key technologies. The effective manager of such projects must understand the essentials of those technologies enough to evaluate whether the work done is of sufficient quality, even if he or she is not as expert as the specialists actually doing the work. Further, when an unfamiliar technology is involved in a problem on the project, the project manager must quickly be able to master the essential technology bearing on the problem from briefings by specialists, so that they can articulate the problem to the client or general management and make effective decisions regarding resolution.

Administrative Skills. Project managers must be experienced in planning, leading, organizing, staffing, and control techniques as they apply to projects. In particular, they should understand the project planning techniques—such as the work breakdown structure, network systems, and others discussed in Chapter 14; design control methods such as design review and configuration/change control; and project cost control methods such as the "earned value" system discussed later in this chapter—especially as they are carried out in that particular company.

Interpersonal Skills. Except in fully projectized organizations, project managers depend heavily on the work of others not under their line control. The ability to inspire, cajole, negotiate, and persuade others therefore becomes very important, and project managers need to be effective with a variety of conflict resolution methods.

Developing Project Management Skills

Managers of large projects typically began in some specialty of engineering or business, learned project planning and control while applying their specialty in a project environment, and were assigned responsibility for a major project only after a series of project and functional assignments of increasing responsibility. However, engineers may find themselves assigned to small projects with little or no preparation (see "Project Management: A Common Path to Engineering Management"). There are many short courses offered by universities, professional societies, and consultants, and many books available on the subject of project management, but none fully substitute for experience. Meetings and publications of the Project Management Institute, of the engineering management divisions of the major engineering societies, and of the American Society for Engineering Management can also help the engineer acquire project management skills.

The Project Manager's Charter

Of vital importance to the project manager is their **charter**, or scope of authority. It is highly desirable that the responsibilities and authority of the project manager be defined in writing in advance to clarify the interfaces between the project manager, functional managers, and others, and to reduce the potential for conflict and confusion. Following are some of the areas a project manager might like to see covered in such a charter:

1. Specification of project priority relative to other activities
2. Designation as the primary contact with the customer
3. Authority to define the work to be performed by supporting departments in terms of cost, schedule, and performance
4. Control over the project budget, with signature authority on all work authorizations
5. Responsibility to schedule and hold design reviews, determine the agenda and representation, and establish responsibility for follow-up action
6. Responsibility for configuration and change control and for approving changes

7. Authority to constitute and chair the make-or-buy and source selection board
8. Responsibility for regular reporting to general management of project status and identification of any factors inhibiting project success
9. Participation in the merit review process for all personnel on loan to the project

Few project managers will be granted all the authorities suggested, but the authority relationships with functional managers and among projects should be clarified where possible for more effective project performance.

Project Management: A Common Path to Engineering Management

Since engineers design things and processes which do not already exist, essentially all engineering work is project based, regardless of whether or not the entire organization is project based. For this reason, engineers are exposed to project-based work from the very beginning of their careers and often have the opportunity to move into a project management role fairly quickly. Based on the experience of my career and observing my students' progress in their own careers, engineers who are interested in assuming formal engineering management and leadership roles should give serious consideration to career opportunities in project management.

At the beginning of my career, I regularly ran projects that involved only a handful or fewer people (including myself). In just a few years, those responsibilities grew rapidly to include growing project budgets and staff, culminating in management of an annual program exceeding $50MM before the age of 30. This growth provided amazing opportunities to understand the broad aspects of a variety of different businesses, work with diverse groups of people around the globe, and learn to manage people in a wide range of circumstances. It provided the development needed for a career move into general management (before leaving industry for the academy). This experience is not unique. During my career, every engineer or computer scientist I worked with who went on to high levels of management started their management career as a project manager. This includes a number of my past students and coworkers, including the current Chairman and CEO of American Express, Stephen Squeri, a computer science graduate whose early career included project work as part of a consulting firm.

MOTIVATING PROJECT PERFORMANCE

Team Building

Thamhain introduces his excellent chapter on "Team Building" by highlighting its importance to project/program management:

> Building the project team is one of the prime responsibilities of the project or program manager. Team building involves a whole spectrum of management skills required to identify, commit, and integrate various task groups from traditional functional organizations into a single program management system. This process has been known for centuries. However, it becomes more complex and requires more specialized management skills as bureaucratic hierarchies decline and horizontally oriented teams and work units evolve.

The newly formed team begins with considerable lack of clarity about purposes, responsibilities, expectations, and a general lack of communication, commitment, and team spirit. Thamhain suggests specific measures for preventing such problems from developing in the first place:

1. The importance of the project to the organization, including its principal goals and objectives, should be [made] clear to all personnel who get involved with the project....

2. Project leadership positions should be carefully defined and staffed at the beginning of the team formation stage. [...] The capabilities, interests, and commitments to the project should be assured before any of the lead personnel are signed up. One-on-one interviews are recommended for explaining the scope and project requirements, as well as the management philosophy, organizational structure, and rewards.

3. Members of the newly formed team should be closely located to facilitate communications and the development of a team spirit. Locating the project team in one office area is the ideal situation. However, this may be impractical, especially if team members share their time with assignments on other projects or the assignment is only for a short period of time. Regularly scheduled meetings are recommended as soon as the new project team is being formed. These meetings are particularly important where team members are geographically separated and do not see each other on a day-to-day basis.

4. All project assignments should be negotiated individually with each prospective team member. [...] The assignment interview should include a clear discussion of the specific task, the outcome, timing, responsibilities, reporting relation, potential rewards, and importance of the project to the company.

5. Management must define the basic team structure and operating concepts during the project formation stage....

6. The project manager should involve at least all key personnel in the project definition and requirements analysis....

7. The project manager should conduct team-building sessions throughout the project life cycle. An especially intense effort might be needed during the team formation stage....

8. Project leaders should try to determine lack of team member commitment early in the life of the project and attempt to change possible negative views toward the project.... Finally, if a team member's professional interests lie elsewhere, the project leader should examine ways to satisfy part of the team member's interests or consider replacement.

9. It is critical for senior management to provide the proper environment for the project to function effectively....

10. Project managers must understand the various barriers to team development and build a work environment conducive to the team's motivational needs. Specifically, management should watch out for the following barriers: (1) disinterested team members, (2) uninvolved management, (3) unclear goals and priorities, (4) funding uncertainty, (5) role conflict and power struggle, (6) incompetent project leadership, (7) lack of project charter, (8) insufficient planning and project definition, (9) poor communication, and (10) excessive conflict, especially personal conflict.

11. Project leaders should watch for changes in performance on an ongoing basis. If performance problems are observed, they should be dealt with quickly by the team [with the help of organizational development specialists if available].

Managing Conflict

Sources of Conflict.　　Conflict is inevitable in any organization, just as it is in any relationship. Conflict that is due to pettiness, lack of understanding of the other person or other group, or intolerance should be avoided. However, much conflict in organizations is natural, and stems from honest disagreement on priorities and the use of scarce resources. Current emphasis on *conflict management* looks for ways to resolve such conflict positively to the overall benefit of the organization.

Thamhain describes seven potential conflict sources, listed here in order of decreasing perceived intensity over the project life cycle:

1. *Conflict over schedules*
2. *Conflict over project priorities*, including conflict over the sequencing of activities and tasks to be undertaken
3. *Conflict over work force resources*, especially in obtaining the desired quality and quantity of personnel from other functional and staff support areas
4. *Conflict over technical opinions and performance trade-offs*
5. *Conflict over administrative procedures* that define how the project will be managed, especially the project manager's reporting relationships, responsibilities, and authority
6. *Personality conflict*
7. *Conflict over cost* and the funds allocated to functional support groups

The relative importance of these seven conflict sources varies over the project life cycle. As the project progresses into the early and main project phases, the overall level of conflict increases, first with staffing concerns and then with priorities, schedules, costs, and the technical issues that surface as subsystems are developed. Toward the end of the project, the overall level of conflict is reduced, and the most likely forms of conflict relate to schedule slippages that delay project completion and project costs.

Methods of Conflict Management.　　One obvious method of resolving conflict between two individuals is to appeal the matter up the chain of command to the level having authority over both individuals and in some cases this will be necessary. However, higher executives simply do not have the time to solve everyone's squabbles, and they expect their subordinate managers and professionals to be mature enough to solve their own problems most of the time. Blake and Mouton identify five methods for dealing with conflict:

1. *Withdrawal*, or retreat from actual or potential conflict
2. *Forcing* one's viewpoint at the potential expense of the other party
3. *Smoothing*, or emphasizing the points of agreement and deemphasizing areas of conflict
4. *Compromising* or negotiating, in which each party must give up something, but each walks away partly satisfied
5. *Confronting* or *problem solving*, in which the parties focus on the issues, consider alternatives, and look for the best overall solution

No single method will work in every situation and it is common for project managers to favor confrontation/problem solving with the goal of moving the project forward in the most advantageous way possible. That said, this type of constructive conflict is not always possible and there will be times that compromise or one of the other approaches will be needed for project success.

Keys to Project Success

Baker, Murphy, and Fisher deduce the following definition of success from research conducted on 650 projects:

> If a project meets the technical performance specifications and/or mission to be performed, and if there is a high level of satisfaction concerning the project outcome among: key people in the parent organization [in which the project is carried out], key people in the client organization, key people on the project team, and key users or clientele of the project effort, the project is considered an overall success.

Characteristics that strongly affect perceived failure of projects were found to include the following:

- Inadequate project manager skills, influence, and authority
- Poor coordination and rapport with the client
- Poor coordination and rapport with the parent organization
- Lack of project team participation and team spirit
- Poor project control: inability to freeze design or close out the project, unrealistic schedules, inadequate change procedures, and/or inadequate status/progress reports
- Project of different type or more complex than handled previously and/or initially underfunded
- Poor relations with public officials or unfavorable public opinion

Characteristics associated with project success emphasized the *commitment* of the parent organization, project manager, and client to established schedules, budget, and technical performance. Frequent feedback from the parent organization and the client, adequate control procedures (especially change control), public support, and lack of excessive government red tape were also among these conditions deemed necessary, but not sufficient, for perceived success.

Baker et al. also made the following observations:

- Cost and schedule overruns were not among the characteristics significantly related to perceived failure, and meeting these targets was not significantly related to perceived success of past projects [but they may certainly affect the survival of managers *during* a project]. In the long run, what really matters is whether the parties associated with, and affected by, a project are satisfied.
- While judicious use of PERT/CPM systems contributes to better cost and schedule performance, their importance is far outweighed by the use of system management tools such as "work breakdown structures, life cycle planning, systems engineering, configuration management, and status reports."
- While each organizational structure discussed at the beginning of this chapter was associated with perceived success in certain situations, the projectized organization structure was most often so associated. This is consistent with the findings of Larson and Gobeli discussed earlier in this chapter.
- The client and parent organizations need to agree on definite goals for the project and to develop close and supportive relationships with the project team, but then need to avoid meddling or interfering with the project team's decision-making process.

- Participative decision making within the project team is highly correlated with success, but public participation in projects affecting the public interest often delays and hampers projects and reduces the probability of success.
- The most important skills in the project manager are technical skills, followed by human and then administrative skills.

Katz and Allen have found that project performance reaches its highest level "when organizational influence is centered in the project manager and influence over technical details of the work is centered in the functional manager." In particular, they reported higher performance "when influence over salaries and promotions is perceived as balanced between project and functional managers" rather than being concentrated in the functional manager.

Customer Communications

This research of Baker et al. makes clear the need for good coordination and rapport with the customer, and this requires good communication. Cleland and Kerzner point out that this communication takes place at three levels, as shown in Figure 15-5. Formal communications are normally between the project manager responsible for project performance in the contractor organization and the project manager or other person designated to monitor project performance in the customer organization. Usually, a separate contracts officer in one or both organizations will be involved when actual changes to the legal contract are discussed.

Often there is a senior management level of informal contact as well, providing checks and balances to assure that correct information has been transferred between contractor and customer and a channel to resolve any conflicts generated at the project manager level. This makes effective vertical communication essential as well, so that executives on each side are apprised of project progress and problems, and the executives themselves pass on to their subordinate project managers any insight or agreements resulting from higher-level communications.

A final, informal level of communication in any large project is between employees of the customer and contractor employee specialists. The customer will often demand this, feeling that essential information is filtered out in the summary that comes from the project manager. The danger of this channel is the temptation at the specialist level to propose and agree to "improvements" that change the scope of the

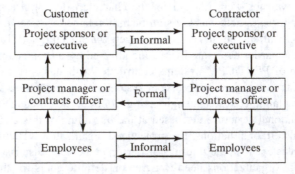

Figure 15-5   Customer (client) communications. (From David I. Cleland and Harold Kerzner, *Engineering Team Management*, Van Nostrand Reinhold Company, Inc., New York, 1986, pp. 63–64.)

project and require additional resources. Engineers and other specialists on both sides must learn, that while they can agree that a change might be "nice to have," no work can be done on such a change until it has been negotiated through formal channels.

Early Learning Path

Balderston et al. emphasize, in their "Strategic Criteria for R&D Project Success," "that, in any development process, *the key management task is strategic creation of an early learning path*." We discussed the phases of the engineering design process at the beginning of Chapter 10, emphasizing the need to reduce uncertainty in the early phases of the process, so that the later, more expensive phases are attempted only when as much early uncertainty as possible has been eliminated. Balderston et al. support this with their focus on the "early" learning path, as shown in Figure 15-6a, in which project activities have been scheduled to gain maximum learning in the early period when cumulative project expenditure (Figure 15-6b) is still low.

These authors urge the project manager to focus on the specific technological approach used in each key area or subsystem:

- Where the *knowledge level* regarding the technological approach is low (uncertainty is high), the risk of project schedule and cost overruns due to the new approach is high. In this situation, they suggest adopting a contingency strategy by bringing along a more conventional, older strategy as a parallel development, with a system design that would accept either. This might also be a time to consider Agile Project Management, as discussed in Chapter 14.
- Where the *analytic potential* of the approach is such that its performance cannot confidently be predicted by analysis, trial-and-error test and evaluation processes should be scheduled early.
- Where the *experience level* of the project staff with the technological approach is low, if the desired experience exists outside the organization, its use should be considered and/or prototypes involving the proposed approach should be developed and tested early.

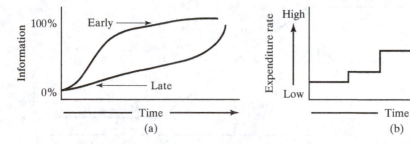

Figure 15-6 (a) Learning paths; (b) expenditure rate versus time. (From Jack Balderston, Philip Birnbaum, Richard Goodman, and Michael Stahl, *Modern Management Techniques in Engineering and R&D*, Van Nostrand Reinhold Company, Inc., New York, 1984, pp. 135–149.)

TYPES OF CONTRACTS

Contracts can be classified by the manner in which contract costs are borne. Two broad classes are fixed-price contracts and cost contracts; each has a number of variations. We discuss contract types in approximate order of decreasing contractor risk and increasing buyer risk.

Fixed-Price Contracts

Firm fixed-price contracts require the contractor to provide an agreed product and/or service for a specified price. The contractor realizes the total profit or loss and thus has maximum motivation to eliminate waste. The buyer must assure that the specifications define clearly what has been purchased for the fixed price.

   Fixed-price with escalation contracts are firm fixed-price, except that the contractor's price can be adjusted to incorporate any increases in specified labor and material rates. This is especially useful in long-term construction work where wage and material rates are susceptible to changes in the market.

   Fixed-price, redeterminable contracts can be adjusted later to reflect actual costs. The contractor therefore has less motivation to keep costs to a minimum before that adjustment is made.

   Fixed-price incentive contracts provide that the contractor and buyer share savings within a certain range, but establish a maximum cost for the buyer above which the *contractor* bears the total risk.

Cost Type Contracts

Cost plus incentive fee contracts establish an estimated target cost and target fee (profit). Within a specified range about the target, contractor and buyer share added costs or savings in an agreed ratio, but outside that range the *buyer* bears the total risk. In Figure 15-7, for example, the target (estimated) cost was $100 million, the target fee 8 percent or $8 million, and the customer/contractor sharing ratio 70:30 within a variation of ± 20 percent of target cost. If the work actually cost $80 million or less, the contractor would

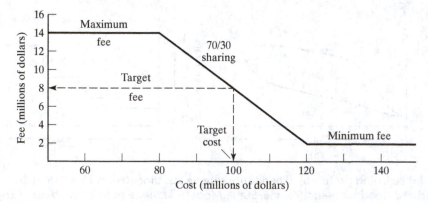

Figure 15-7   Example of cost plus incentive fee contract.

gain 30 percent of the first $20 million saved, increasing his fee to $14 million on less work. If the cost overran to $150 million, the contractor would only have to share 30 percent of the first $20 million overrun and would still receive a fee of $2 million regardless of cost.

Cost plus fixed-fee contracts require the buyer to pay all costs, plus an agreed-upon fee. In this example, if the initial estimate of costs were $100 million and the fee rate agreed were 8 percent, as in the preceding example, the contractor would receive an $8 million fee for services regardless of cost overruns. This arrangement is not uncommon in R&D contracts, where the scope of work is difficult to estimate in advance. Obviously, the contractor's only motivation for reducing waste is the effect of costs on future contracts. Indeed, there is every motivation to load onto the contract new employees for training or improper overhead charges; both contractor and customer have a considerable accounting burden regarding allowable costs.

Time and materials contracts are common in repair, maintenance, and emergency situations, and they involve payment at agreed-upon rates (high enough to include profit) for hours worked, plus reimbursement of invoices for actual materials used. The only motivation for efficiency is the hope of future work; the advantage is its speed and economy for small jobs not worth detailed estimating.

A **letter contract** is a preliminary contract authorizing the contractor to proceed with specified work at customer cost in the interim until a formal contract is negotiated.

DISCUSSION QUESTIONS

15-1. The four categories or "elements of the project-driven organization" were written largely for projects in which systems were designed and manufactured. How would the categories and their contents differ for an architecture/engineering (A/E) firm whose projects involved the design of structures and management of their construction?

15-2. What types of projects are managed most effectively by using the (fully or partially) projectized organization? What types are managed most effectively as part of the normal functional organization?

15-3. Discuss the ways in which the matrix management structure tends to reduce the problems of project management under (**a**) a fully projectized organization and (**b**) a normal functional organization. (**c**) What are the disadvantages of going to the matrix structure?

15-4. Project managers have a variety of professional certifications available to them through the Project Management Institute. Research these certifications and identify any that might be valuable to you at this stage in your career. What is this value?

15-5. In the text we discuss the technical, administrative, and interpersonal skills desirable in a project manager. What attributes do you believe are desirable in an engineering specialist working on a project in a matrix organization?

15-6. If you were a project manager, which three or four responsibility areas of the nine listed under "The Project Manager's Charter" (or others you think important) would you *most* want to have included in a charter granted to you by management? State your reasons.

15-7. Discuss the precautions you might take as manager of a new project in building a project team to assure that the project does not get off "on the wrong foot."

15-8. Explain why informal communication channels are important in matrix organizations.

15-9. Describe what types of products or services each of the following contracts would be appropriate for, and give an example for each: (**a**) firm fixed-price, (**b**) cost plus fixed-fee, and (**c**) time and materials.

SOURCES

Allen, Thomas J., *Managing the Flow of Technology: Technology Transfer and the Dissemination of Technological Information Within the R&D Organization* (Cambridge, MA: The MIT Press, 1977), pp. 110–113.

Baker, Bruce N., Murphy, David C., and Fisher, Dalmar, "Factors Affecting Project Success," in David I. Cleland and William R. King, eds., *Project Management Handbook* (New York: Van Nostrand Reinhold Company, Inc., 1983), Chapter 33, pp. 669–685.

Balderston, Jack, Birnbaum, Philip, Goodman, Richard, and Stahl, Michael, *Modern Management Techniques in Engineering and R&D* (New York: Van Nostrand Reinhold Company, Inc., 1984), pp. 135–149.

Blake, Robert R. and Mouton, Jane S., *The Managerial Grid III: The Key to Leadership Excellence* (Houston, TX: Gulf Publishing Company, Book Division, 1985).

Cleland, David I. and Kerzner, Harold, *Engineering Team Management* (New York: Van Nostrand Reinhold Company, Inc., 1986), pp. 63–64.

Cleland, David and King, William R., *Systems Analysis and Project Management*, 3rd ed. (New York: McGraw-Hill Book Company, 1983), pp. 337–341.

Katz, Ralph and Allen, Thomas J., "Project Performance and the Locus of Influence in the R&D Matrix," *Academy of Management Journal*, 28:1, 1985, pp. 67–87.

Kerzner, Harold, *Project Management: A Systems Approach to Planning, Scheduling, and Controlling*, 12th ed. (Hoboken, NJ: John Wiley and Sons, Inc., 2017), p. 22.

Larson, Erik W. and Gobeli, David H., "Significance of Project Management Structure on Development Success," *IEEE Transactions on Engineering Management*, May 1989, p. 124.

Lawrence, P. R. and Lorsch, J. W., *Organization and Environment* (Boston: Harvard Business School, Division of Research, 1967).

Newman, William H., Warren, E. Kirby, and McGill, Andrew R., *The Process of Management: Strategy, Action, Results*, 6th ed. (Englewood Cliffs, NJ: Prentice Hall, Inc., 1987), p. 268.

Thamhain, Hans J., *Engineering Program Management* (New York: John Wiley & Sons, Inc., 1984), p. 178. Reprinted by permission.

Thamhain, Hans and Wilemon, David L., "The Effective Management of Conflict in Project-Oriented Work Environments," *Defense Management Journal*, July 1975, pp. 29–40.

Young, Edmund J., "Project Organisation," in Dennis Lock, ed., *Project Management Handbook* (Aldershot, Hampshire, England: Gower Publishing Co. Ltd., 1987), p. 27.

Part V

Managing Your Engineering Career

16

Engineering Ethics

PREVIEW

Part V of the text, Managing Your Engineering Career, includes a series of topics important to engineers and engineering managers. The first chapter in this section considers engineering ethics. The remaining chapters examine time management, diversity within the engineering profession, and global opportunities. This chapter is devoted to professional ethics and conduct, beginning with some definitions and a look at engineering codes of ethics. Ethical problems that are faced by the engineer in industrial employment, consulting, and contracting are presented.

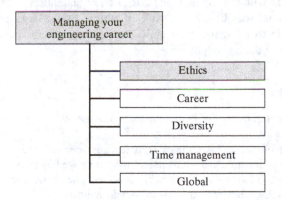

LEARNING OBJECTIVES

When you have finished studying this chapter, you should be able to do the following:

- Explain the importance of ethics in engineering.
- Describe what is meant by whistle-blowing.

- Describe the need for ethics in design and construction.
- Describe situations where conflict of interest may arise.
- Apply guidelines for facilitating solutions to ethical dilemmas.

PROFESSIONAL ETHICS AND CONDUCT

The emphasis of this chapter is on professional ethics, not personal ethics. However, it is hard to separate the two, since personal ethics are the foundation for professional ethics. Ethics is about behavior, doing the right thing in the face of dilemma. Ethical people take the "right" and "good" path when they come to decisions.

Our values ultimately drive our behavior, since values exert influence over our attitudes, and attitudes influence our behavior. Values are integral to attitude formation and to how we respond to people and situations. No one person is quite like any other person, but a handful of core values and beliefs do underlie and permeate cultures. These values and beliefs do not apply across the board in every situation, but they are at the heart of our culture.

Established behavioral standards and written codes of ethical conduct can help bolster virtuous values and promote ethical organizational behavior. Behavioral standards usually incorporate specific guidelines for acting within specific functional workplace areas. The greatest difference between personal ethics and professional ethics has to do with the ethical standards of an organization or the engineering community. Most engineering professional societies have a code of ethics. As the engineering community becomes more global, it is important to recognize how ethical norms can change based on culture. Chapter 18 includes greater discussion of these changes.

Some Definitions

Koestenbaum describes the importance of being ethical in business:

> Little else is as distinctively human as our ethical conscience and our moral sense. To be ethical means to live by the stern demands of reason and not to be governed or swayed by the seduction of the emotions. To be ethical is to be just, consistent, and predictable. It is preeminently our capacity to act ethically and our possession of a moral sense, which set us aside from the animals.

Gluck distinguishes between morality and ethics as follows:

> *Morality* is concerned with *conduct* and *motives*, right and wrong, and good and bad *character*. *Ethics* is the philosophical study of morality; it is *moral philosophy*. When we exercise moral philosophy—that is, when we practice the philosopher's craft—we are subjecting the questions *of* morality to other critical and analytical questions *about* morality.

> Morals is a set of rules of conduct and standards of evaluation that a culture uses to guide its individual and collective behavior and direct its judgments. Codes of professional moral conduct (codes of ethics) are specialized subsets of these rules and standards.

The study of *moral philosophy* used to be a formal part of every liberal education, preparing the graduate for ethical analysis of future problems occurring in his or her life and career. This study is less common in the crowded curricula of the twenty-first century, especially in engineering programs.

Ethicists, after lifetimes of thought, are unable to agree on a simple definition of ethics. Instead, their definitions fall into several categories:

1. **Utilitarian** ethics is a goal-based approach in which we seek to obey those rules or choose those acts that will result in the greatest good for the greatest number of people. This involves value judgments in the weighting of different "goods," and raises the question of the rights of those with less voice in the decision.
2. **Ethical egoism** or **Individualism** ethics is a goal-based theory of "rational" self-interest. Under this theory, individuals make decisions to promote their long-term self-interests. Adam Smith developed the theory that, if everyone acts in his own self-interest, the "invisible hand" of the marketplace will transform this into social good.
3. **Duty-based, deontological, normative, or Kantian** ethics asserts that there are moral imperatives that we must obey, regardless of the consequences. This view stems from the *categorical imperatives* of the German philosopher Immanuel Kant, who believed that to steal, lie, or break promises is universally immoral, regardless of the consequences.
4. **Rights-based** ethical theories are based on the belief that there are certain fundamental human rights, and that moral obligations arise in the context of these rights. The English philosopher John Locke (1632–1704) believed that these rights included life, liberty, and property; his writings inspired the framers of the Declaration of Independence of the United States to declare it "self-evident" that all men are endowed with "certain inalienable rights," including those of "life, liberty, and the pursuit of happiness."
5. **Environmental** ethics broadens the moral community to whom we owe ethical responsibility to include animals, plants, and even inanimate objects. Environmental ethics may be either goal based (utilitarian) or duty based (deontological). They should be of particular interest to civil and chemical engineers.

No single one of these views provides us with a simple, reliable guide to resolving the ethical problems we are sure to encounter in our lives and careers. In the end, we must fall back on our personal set of *values* to guide our final decisions. Table 16-1 provides an extensive "compilation of values," divided into four categories: individual, professional, societal, and human values.

How our values influence our ethical decision making can be represented in the three levels of Kohlberg's moral development. As shown in Figure 16-1, this framework explains moral development as happening over six stages across three levels. These levels begin with the pre-conventional level of self-centered behavior, progress to the conventional level of social-centered behavior, and culminate in the post conventional level of principle-centered behavior.

Table 16-1   Compilation of Values

Four categories of values are highlighted in the following listing: individual, professional, societal, and human. The values in any category are not mutually exclusive, nor is the listing complete.

Individual Values	Civic consciousness	Prestige	Privacy
Curiosity	Collegiality	Pride in work	Progress
Endurance	Communication	Problem-solving ability	Public service
Family	Compassion	Professionalism	Social justice
Flexibility	Competence	Prudence	Societal harmony
Friendship	Conformity	Rationality	Survival of society
Hard work	Conscientiousness	Realism	Tradition
Honor	Cooperation	Recognition of	
Independence	Courtesy	accomplishments	**Human Values**
Initiative	Creativity	Self-education	Autonomy
Intellectual stimulation	Curiosity	Selflessness	Beauty
Intelligence	Decisiveness	Service to others	Beneficence
Leisure	Devotion to principle	Tolerance	Bravery
Optimism	Duty	Trustworthiness	Fairness
Personal liberty	Economy		Faith
Personal morality	Effectiveness	**Societal Values**	Freedom
Personal power	Efficiency	Capitalism	Friendship
Personal security	Fair play	Centralization	Happiness
Privacy	Flexibility	Change	Health
Property	Forthrightness	Competition	Hope
Quality of life	Freedom of inquiry	Culture	Human dignity
Self-advancement	Honesty	Democracy	Humility
Self-control	Idealism	Education	Idealism
Self-fulfillment	Imagination	Equality	Justice
Self-reliance	Informedness	Equality of opportunity	Love
Self-respect	Initiative	Freedom of religion	Morality
Self-worth	Innovativeness	Freedom of thought	Pleasure
Strength	Integrity	Governance by law	Prevention of evil
Success	Leadership	Improvement of society	Progress
Wealth	Literacy	Individual rights	Promotion of goodness
Wit	Loyalty	Individualism	Prudence
	Obedience	Liberty	Reason
Professional Values	Openness	National pride	Reverence for life
Ability to analyze	Patience	National prosperity	Self-sacrifice
Ability to synthesize	Perseverance	Order	Truth

Source: Heinz C. Luegenbiehl and Don L. Dekker, "The Role of Values in the Teaching of Design," *Engineering Education*, January 1987, p. 245.

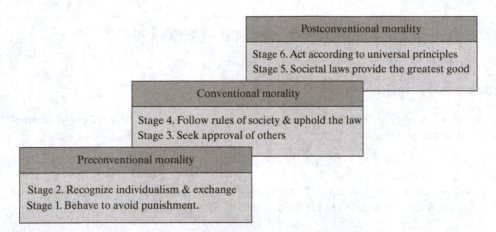

Figure 16-1   The three stages of Kohlberg's moral development.

ENGINEERING CODES OF ETHICS

The profession's first code of ethics was written by Isham Randolph and adopted by the American Association of Engineers in 1918. Soon thereafter, five engineering societies collaborated in preparing a uniform code for engineers, but the American Society of Mechanical Engineers was the only one to adopt it (in 1922). Oldenquist and Slowter trace the history of more recent attempts to achieve a universal code:

> By 1947, ECPD [the Engineers' Council for Professional Development] developed Canons of Ethics which eventually had acceptance by a significant number, but not a preponderance, of engineering societies; these canons were updated in 1963, and again some acceptance was obtained; more recently, a major effort by ECPD produced a Code of Ethics for Engineers in 1974 which contained three levels of specificity: Fundamental Principles; Fundamental Canons; and Suggested Guidelines.

> Although a number of societies participated in the preparation of this three-level code and it seemed to offer opportunity for universal acceptance of at least the Fundamental Principals, it has not secured the support of a majority of the professional societies, and the goal of a universal code continues to elude the profession.

Center for Engineering Ethics and Society (CEES)

The Center for Engineering Ethics and Society (CEES) focuses the talents of the nation on addressing the ethical and social dimensions of engineering, as both a profession and an agent of innovation. It achieves these objectives by 1) addressing ethically significant issues that arise in engineering and scientific research, education, and practice; 2) sharing knowledge through the Online Ethics Center, organizing workshops, and publishing informative materials; and 3) including diverse perspectives in ethics discussions to promote better understanding of ethical challenges.

Source: https://www.nae.edu/Activities/Projects/CEES.aspx, September, 2018.

The American Society for Engineering Management Code of Ethics

Adopted 10/17/2012

Fundamental Principles

The American Society for Engineering Management (ASEM) requires its members to conduct themselves in an ethical manner by:

- Being honest and fair in serving the stakeholders in the Society including members, other organizations, clients, and the public
- Striving to increase the competence and prestige of the engineering management profession
- Using their knowledge and skill for the protection of the public health, safety, and welfare

Fundamental Canons

Members of ASEM are required to observe the following canons:

1. Hold paramount the safety, health, and welfare of the public
2. Undertake assignments in the profession of engineering management only when qualified by education or experience
3. Maintain professional competence, strive to advance the engineering management body of knowledge, and provide opportunities for the professional development of fellow educators and practitioners in the profession
4. Act in a professional manner towards colleagues, ASEM staff, customers and clients of ASEM, and the public
5. Give proper credit for the work of others and accept and give honest and fair professional criticism
6. Abide by the Constitution, By-Laws, and policies of ASEM
7. Disclose to the ASEM Executive Committee any alleged violation by a member of this Code of Ethics in a prompt, complete, and truthful manner

Source: www.asem.org, April 2019.

Oldenquist and Slowter have analyzed numerous codes, and they identify 20 basic concepts that pervade them, divided into three groups: (1) the public interest; (2) truth, honesty, and fairness; and (3) professional performance (see Table 16-2). They add that "mixed with these ethical principles [in the existing codes] are a number of rules and customs concerning business practices and political conventions which, at least in the eyes of much of the public, seem more designed to protect the interests of engineers than to serve the general good." A prime example was the provision of the former Section 11(c) of the National Society of Professional Engineers (NSPE) code, which held that the engineer "shall not solicit

Table 16-2 Core Concepts in Engineering Ethics

I. The public interest
   A. Paramount responsibility to the public health, safety, and welfare, including that of future generations
   B. Call attention to threats to the public health, safety, and welfare, and act to eliminate them
   C. Work through professional societies to encourage and support engineers who follow these concepts
   D. Apply knowledge, skill, and imagination to enhance human welfare and the quality of life for all
   E. Work only with those who follow these concepts
II. Qualities of truth, honesty, and fairness
   A. Be honest and impartial
   B. Advise employer, client, or public of all consequences of work
   C. Maintain confidences; act as faithful agent or trustee
   D. Avoid conflicts of interest
   E. Give fair and equitable treatment to all others
   F. Base decisions and actions on merit, competence, and knowledge, and without bias because of race, religion, sex, age, or national origin
   G. Neither pay nor accept bribes, gifts, or gratuities
   H. Be objective and truthful in discussions, reports, and actions
III. Professional performance
   A. Competence for work undertaken
   B. Strive to improve competence and assist others in so doing
   C. Extend public and professional knowledge of technical projects and their results
   D. Accept responsibility for actions and give appropriate credit to others

Source: Andrew G. Oldenquist and Edward E. Slowter, "Proposed: A Single Code of Ethics for All Engineers," *Professional Engineer*, May 1979, p. 9.

or submit engineering proposals on the basis of competitive bidding...defined as the formal or informal submission, or receipt, of verbal or written estimates or cost...whereby the prospective client may compare engineering services on a price basis" before selecting one engineer or engineering organization for negotiations. This section was deleted by order of the U.S. District Court, as affirmed by the U.S. Supreme Court, on April 25, 1978. At the same time, provisions of Section 3(a) limiting advertising by engineers to the identification of name, address, telephone number, and fields of practice were replaced by statements emphasizing the ethical consideration of avoiding misrepresentation of fact.

CORPORATE CODES OF ETHICS

Just as professions have codes of ethics, a growing number of companies have their own codes of ethics, with Kaptein reporting that over 50 percent of the Fortune 200 had one in place in 2004. These codes, also commonly known as codes of conduct or corporate values, are voluntary commitments by these organizations to guide the behavior of their employees by formally stating the values and ethical principles they are expected to uphold.

According to Ryan, the development of corporate codes of ethics began in the 1960s and 1970s largely in response to corporate scandals. Regrettably, the results of these efforts did not appear overly beneficial during the first two decades of their growth, as companies with codes of ethics are actually cited by federal agencies *more often* than are those that lack such standards. In 1987 Marilynn Mathews found in a survey of ethical codes of 202 Fortune 500 companies that more than three-fourths of these codes stressed relations with the U.S. government (87 percent), customer/supplier relations (86 percent), political contributions (85 percent), conflicts of interest (75 percent), and honest records (75 percent). On the other hand, more than three-fourths *did not* mention personal character matters (94 percent), product safety (91 percent), environmental affairs (87 percent), product quality (79 percent), or civic and community affairs (75 percent).

More recently, Kaptein found that corporate codes include a variety of elements more in line with the things that were missing during the earlier study. These typically include the organization's mission, core values, responsibilities toward key stakeholders such as customers and community, and detailed norms of behavior. However, given the continued scandals of organizations with these codes in place, such as Volkswagen and Wells Fargo, it is not yet clear the codes are having the desired effects.

ETHICAL PROBLEMS IN CONSULTING AND CONSTRUCTION

Significance.   Engineering codes of ethics have traditionally been geared toward the consulting engineer in private practice, especially the civil engineer. Much of the literature on ethical problems stems from the NSPE, whose membership is weighted heavily toward civil engineers and private consultants, since these are the engineers most motivated to become registered professional engineers (PEs). The engineer in private practice and in public works management has a special set of ethical problems, some of which are discussed next.

Political Contributions.   Private consultants must rely on public and private clients for work. In an article entitled "I Gave Up Ethics—To Eat," an anonymous engineer who went into consulting in the public works field after a good education and "a long apprenticeship with an old, established engineering firm" describes how he went two full years without work until he learned about "political engineering." He was introduced to "The Reverend," who "was in pretty good with the public works department and other state agencies." On a 10 percent commission of his fees to The Reverend, and through a practice of large campaign contributions (to both parties, just in case), our engineer began to prosper. The editors of the compendium in which this article appeared conclude the following: "His story may not be true for all branches of private practice nor in all parts of the country, but his story is authentic for his field of practice and also his state. We checked it."

Distribution of Public Services.   Price describes at some length the 1971 case of *Andrew Hawkins* v. *Town of Shaw*, Mississippi. Shaw was a segregated town of 1,500 black inhabitants (including Hawkins) and 1,000 whites. 98 percent of the homes fronting on unpaved streets and 97 percent of those not served by sanitary sewers were occupied by blacks. White neighborhoods had mercury vapor streetlights and six-inch water mains; black neighborhoods had bare bulbs and two-inch mains. The Fifth Circuit Court of Appeals found that this violated the equal protection clause of the 14th Amendment (the same clause that was applied to segregated schools in the famous *Brown* v. *Board of Education* case in 1954). Public works improvements in this case were paid for from the city's general tax funds; improvements reimbursed by special assessments on the property owners who benefited would not be subject to this equal protection clause. This case is just one example of the political environment in which the city engineer or public works director may work.

Guidelines for Facilitating Solutions to Ethical Dilemmas in Professional Practice

Step 1: Determine the facts in the situation. Obtain all of the unbiased facts possible.

Step 2: Define the stakeholders—those with a vested interest in the outcome.

Step 3: Assess the motivations of the stakeholders by using effective communication techniques and personality assessment.

Step 4: Formulate alternative solutions based on most complete information available, using basic ethical core values as a guide.

Step 5: Evaluate proposed alternatives—shortlist ethical solutions only; may be a potential choice between or among two or more totally ethical solutions.

Step 6: Seek additional assistance, as appropriate—engineering codes of ethics, previous cases, peers, reliance on personal experience, even prayer if desired.

Step 7: Select the best course of action—that which satisfies the highest core ethical values.

Step 8: Implement the selected solution. Take action as warranted.

Step 9: Monitor and assess the outcome. Note how to improve the next time.

Source: http://www.depts.ttu.edu/murdoughcenter/products/resources/guidelines-for-facilitating-solutions.php, May 2019.

Nine Basic Steps to Personal Ethical Decision Making

Step 1: Practice ethical behavior actively (initiate a personal ethical awareness training program), including definition of personal worldview and review of core ethical values.

The ethical design professional is consistently ethical!

Step 2: Beware of "new ethics" programs. Very little of true value is "new"; all of the necessary tools are already at your fingertips.

Step 3: Define the ethical problem when it arises. Ignoring the problem doesn't make it go away.

Step 4: Formulate alternatives. Avoid "first impulse" solutions without having extensive ethical awareness training and experience.

Step 5: Evaluate the alternatives. Are they ethical? Am I the sole beneficiary? How would I feel if the roles or circumstances were reversed?

Step 6: Seek additional assistance, as appropriate—previous cases, peers, reliance on personal experience, prayer.

Step 7: Choose best ethical alternative—the one that does the most good for all the right reasons.

Step 8: Implement the best alternative—no initiative, no results.

Step 9: Monitor and assess the outcome—how to improve the next time.

Source: http://www.depts.ttu.edu/murdoughcenter/products/resources/steps-to-personal-ethical-decision-making.php, May 2019.

Construction Safety.   If you seek maximum safety, some suggest you work for a nuclear plant or an explosives manufacturer, for in these occupations safety engineers examine hazards exhaustively and large sums are spent to mitigate them. Avoid unskilled construction jobs such as digging trenches! At least 100 fatal accidents occur in the United States each year from trench cave-ins, most of which could be prevented by prudent engineering and construction supervision. Large contractors, driven by liability insurance costs, OSHA (Occupational Safety and Health Administration) regulations, and the desire to avoid bad publicity, have few such accidents; small contractors are tempted to cut corners and ignore regulations and have a disproportionately greater percentage of accidents. The ethical implication is clear, not only for the contractor's engineer, but also for the client's engineer, who has an ethical obligation "to hold paramount the safety" of the public and to specify and insist that construction be carried out in a safe manner.

Construction Case Study: The Hyatt Regency Disaster

Engineers in construction bear ethical obligations to the public that uses the structures they design. These are at least as important as the obligations to workers who build these structures. The Hyatt Regency Hotel in Kansas City, Missouri, was designed with a high-roofed atrium over the lobby area, crossed by three pedestrian bridges, or "skywalks," of which the one at the fourth-floor level was directly above the one at the second-floor level. These two skywalks, each about 8 feet (2.4 meters) wide and 120 feet (36.6 meters) long, were suspended from the ceiling by three pairs of steel rods about 30 feet (9.1 meters) apart.

On July 17, 1981, not too long after the hotel opened, a tea dance was held, and the skywalks as well as the lobby floor were crowded with people. A structural support for the fourth-floor skywalk gave way, causing it to crash into the lobby. It took the second-floor walk with it, killing 114 people and injuring more than 200.

The original design of the hotel specified that the support rods for these skywalks be made from 1.75-inch (4.44-cm) diameter high-strength steel; the actual construction used 1.25-inch (3.18-cm) diameter mild (A36) steel rods. Moreover, the six support rods were intended to be continuous from the ceiling to the second floor, threaded so that separate nuts would support the two skywalks; in a field change, the rods were terminated at the fourth floor, with additional rods slightly offset beginning there and going down to the second floor, adding some torsional stress.

In the original design, the nuts on the fourth-floor walkway had to support only the weight of the fourth-floor walkway itself and were sized according to that requirement. In the revised design, however, the fourth-floor nuts were required to support both the fourth-floor walkway and the second-floor walkway hanging from it. This change in requirement was not noticed, and so the same nuts as in the original design were used—now holding up twice the weight they should have been. When the walkways became heavily loaded, the nuts and washers on the fourth-floor walkway pulled through the walkway's support beams, and both walkways collapsed.

A structural engineering consulting firm, GCE International Inc., was retained by the architects and "sealed" the drawings; no one at GCE did a detailed analysis of the support of the skywalks, despite their promise to do so, maintaining instead that this was the customary responsibility of the steel fabricators providing the actual steel components.

In February 1984 the Missouri Board for Architects, Professional Engineers, and Professional Land Surveyors and Landscape Architects (Missouri's licensing agency) filed a complaint against Daniel M.

Duncan and Jack D. Gillum, principals in GCE, and a 26-day hearing was held before Administrative Hearing Commissioner James B. Deutsch. In a carefully written decision, Judge Deutsch supported the complaint, and the Missouri Board stripped Duncan and Gillum of their professional engineer licenses, and GCE of its engineering license. The action was upheld in appeals to the City of St. Louis Circuit Court in December 1986 and the Missouri Court of Appeals on January 26, 1988. In discussing the case later before a civil engineering class at the University of Missouri–Rolla (now the Missouri University of Science and Technology), Judge Deutsch identified three sources of professional responsibility:

1. The statutory responsibility "to protect the public interest" and the regulatory law amplifying this
2. Contractual responsibility (in this case, to perform all structural engineering and analysis on the design)
3. The customs and practices of the engineering profession, including the common-law requirement to act as a "reasonable man" would and its modern extension to act as a reasonably skilled and prudent professional should

In the Hyatt affair, Deutsch told the class, none of these three were observed. The engineer who imprinted his PE seal on the applicable drawing did not design it, supervise it, or even check it. Deutsch rejected the defense that "everybody does it this way," quoting a prior opinion that an unsafe practice does not become acceptable even if uniformly adopted. As a result, professional engineers can have no doubt of their personal responsibility for designs that bear their seals. (Following the Hyatt action, liability insurance for Missouri structural engineers soared to 9.3 percent of billings, three times the level of 1983.)

Source: AP/Shutterstock

Bridge Failure and Engineers' Responsibility

As a student reading the case study of the Hyatt Regency disaster, it might be easy to dismiss this disaster as the errors of uninformed engineers from decades ago. Unfortunately, a number of far more recent examples illustrate how this issue continues to plague the profession.

On August 1, 2007, at the peak of the evening rush hour, the I-35W bridge over the Mississippi River just east of downtown Minneapolis experienced catastrophic failure. This failure included 1,000 feet of deck truss collapsing, with nearly 500 feet of the main span falling over 100 feet into the river. At the time of failure, 111 vehicles were on the collapsed section of the bridge and 17 had to be recovered from the river. As a result of the failure, 13 people died and 145 were injured.[1] The collapse of the bridge was captured live by a U.S. Army Corp of Engineers motion-activated camera and can be found via internet search today. Following a complete investigation by the National Transportation Safety Board (NTSB), a number of engineering failures were found to be the cause of the accident. The primary cause of the failure was inadequate load capacity due to a design error regarding gusset plates by the company responsible for structural engineering design. Similar to the Hyatt Regency case, this error was allowed to be incorporated into the structure due to subsequent failures of the engineering review process at the company and by state and federal transportation engineers. Subsequent failures by these same state and federal agencies with regard to inspection procedures and load rating calculations allowed additional structural degradations to remain in place without correction. This series of failures over time and across numerous parties led to the disaster.

In a more recent example, in the early afternoon of March 15, 2018, a new pedestrian bridge being constructed to improve the safety of students at Florida International University in Miami collapsed, crushing eight vehicles and killing six people. While the accident is too recent to allow for a full NTSB report to be completed, the agency has confirmed that they are focusing the investigation on bridge design, the construction process, and the construction materials as well as the emergence of cracks in the structure that were reported two days before the collapse.[2] Not only are all of the areas under investigation the responsibility of the bridge's design engineers, it was an engineer with the company responsible for construction who reported the cracks, stating that the company wasn't concerned about the cracks from a safety perspective.[3] While the final reasons behind the collapse are not yet determined, based on reporting and expert opinions shortly after the tragedy, it appears that economic pressures due to construction being overbudget and behind schedule may have led to the use of construction techniques that had not been adequately tested, increasing the risks of future failure and leading to multiple federal criminal investigations surrounding the collapse.

In all of these cases, it seems that failures of engineering and engineers led to destruction of property and death. This represents a failure of the first canon of engineering ethics to hold paramount the safety, health, and welfare of the public. As this edition is being written, yet another bridge collapse led to loss of life—as 43 people were killed when a bridge in Genoa, Italy collapsed on August 14, 2018. This tragedy illustrates that these issues are not limited to U.S. engineering practices.

Source: National Transportation Safety Board

Sources

1. National Transportation Safety Board, *Collapse of I-35W highway bridge Minneapolis, Minnesota, August 1, 2007*. 2008: Washington, DC.
2. National Transportation Safety Board, *Preliminary Report- Highway: Collapse of Pedestrian Bridge Under Construction Miami, Florida (HWY18MH009)*. 2018: Washington, DC.
3. Klas, M.E., D. Smiley, and D. Chang, *Bridge designer left state voice mail about cracks days before FIU bridge collapsed*, in *Miami Herald*. 2018: Miami, FL.

ETHICAL PROBLEMS IN INDUSTRIAL PRACTICE

Significance. The examples of ethical problems in the prior section refer to engineers in private consulting practice or in public works, areas to which many engineering codes of ethics have been directed. But Wilcox points out the following:

> First, engineers are, for the most part, employees of large corporations. They face the dilemmas of many professionals whose work is situated in the context of complex organizational structures. How are engineers as professionals to understand themselves as employees of institutions which do not have the same ethical codes as the engineers?

Four examples of problems engineers may face in corporate employment appear in the following paragraphs, including environmental responsibilities, conflicts of interest, non-compete agreements, and whistle-blowing. This section culminates in three case studies: a design engineering study of the Titanic, a whistle-blowing case involving Boeing and what is perhaps the most studied ethical failure of engineering managers, those that led to the *Challenger* disaster.

Environmental Responsibilities.   Engineers and engineering managers are intimately involved in decisions about effluent discharge from their plants and the effect it has on the air we breathe and the water we drink. For example, oil- or coal-burning electric power plants are among the largest sources of sulfur dioxide in our atmosphere, a prime cause of smog, acid rain, and a contributor to climate change. Reducing the sulfur dioxide discharge requires using much more expensive low-sulfur fuel or construction of very expensive "scrubbing" equipment. These costs *may* be recoverable by increasing electrical rates, but achieving rate increases requires lengthy and uncertain pleadings before regulatory agencies, and any costs not recovered fall heavily on stockholders.

The decision may be even more difficult in unregulated industries. Ideally, one might want a chemical plant or a metal smelter to return water to our streams as clean as it was received. To accomplish this, however, requires substantial expenditure. If industrial competitors, whether in the United States or abroad, ignore their environmental responsibilities, it may not be economically feasible for a single firm to modify an old plant to meet the new standards completely, and the alternative may be to close the plant and lay off all the workers. In such situations the ethical answer is not always clear-cut.

Conflict of Interest.   There are a number of situations where conflict of interest may arise in the work of an engineering employee of a corporation. Three of the more obvious are as follows:

1. *Gifts.* Engineers typically provide the specifications for technical products and recommend which among alternative offerings should be accepted. Salespersons (including sales engineers) for suppliers attempt to build goodwill and capture business from potential customers and clients. Part of the custom in this relationship is the offering of tokens of appreciation (gifts). Each person (and each organization) needs to establish what is an acceptable gift, and at what level a gift carries a connotation of undue influence. Many organizations establish an approximate dollar limit. Federal government employees may have almost a zero limit, down to not accepting a free meal. No hard and fast rule can be provided—each engineer must develop a feeling for what is proper for the industry and his or her employer, and the sensitivity involved in the job held with that employer.

2. *Moonlighting.* Most engineers would agree that they should not compete with their employer in bidding on a project. However, if the employer declines to bid, may the engineer offer to do the job privately on nights and weekends? If that brings the client back to the engineer rather than to his or her employer on the next offering, must the employer again be offered the first opportunity?

3. *Inside information.* Engineers who have a significant percentage ownership in a small firm are limited in their right to buy and sell large blocks of stock based on "inside information" of favorable and unfavorable information that will affect stock purchases. Is the engineer who owns an insignificant amount ($10,000 stock in a billion-dollar corporation) similarly constrained?

Postemployment Limitations/Non-Compete Agreements.   What are reasonable limitations to the contracts required by some employers of engineers and other professionals that they will not work for a competitor for a specified time after leaving their current employer? Certainly, it is reasonable to require an engineer working on an unusually important confidential development not to move to another firm and use this proprietary information (i.e., trade secrets) to create a product that competes directly with that of the original employer. On the other hand, knowledge and skill in a specific specialty is all that a knowledge worker has to offer a potential new employer, and to deny an engineer the right to use these skills (as opposed to proprietary information) elsewhere would also be unprofessional. Where the ethical divider between these extremes lies must fall to individual consciences—and, failing that, the courts—to decide.

Whistle-blowing. Engineers and other professionals may well come across an example where their employer is doing something that they believe is dishonest, illegal, or damaging to the public, and the employer is unwilling to change. Here, the engineer's professional ethics may come in conflict with loyalty to the employer (and "business ethics," if these are perceived differently). DeGeorge provides some guidelines for such occasions:

> Whistle-blowing is morally justifiable when there is impending danger and a concerned employee "has made his moral concern known" to his immediate superior who has subsequently failed to act. When this happens, a concerned employee should take his or her complaint upward through company channels, if necessary, to top management. After all internal efforts have failed, public disclosure is justifiable. For whistle-blowing to be obligatory as well as justifiable, two more conditions must be met: first, that the employee have documentation or other hard evidence; and second, that they "must have good reason to believe that by going public they will be able to bring about the necessary changes."

Construction Case Study: Titanic

The Titanic sank into the North Atlantic in 1912. Since then, scientists and Titanic buffs have debated what really caused the biggest passenger ship of her day to sink just 2 hours and 40 minutes after hitting an iceberg, carrying 1,522 people to their deaths.

Those questions were answered long ago, in a confidential investigation by the ship's builders. To date, experts have amassed enough evidence to demonstrate that the ship broke into three pieces, not two—before sinking, not after—and she went down faster and at a much lower angle—all thanks to skimpy rivets and a flimsy hull. But a trove of documents from Harland and Wolff—the Belfast, Ireland, shipyard where the Titanic and her sisters were born—reveal that the problem was not just one of incompetence and poor construction. It was negligence: the ship's builders suspected that the ship's hull was too flimsy, but they overrode the concerns of their engineer in a bid to get the Titanic on the seas in time. An investigation held after the ship sank was not made public; the heads of Harland and Wolff allowed two formal government inquiries to lay blame for the wreck on the shoulders of the ship's captain. The lawsuits of so many victims would have bankrupted the Titanic's owners—J. P. Morgan among them.

Not until 2005 did divers find two large sections of the ship's bottom—enough for forensic scientists to determine that the flimsy hull and skimpy rivets were, in fact, responsible for the ship's fate.

When the safety of the rivets was first questioned, the builder ignored the accusation and said it did not have an archivist who could address the issue.

Now, historians say new evidence uncovered in the archive of the builder, Harland and Wolff, in Belfast, Northern Ireland, settles the argument and finally solves the riddle of one of the most famous sinkings of all time. The company says the findings are deeply flawed.

A team of divers and scientists and Harland and Wolff's engineers concluded that a stronger hull and rivets would have kept the ship afloat much longer, resulting in a dramatically lower death toll.

(Harland and Wolff then retrofitted the hull of Titanic's older sister with extra steel. They also built Britannic—the sister ship that was under construction when the Titanic sank—to the original specifications.)

The rivets holding the hull together were much more fragile than once thought. From 48 rivets recovered from the hulk of the Titanic, scientists found many to be riddled with high concentrations of slag. A glassy residue of smelting slag can make rivets brittle and prone to fracture. Records from the archive of the builder show that the ship's builder ordered No. 3 iron bar, known as "best"—not No. 4, known as "best-best," for its rivets, although shipbuilders at that time typically used No. 4 iron for rivets. The company also had shortages of skilled riveters, particularly important for hand riveting, which took great skill: the iron had to be heated to a precise color and shaped by the right combination of hammer blows.

The Titanic had been the product of a colossal rivalry spurred by the growth in shipping profits from the Spanish American War. In the hopes of controlling the North Atlantic, J. P. Morgan bought controlling interests in a handful of British and American shipping companies.

Making the hull plating a quarter of an inch thinner and the rivets an eighth of an inch thinner and made of lower grade iron than the original designs called for would reduce the ship's weight by 2,500 tons, enabling her to cross the English Channel faster than the competition. Because shipbuilding regulations had not kept pace with the push toward larger vessels, the thinner specifications still met the standards of the day.

Source: *New York Times*, April 15, 2008

Whistle-blowing Case Study: Boeing

In 1996, Boeing Company was locked in a fierce competition with Lockheed Martin Corporation to become the government's primary maker of rockets for launching spy, communications, and other satellites. With Boeing as the underdog, and the future of its space-launch business at stake, company officials were seeking any advantage they could get over their rival. In 1997, Kimberly Tran saw a Boeing senior engineering scientist and a former Lockheed Martin employee with access to its Atlas V secrets, carrying a binder at Boeing's Huntington Beach, California facilities that appeared to be filled with those secrets. She reported the infraction to her immediate manager, senior manager, and the leader of the Delta IV senior infrastructure team. Instead of acting on her efforts, it appears Boeing attempted to "keep a lid" on the issue, according to Lockheed Martin's lawsuit. In Florida, two years later, the supervisor of that former Lockheed Martin employee bragged to Steven Griffin about hiring this man to obtain Lockheed Martin's proprietary data. Griffin reported the comments to Boeing's human resources department and his supervisor. In contrast to Tran's experience, he was interviewed promptly. "Boeing did what it was supposed to do," Griffin said to *Aviation Week & Space Technology*. However, the launch community is tight-knit and Griffin had somebody else to talk to—his wife. Bridget Griffin, lead engineer on payload operations for Lockheed Martin's Atlas program, reported the matter to her supervisors, and immediately alarm bells began ringing. Mrs. Griffin's report has earned widespread recognition from Lockheed Martin and shined a bigger light on the good conduct that her husband and Tran demonstrated.

A Boeing attorney discovered documents marked "Lockheed Proprietary" in both Bill Erskine's and Ken Branch's office. Calls were then made to Lockheed and the Air Force informing them that seven pages of "harmless documents" had been discovered and they had only been seen by Erskine and Branch. Later depositions revealed that a five- to six-inch stack of documents had been found in Branch's office and a box of documents had been found in Erskine's office during a later search. After the Air Force completed its investigation it was discovered that eight boxes containing 24,500 pages of Lockheed proprietary documents had been found!

At first Boeing put forth a story that an "exhaustive investigation" had revealed only 197 pages of Lockheed proprietary data and reiterated that only Branch and Erskine had seen the documents. They stuck to this story for two years. Unfortunately for Boeing, when Branch and Erskine filed wrongful termination suits the truth came out. The two lawyers hired to defend Boeing acknowledged the box of documents discovered by the Boeing attorney and returned 2,700 documents to Lockheed, many more than the initial 204 returned as proprietary.

Now aware that more documents existed, Lockheed asked for full disclosure from the two attorneys representing Boeing. The attorneys tried to explain away the extra documents stating they believed only the original documents were proprietary and were under no obligation to return the remaining documents. But it also came to light that the Air Force had never been notified of any documents beyond the original seven "harmless documents." The Air Force subsequently demanded to see all material found in Erskine's and Branch's offices. The huge amount of documents turned over to the Air Force by Boeing resulted in charges being made that Boeing made "false and misleading" statements to the government.

The Air Force stripped Boeing of seven launches worth $1 billion and handed them over to Lockheed. Boeing was also suspended from bidding on government launch contracts until the investigation was complete. The U.S. Attorney in Los Angeles charged Branch and Erskine with conspiracy to steal trade secrets. Lockheed in turn filed a civil racketeering suit against Boeing indicating that Branch turning over the documents showed intent to "engage in economic espionage."

Since the racketeering suit was filed in 2003, Boeing has conducted more company searches for Lockheed documents. These searches revealed another 35,000 more Lockheed documents in the possession of just two engineers. Boeing defends these document discoveries as just "harmless work papers" again.

Sources: http://seattletimes.nwsource.com/html/businesstechnology/2002146025_boeinglockheed09.html; http://www.nlpc.org/cip/030505bg.html; *Aviation Week & Space Technology*, April 19, 2004, p. 45; *Wall Street Journal*, May 5, 2003;

Whistle-blowing Case Study: The *Challenger* Disaster

On January 24, 1985, Roger Boisjoly, Senior Scientist at Morton Thiokol, Inc. (MTI), watched the launch of Flight 51-C of the space shuttle program and remained to inspect the solid rocket boosters after their recovery from the Atlantic Ocean. These immense boosters are too large to transport, so they are manufactured in cylindrical sections and fastened together with "field joints" before launch. At each

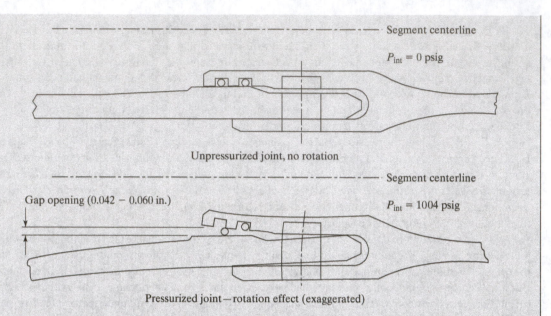

Figure 16-2 Cross section of *Challenger* booster flange. (From Russell J. Boisjoly and Ellen Foster Curtis, "Roger Boisjoly and the *Challenger* Disaster: A Case Study in Engineering Management, Corporate Loyalty, and Ethics," *Proceedings of the Eighth Annual Meeting, American Society for Engineering Management*, St. Louis, MO, October 11–13, 1987, p. 10.)

joint, the straight terminal ring of one segment (the "tang") slid into a clevis ring (with a Y-shaped cross section) on the mating segment, and this joint was sealed with two O-rings (see Figure 16-2). Boisjoly, then "considered the leading expert in the United States on O-rings and rocket joint seals," was dismayed to find "that both the primary and secondary O-ring seals on a field joint had been compromised by hot combustion gases (i.e., hot gas blow-by had occurred), which had also eroded part of the primary O-ring," although the temperature of the field joint at launch was believed to be a comfortable 53°F (12°C). (See Figure 16-3.)

Since rocket motor pressurization following ignition causes some rotation in the field joint, opening the annulus sealed by the O-rings, Boisjoly sponsored a series of subscale laboratory tests in March 1985 of the effect of temperature on O-ring resiliency. In these tests O-rings were squeezed, the pressure removed, and the time for the O-ring to regain shape measured. At 100°F (38°C) recovery was immediate, at 75°F (24°C) it took 2.4 seconds, but at 50°F (10°C) the seal had not recovered even after 10 minutes (600 seconds). In the ensuing months, Boisjoly emphasized in the strongest terms the need to redesign the field joint. On August 20, 1985, Robert K. Lund, MTI Vice President, Engineering, announced formation of a Seal Erosion Task Team, but little progress was made on solving the problem—despite further blow-by on a flight on October 30, 1985, when the field joint temperature was estimated at a balmy 75°F (24°C).

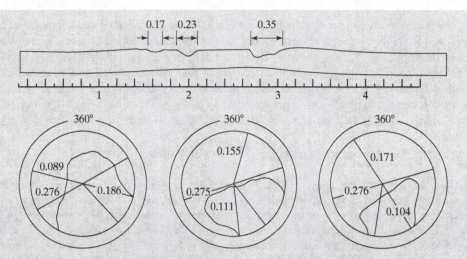

Figure 16-3 Multiple burn-through of *Challenger* nozzle joint primary O-ring. (From Russell J. Boisjoly and Ellen Foster Curtis, "Roger Boisjoly and the *Challenger* Disaster: A Case Study in Engineering Management, Corporate Loyalty, and Ethics", *Proceedings of the Eighth Annual Meeting, American Society for Engineering Management*, St. Louis, MO, October 11–13, 1987, p. 8.)

The stage was now set for the eve of the *Challenger* tragedy:

> According to Boisjoly's account at 10 A.M. on January 27, 1986, Arnie Thompson (MTI Supervisor of Rocket Motor Cases) received a phone call from Thiokol's Manager of Project Engineering at MSFC (Marshall Space Flight Center), relaying the concerns of NASA's Larry Wear, also at MSFC, about the 46°F (8°C) temperature forecast for the launch of Flight 51-L, the *Challenger*, scheduled for the next day. This phone call precipitated a series of meetings within Morton Thiokol; at the Marshall Space Flight Center; and at the Kennedy Space Center (KSC) that culminated in a three-way telecon, involving three teams of engineers and managers, that began at 8:15 P.M. EST.

Present on the telephone were 14 managers and engineers at Thiokol's Wasatch (UT) Division Management Information Center, 15 at MSFC, and 5 at KSC. Boisjoly and Thompson began by detailing the flight and laboratory experience previously outlined. Lund presented the final chart, recommending against launch unless the O-ring seal temperature exceeded 53°F (12°C); Joe Kilminster, MTI Vice President, Space Booster Programs, supported his engineers and would not recommend a launch below 53°F.

George Hardy, Deputy Director of Science and Engineering at MSFC, was "appalled at that recommendation," but would not recommend to launch if the contractor was against it. Lawrence Mulloy, Manager of Booster Projects at KSC, also strenuously objected, saying "My God, Thiokol, when do you want me to launch? Next April?" Boisjoly continued to object to a launch, but finally Kilminster asked for a five-minute caucus of Thiokol's people. Jerry Mason, MTI Senior Vice President of Wasatch Operations, began the caucus by saying that "a management decision was necessary" (influenced, very likely, by the fact that MTI was at that time negotiating a billion-dollar follow-on contract with NASA). Thompson and Boisjoly re-reviewed their reasons for not launching, but quit when it was obvious that no one was listening; Mason then turned to Bob Lund and, in a memorable statement, asked Lund to "take off his engineering hat and put on his management hat." At that point Lund, Mason, Kilminster, and Calvin Wiggins (MTI Vice President and General Manager of the Space Flight Division) held a brief discussion and voted unanimously to recommend *Challenger*'s launch. The following day, about 73 seconds into launch, the *Challenger* exploded in a ball of flame on the television screens of the entire world.

In discussing this "management" decision, Florman concludes as follows:

> The four so-called Thiokol "managers" are, in fact, engineers. Mason has a degree in aeronautical engineering; Lund in mechanical; Wiggins has a degree in chemistry; and Kilminster, a master's in mechanical engineering on top of an undergraduate degree in mathematics. The two NASA "officials," Hardy and Mulloy, who urged that Thiokol approve the launch, are also engineers, as are the key NASA people above them. These men were educated as engineers and had worked as engineers, eventually moving into positions of executive responsibility. They did not thereupon cease being engineers, any more than a doctor who becomes director of a hospital stops being a doctor.

> Were these engineer–executives under pressure to meet a launching schedule? Of course. But pressure is inherent in engineering…Pressure goes with the job like the proverbial heat in the kitchen. It may help explain, but it cannot excuse, an engineering mistake.

Boisjoly's testimony before the Rogers Commission regarding the foregoing events led to increasing friction with MTI management. "Although given the title of Seal Coordinator for the redesign effort, he was isolated from NASA and the seal redesign effort. His design information had been changed without his knowledge and presented without his feedback." As Boisjoly later concluded, "The research on [whistle-blowing] leads to two conclusions. First, all whistle-blowers attempt to achieve problem resolution through their organizational chain of command; and, second, they are all punished by the organization after whistle-blowing outside the organization." Boisjoly cited as "timeless" the advice of Adolph J. Ackerman in a June 1967 IEEE article:

> Engineers have a responsibility that goes far beyond the building of machines and systems. We cannot leave it to the technical illiterates, or even to literate and overloaded technical administrators to decide what is safe and for the public good. We must tell what we know, first through normal administrative channels, but when these fail, through whatever avenues we can find. Many claim that it is disloyal to protest. Sometimes the penalty—disapproval, loss of status, even vilification, can be severe.

Boisjoly understands the last sentence well. His position at MTI became untenable, and he requested extended sick leave on July 21, 1986, with a case of post-traumatic stress syndrome. More than two years later, when the redesigned shuttle put America back in space, it was clear that Boisjoly would never return to work.

In a later analysis of testimony before the Rogers Commission on the correlation between temperature and O-ring erosion, Lighthall quotes testimony of participants Boisjoly ("I couldn't quantify it"),

MTI engineer Jerry Burn ("it is speculation"), NASA's Mulloy ("I can't get a correlation"), and NASA's Hardy ("obviously not conclusive"), then shows by statistical analysis of data available before the flight a better than 99.5 percent probability of just such a correlation. He concludes "that none of the participants had ever learned, or had long since forgotten, elementary ideas and methods of statistical analysis and inference." This conclusion is of obvious significance for engineering education, with similar issues discussed in Chapter 12.

SUMMARY: MAKING ETHICAL DECISIONS

This chapter has shown that engineers, managers, and other professionals have many occasions in which corporate, professional, and personal objectives and values may conflict. In some of these there will be real conflicts between positions where each position has an ethical justification. Questions such as the following can lead to a solution you can adopt with self-respect and live with:

- Does the action I am considering make good sense?
- Does this action fit my best concept of a dedicated professional engineer?
- Will my action unnecessarily harm others? Is there some way that I can compensate them?
- Would my action stand up to close public scrutiny? Would I have difficulty explaining it to a reporter? To a judge and jury? To my colleagues? To my own family?
- Am I hiding behind a superior's judgment or wish, or can I justify it based on my own values?

A System Dynamics View of Engineering Ethics

This chapter has described the place of ethical principles, canons, and codes within the professional engineering societies—and the societies' inability to agree on the details. There are examples of engineering failures, such as the Hyatt Regency balcony and I-35 bridge collapses, due to ethical lapses. There are examples of engineers who gave up or damaged their careers to expose or prevent unethical and/or unsafe corporate actions—and of engineers who received awards of recognition or money for speaking up.

Some of these represent choices by the engineer and the organization between ethics and expediency. Others represent ethical dilemmas where different ethical values are conflicting. The *New Oxford American Dictionary* definition of integrity includes "the state of being whole and undivided." In a situation where expediency issues exert significant force or there are ethical dilemmas the engineer's focus is "divided," and even when there is no ethical compromise it is difficult to concentrate on the design process.

We suggest that a systems view of ethics can help us understand the complexity of ethical situations. This systems view focuses on forces that tend to increase or decrease ethical behavior by the individual engineer. For example:

- The ethics code is well-meaning but ambiguous; the cost cutting pressure is severe!
- For a previous employer you created an optimal design, which still seems optimal for a design requirement of your current employer. Are you about to misappropriate a trade secret? Which best serves the public welfare, using or not using the optimal design?

- You believe a design may be infringing on a patent held by another firm, but your boss tells you to go ahead because "they cannot afford to fight us in court."
- What if engineering firms are in a recession; should you adjust your ethical standards to protect your job and your colleagues' jobs?

This systems approach considers social, cultural, legal, economic, technological, and organizational systems and their interactions. The system behavior is dynamic, and it evolves from previous to future states. These system interactions contain multiple (possibly non-linear) feedback loops that may combine to produce counterintuitive behavior. As an example of counter-intuitive behavior, an ethics code can actually increase the search for ethics "loopholes."

Figure 16-4 is a systems dynamic model that uses causal-loops to summarize the interactions of the many forces. Variables are connected by directional arrows; if an increase in an originating variable increases the destination variable, the arrowhead at the destination is positive. This model was created using *Vensim®* software. Our goal is a better understanding of the complexity inherent in the engineering ethics situation, and the *model is intended to be exploratory and not in any sense definitive.*

The approach also is based on two value propositions: First, no engineer is justified in violating the law, but obeying the law is at best the floor for ethical behavior and not the ceiling. Second, the natural level of engineering ethics in the United States is fairly high and the problem is to maintain this high level or even enhance it, but the more important issue is reducing the frequency and severity of deviations.

The forces shown in Figure 16-4 can be categorized with some overlap as follows (listed clockwise from top left):

- The engineer's personal ethical development growing up
- The engineering profession's values and pressures
- The ethical climate of the engineer's organization
- The competitive pressures on the organization
- The oversight and sanctions (including licensing) from outside the organization
- The extent the engineer is moving from engineering to managerial responsibilities
- The personal risk-taking ability of the engineer

This model is a basic system dynamics model, with positive and negative influences between variables included, but not their strength or rate of change. If these can be identified from empirical research for specific situations, then the possibility exists for simulating the behavior of the ethics systems.

How can managers, including engineering managers, benefit from the systems model of ethics? *We suggest the most important benefit is recognizing the complexity of the ethics system.* Often there is a managerial tendency to visualize simplistic relationships. For example, if engineering ethics is a problem, we will solve it by ethics training for our engineers. Or by writing an ethics code. Or perhaps the engineering profession as a whole will try to improve ethics by strengthening or clarifying its ethics canons.

But these actions, whether at the managerial or profession-as-a-whole level, may miss the actual causal factors leading to ethical expediency. Experience with complex systems of all kinds reveals both

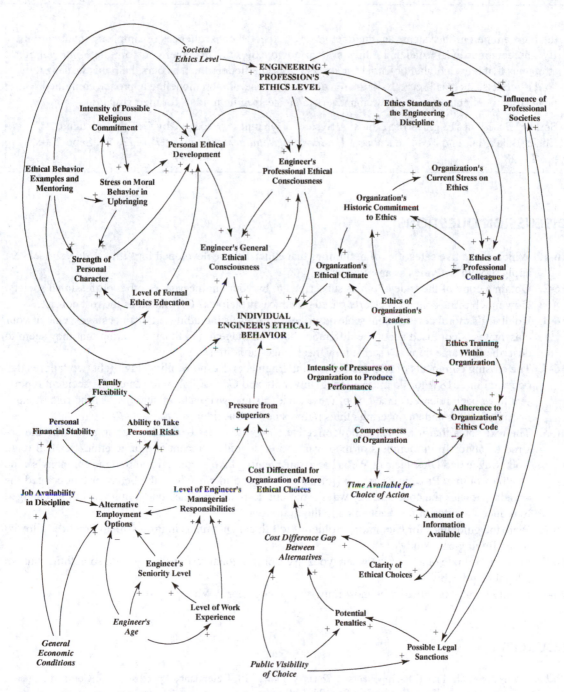

Figure 16-4 Possible forces shaping an individual engineer's ethics choices.

357

the inertia inherent in systems and ineffective or unexpected consequences of simplistic interventions. By appreciating and visualizing ethics system complexity, the manager can formulate intervention strategies that have a higher potential for success. System models, such as ours, help this visualization. And if models, such as Figure 16-4, are too generic, the causal loop modeling approach combined with powerful modeling software allows managers to create specific models for their situation.

Source: Based on "Engineering Ethics: A Systems Dynamic Approach," by George Geistauts, Elisha Baker, IV, and Ted Eschenbach, *Engineering Management Journal*, Vol. 20 #3, Sept. 2008, pp. 21–28.

DISCUSSION QUESTIONS

16-1. Which of the five categories of definitions that ethicists have developed for "ethics" do you feel best apply to your activities as an engineer?

16-2. Obtain a copy of the code of ethics subscribed to by your engineering or other professional society. How much of its content is clearly based on ethical principles? Of what does the rest consist?

16-3. Kohlberg's moral development scale describes three levels including six distinct stages. One of your classmates is confused about the difference between Stage 4 and Stage 5, stating that they seem to describe the same thing. Describe how these stages are different.

16-4. The closing comments of "Bridge Failure and Engineers' Responsibility" highlight two failures that occurred in 2018 (Florida International University and Genoa, Italy). Research the accident reports that have been released since one of these accidents and write a brief summary of the role of engineering failures and engineering ethics failures in these accidents.

16-5. The text notes that it is unclear how effective corporate codes or conduct are in achieving the desired behavior from members of the organization. Identify a recent corporate ethics scandal (e.g., Volkswagen and Dieselgate or Wells Fargo and creating fake accounts) and find the company's code of ethics online. Provide a summary of the scandal and identify how the behavior that created the scandal violates the code and any ways it did not. Explain how the code might need to be changed to reduce the odds of this happening in the future.

16-6. Were the guidelines for Facilitating Solutions to Ethical Dilemmas in Professional Practice followed in the Boeing case study?

16-7. Give an example of a situation where you have a professional responsibility to do something, but not a legal responsibility.

16-8. Should loyalty or integrity be most important to engineers? Why?

SOURCES

Andrew Hawkins et al., Plaintiffs-Appellants v. *Town of Shaw*, MS, Defendants-Appellees, U.S. Court of Appeals, 5th Circuit, January 23, 1971, 437 Fed. Rep. 2d, 1286.
Anonymous, "I Gave Up Ethics—To Eat," in Schaub and Pavlovic, *Engineering Professionalism and Ethics*, pp. 233–238.

Bagley, Fenton, "Ethics, Unethical Engineers, and ASME," *Mechanical Engineering*, July 1977, p. 42.

Baum, Robert J. and Flores, Albert, *Ethical Problems in Engineering* (Troy, NY: Rensselaer Polytechnic Institute, Center for the Study of the Human Dimensions of Science and Technology, 1978), pp. 33–52.

Boisjoly, Russell J. and Curtis, Ellen Foster, "Roger Boisjoly and the *Challenger* Disaster: A Case Study in Engineering Management, Corporate Loyalty, and Ethics," *Proceedings of the Eighth Annual Meeting, American Society for Engineering Management*, St. Louis, MO, October 11–13, 1987, pp. 6–7.

"Builders Fined $5 Million in Fatal Collapse," *St. Louis Post-Dispatch*, October 23, 1987, p. 1F.

Byrne, John A., "Businesses are Signing Up for Ethics 101," *BusinessWeek*, February 15, 1988, pp. 56–57.

Davenport, Manuel M., "Ethical Issues in Excavation Safety," *Business and Professional Ethics*, quarterly newsletter of the Center for the Study of the Human Dimensions of Science and Technology (Troy, NY: Rensselaer Polytechnic Institute, n.d.), p. 11.

DeGeorge, Richard T., *Business Ethics* (New York: Macmillan, 1982), quoted in Gluck, "Ethical Engineering," pp. 185–186.

Deutsch, James B., talk before civil engineering class, University of Missouri–Rolla, November 3, 1988.

Florman, Samuel C., *The Civilized Engineer* (New York: St. Martin's Press, 1987), p. 163.

Gluck, Samuel E. "Ethical Engineering," in John E. Ullmann, ed., *Handbook of Engineering Management* (New York: John Wiley & Sons, Inc., 1986), p. 176.

http://whistleblowerlaws.com, December 2012.

Kaptein, Muel. "Business codes of multinational firms: What do they say?," *Journal of Business Ethics, vol.* 50, iss.1, 2004, pp. 13–31.

Lighthall, Frederick F., "Engineering Management, Engineering Reasoning, and Engineering Education: Lessons from the Space Shuttle *Challenger*," *Proceedings of the International Engineering Management Conference*, Santa Clara, CA, October 21–24, 1990, pp. 369–377.

Mathews, Marilyn C. "Codes of ethics: Organizational behavior and misbehavior," *Research in corporate social performance and policy* 9, 1987, pp. 107–130.

Munger, Paul (Professor of Civil Engineering, University of Missouri–Rolla and chairman (1981–1984), Missouri Board for Architects, Professional Engineers and Land Surveyors), personal communication.

Naj, Amal Kumar, "Federal Judge Awards Ex-GE Staffer Record Amount in Whistle-Blower Case," *Wall Street Journal*, December 7, 1992, p. A5.

Oldenquist, Andrew G. and Slowter, Edward E., "Proposed: A Single Code of Ethics for All Engineers," in James H. Schaub and Karl Pavlovic, eds., *Engineering Professionalism and Ethics* (New York: John Wiley & Sons, Inc., 1983), pp. 446–447.

Price, Willard, "Values in Public Works Decision Making: The Distribution of Services," in Daniel L. Babcock and Carol A. Smith, eds., *Values and the Public Works Professional* (Rolla, MO: University of Missouri–Rolla, 1980), pp. 23–35.

Ryan, Brother Leo V. "Conflicts Inherent in Corporate Codes," *International Journal of Value-Based Management, vol.* 4, iss. 1, 1991, pp. 119–136.

Wartzman, Rick, "Nature or Nurture? Study Blames Ethical Lapses on Corporate Goals," Marilynn C. Mathews, quote in *Wall Street Journal*, October 9, 1987, p. 21.

Wilcox, John R., "The Teaching of Engineering Ethics," *Chemical Engineering Progress*, May 1983, pp. 15–20.

CASE STUDY WEBSITES

The following are useful websites for ethics case studies. (September, 2018)
http://ethics.tamu.edu/
https://www.nae.edu/Activities/Projects/CEES.aspx
http://www.onlineethics.org/Topics/ProfPractice/PPCases.aspx
https://www.scu.edu/ethics/about-the-center/

17

Achieving Effectiveness as an Engineer

PREVIEW

With the expansion of global economies, the world we live in has become more diverse in needs and production of products. Consequently, companies require diverse intellectual capital that can bring ideas and knowledge to the workplace to improve their competitive advantage in the global market. These evolving trends in industry must be applied back into the engineering curriculum. Industry needs universities to respond with the following changes:

1. Retain strengths in math and science fundamentals, plus enhanced information technology emphasis.
2. Increase emphasis on design and manufacturing skills.
3. New emphasis on breadth, developing professional skills, and process-related issues—for example, engineering economics, business fundamentals, project management, environmental and social issues, teamwork and communication skills, and life-long learning are some of the attributes the graduating engineers of today and tomorrow need in their future careers.

This chapter is concerned with some of the areas in which many basic engineering education programs are weak. This includes the advice of some successful engineers, past and present, on "getting off to the right start" in early professional first assignments. Then choices in career field and the sequence of stages in a career are considered. The need to communicate your ideas effectively is discussed, with some techniques for effective oral and written communication. Fourth is a discussion of how to stay technically competent in an age of exploding information. Next, the areas of professional society activity, registration, and certifications are discussed.

The chapter next looks at the changing positions of women and of under-represented minorities in engineering and management. Particular attention is paid to the reasons for the disproportionately small numbers of women and minorities entering engineering schools and graduating from them, and the barriers they face as engineering professionals and managers.

This chapter closes the section on achieving effectiveness as an engineer by examining the relation of management to the engineer's career, and the need for engineers in top management. The reasons technologists give for moving into management are discussed, as well as the effectiveness of parallel career ladders in technology and management. Lastly, some of the methods of preparing for management responsibilities, including formal and informal courses and job experience, are presented. The chapter concludes with a discussion of an item all professionals need—time management.

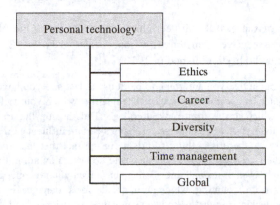

LEARNING OBJECTIVES

When you have finished studying this chapter, you should be able to do the following:

- Discuss the steps in getting off to a right start in a new job.
- Explain why communication skills are important to an engineer.
- Discuss ways to remain technically competent in the engineering field.
- Discuss the position of women and under-represented minorities in engineering.
- Discuss the responsibilities of engineers in management.
- Discuss techniques to manage time.

GETTING OFF TO THE RIGHT START

Background

In engineering education you have worked hard to survive a demanding curriculum and to build an academic record you can be proud of. Many engineers land their first job because the employer came to them: Although less than 10 percent of college graduates are engineers, they receive about half of all on-campus job offers. Despite new engineering graduates being sought after employees, very few positions can be considered "secure jobs" in current employment practices, due to rapidly changing economic conditions, new competitive forces, etc.

In addition, merely having a secure job is not satisfying for most professionals. For a rewarding professional life, you must begin building for yourself a personal reputation on which your future career success will depend. Several decades later you may come to reflect on the actions and decisions in your early career that made you successful, or that might have made you more successful; perhaps then you will be willing to share your hard-earned wisdom with young engineers who are following in your footsteps.

Fortunately, a number of successful engineers have done just that. One of the first was the famous French mining engineer and executive Henri Fayol, whose *Advice to Future Engineers* is as fresh and appropriate today as it was when he first published in 1916.

> You are happy in the thought that you are going to be of use at last and you have the legitimate desire to win an honourable place by giving your service. The qualities which you will have to call into play are not precisely those which confer front rank at college. Thus health, the art of handling [people], and bearing, which are not assessed in examinations, have a certain influence on an engineer's success. Circumstances, too, vary, so there is nothing surprising in the fact that the first class ["*A* students"] or even the highest of their year are not always those who do best. You are not ready to take over the management of a business, even of a small one. College has given you no conceptions of management, nor of commerce, nor yet of accounting, which are requisite for a manager. Even if it had given you them, you would still be lacking in what is known as practical experience, and which is acquired only by contact with [people] and with things....
>
> Your future will rest much on your technical ability, but much more on your managerial ability. Even for a beginner knowledge of how to command, plan, organize, and control is the indispensable complement of technical knowledge. You will be judged not on what you know but on what you do and the engineer accomplishes but little without other people's assistance, even when [they] start out. To learn how to handle [people] is a pressing necessity.

Regarding Your Work

The new graduate makes their mark within the first few years in the organization. This makes it essential to give your best efforts to your early assignments, regardless of how trivial they may appear. Doing an exceptional job on a minor assignment is the best way to be recognized as a reliable contributor and assigned more important, more challenging, and more satisfying work. This applies to cooperative (co-op) work assignments of undergraduate engineering students in industry as well. Lee found that higher-performance co-op students were those who "seemed to work harder and they also found their work assignments to be more challenging"; this work experience "also had an indirect effect on subsequent professional job performance" evaluated two years after graduation.

In the occasional situation where you have given simple assignments your best for a reasonable period and have not been given anything more challenging, discuss with your manager your interest in getting more varied experience. If they cannot (or will not) do anything, ask permission to make an appointment with another manager who controls jobs you think you might like—or inquire of the personnel manager where there might be other opportunities in the organization. If this still does not work, you may need to find a more promising organization!

Do Not Wait for Others—Get Things Done.   Just because you have asked a manager, a vendor, or a colleague to provide something you need does not mean that it is going to happen in a timely fashion. Keep a "tickler" file and call (and call again if needed) to check on progress (calling is better than email). Find another way to get it done or work two techniques in parallel if necessary. Be understanding, but persistent, and learn to know the difference.

Go the Extra Mile—and Hour.   Only in the rare organization is a reputation made in a 40-hour week. To continue to be an effective professional you will at least have to do your professional reading largely on your own time; as you increase in responsibility you will also find that you need uninter-rupted blocks of time that never seem to be available during the day, for planning and thinking problems through. The fastest promotions generally go to those who put forth the extra effort and *meet deadlines*. This must be balanced against our other values—time spent raising our families, recreation to keep us whole and renewed, service to our community, and other investments of time that are important to us. Each of us must reach a balance (or make our compromises) customized to our individual needs. Fortunately, success in most organizations is measured in effectiveness, not just hours, as we will see under "Managing Your Time."

Look for Visibility.   You can do a good job every day, but you need to be seen to be recognized as a "rising star." Look for chances to make a presentation, to take leadership in a professional society chapter, etc. Sometimes a careful choice of car pool, lunch time and location, or exercise site will put you in touch with established professionals who can give you more insight into the forces driving the organization. Learning the dividing line between making your capabilities visible and being overly self-promoting takes maturity, but it is maturity that leads to greater responsibility.

What Employers Are Looking For

Besides your technical degree there are several skills desired by employers according to Bill Coplin. It is more than an engineering degree. The three most important skills are:

Work well with others. Social skills are important, but you need to be a team player. Teamwork in the classroom is training for this skill.

Know your numbers. Engineers know their numbers, but make sure you understand your statistics. You need to understand your financials—how to manage a spreadsheet.

Be responsible for yourself. To get ahead you need motivation and time management skills. Take responsibility for yourself and your career.

Source: http://www.billcoplin.org/, December 2019.

Learn the Corporate Culture.   Keep your eyes and ears open. Notice how successful engineers dress, and do likewise: Save your expressions of independence for important things. Notice how your more effective colleagues interact and how they get things accomplished. If you cannot be comfortable and effective in your company's culture, perhaps it is time to start looking for a better fit!

Regarding Your Manager

Be as Careful as You Can in Selecting Your Manager.   King believes that "this is second in importance only to the selection of proper parents....Long before the days of universities and textbooks, craftsmen in all the arts absorbed their skills by apprenticeship to master craftsmen." By observing the master engineer (or engineering manager) you can learn much more quickly the art of being an effective engineering professional.

Keep Your Manager Informed.   Ask yourself, "What does my manager need to know to do their job effectively?" In particular, never let your manager be caught by surprise. If something is going wrong or an assignment will be late, let your manager know. They may be able to reduce the consequences to both of you if they learn of it before *their* manager does. Again, if circumstances force you into a commitment to a higher executive or outsider, let your manager know the situation and to what you have committed as soon as feasible. Finally, if you are given a job to do, complete it or, if your initial effort convinces you it is not worth doing, tell your manager what you have found and get their agreement before dropping it; do not let your manager continue to think you are working on it. None of this implies that you should deluge your manager with unwanted trivia: More new professionals err by communicating too much than by communicating too little.

Make Your Manager's Job Easy.   Your primary job is to help your manager carry out their responsibilities, so give top priority to whatever the manager wants done or ask the manager for guidance on conflicting priorities. Learn to do *completed staff work*: Do not just go to your manager with a problem—state the problem, the alternatives you have considered, and your recommended action.

Regarding Associates and Outsiders

Network—Keep up the *old school ties* by staying in touch with past school friends, professors, old colleagues, and past managers. Someday you may need help in finding a new job, getting a recommendation for graduate school, or some other venture. Also, you may need outside sources of information on people or other resources needed to solve a company problem; you will be measured not just by what you know, but by what you can find out when needed.

In Dealing with Customers and Other Outsiders, Remember that You Represent Your Organization.   Do not bad-mouth your organization or anyone in it—instead, put them in the best possible light. (Then take any problem back to the person who is responsible, so that it can be handled internally.) Moreover, realize that the outsider is likely to regard you as being the technical, legal, and financial agent of your company, even if you have just recently joined it, and so be very careful of your commitments.

In the Apollo program, young engineers from NASA and from contractor organizations met many times to discuss technical problems; considerable confusion (and some litigation) took place before engineers on both sides learned that, although they might come to an agreement as to what change *should* be made, they could not implement it until the contract representatives who had authority to commit each party negotiated the change.

CHARTING YOUR CAREER

Defining Career Success

Morrison and Vosburgh point out that a career is defined differently for different people. For the *pure* professional the occupation is usually paramount while the link to the organization is tenuous, and the professional will go elsewhere if the organization does not provide the technical challenge, collegial relationships, or the values and rewards expected from the occupation; career and occupation become one. Other individuals become totally committed to an organization and transfer from design to project management to production to corporate staff within that organization; career and organization become one; large Japanese firms promote this orientation. Most engineering careers lie somewhere between these two extremes. In setting out to manage your own career you need to define what success in a career means to you, because this in turn influences what actions are appropriate.

Career Fields

There are a variety of broad fields of career endeavor for which your engineering education has given you a basic preparation. Some are entry-level opportunities for a new graduate, yet broad enough for a lifetime of challenges; others are best entered after you have gained some professional experience and, perhaps, graduate education. Leonard Smith identifies the following career fields:

1. *Operational careers.* Many engineers begin their careers in operating areas such as manufacturing, purchasing, planning, customer service, and sales. Each of these areas will have applications of your engineering knowledge and skills. Operational assignments are very likely to lead into operational management and, if you prove successful, general management positions. Chapters 11, 12, and 13 discussed these areas.
2. *Research and design careers.* These careers include research in new engineering technology and the advanced design and preliminary development of sophisticated new systems. Engineers in this career field must exert a continuing effort to stay at the top of their technology, using some of the methods discussed later in this chapter. Chapters 9 and 10 and, from an advanced development point of view, Chapters 14 and 15 (project management) sought to give you some feeling for these careers.
3. *Engineering management careers.* As suggested in Chapter 1, about two-thirds of engineers traditionally have spent about the last two-thirds of their careers with some level of management responsibility. The first half of this chapter discusses the engineer's transition to manager.

4. *Entrepreneurial careers.* Quite a few engineers opt at some point in their careers to form their own company, either alone or with selected colleagues. This company might be involved in research, design, manufacturing a product, or providing sales or service for a product. The risks of failure are greater, but so is the potential reward.

5. *Consulting careers.* These are careers in which you use your engineering knowledge and experience for the benefit of a variety of other organizations and individuals. For example, you might be asked to provide designs, give advice, solve problems, or provide expert testimony. To be successful, you will have to develop a substantial expertise and reputation in your selected area. You can combine this with an entrepreneurial career by forming your own consulting organization, but may want to develop a specialty by working for another organization first.

6. *Writing careers.* If you enjoy writing, you might find yourself writing for a technical magazine or journal, serving as a technical editor, writing training or maintenance manuals, preparing sales literature for technical products, or writing a text book.

7. *Academic careers.* You may decide you want to teach others. With some courses in education you can fill a great need teaching math or science in a high school, or with a master's degree you can teach these topics or preengineering subjects in teaching colleges. For a career in university teaching in most engineering colleges, a doctorate is mandatory. A career of university teaching can be combined effectively with research, consulting, and writing. Indeed, progress in research and technical publication are necessary for obtaining tenure (keeping a faculty job) at most engineering colleges.

8. *Other careers.* You always have the choice of leaving engineering and going into something entirely different. The training in problem solving you have received as an engineer can give you an advantage in thousands of other positions.

Career Stages

Super proposed in *The Psychology of Careers* the following five-step career sequence, together with fixed age periods that applied to them:

1. The *growth* stage (birth to age 14), from the first awareness of impending career decisions to the initial development of career aspirations, interests, and abilities

2. *Exploration* (ages 15 to 24), involving making and trying out tentative choices, transition involving entering the labor market or advanced training, and trial in a beginning job

3. *Establishment* (ages 24 to 44). "The first five years of this stage are considered to comprise a trial period in which one or two changes in the field of work would be made before a life work would be found or it became clear that life work would be a series of unrelated jobs. The last fifteen years in this stage are classified as a stabilization period in which the individual acknowledges commitment to the life's work and to the organization by becoming socialized, progressing, and making a secure place in the field or organization."

4. The *maintenance* stage (ages 45 to 65), in which the primary concern is to hold on to the place achieved in the world of work, and the person continues along established patterns

5. The *disengagement* stage, in which physical and mental powers decline, and participation in the working world changes and then stops

Super and others later realized that these stages may occur at widely varying ages: Establishment may be interrupted by a change in societal needs or a *midlife crisis*, resulting in reentry into the exploration stage; disengagement may be deferred into the 70s or even 80s, or it may be triggered by early retirement from industry in the early 50s (or the military in the 40s), leading to exploration of a new career and a new cycle.

Dalton and Thompson proposed a sequence of four career stages for professionals: *apprentice, colleague, mentor*, and *sponsor*. Morrison and Vosburgh summarize their findings:

1. The new professional serves as an apprentice and learns to be an effective subordinate who demonstrates willingness to do routine assignments, yet aggressively searches out new and more challenging tasks. By leaving this stage too early, the individual does not learn from the experience of others....

2. The young professional earns the way into the second, collegial stage by building a reputation as a technically competent individual....The individual becomes less dependent and starts to contribute personal ideas about what to do in a given situation. Many professionals stay in this stage for the rest of their careers and have a reasonably successful career although their value to the organization dwindles over time....

3. Movement into stage 3, mentor, takes place because the individual is able to take increased responsibility for influencing, guiding, directing, and developing other people....The individual in this stage may serve with one or a combination of three roles: an informal mentor, a manager, or the *idea person* [gatekeeper]....80 percent of those who make it to this stage were perceived by the organization to be above-average performers after age 40. [Mentors are busy professionals; the young engineer is well advised to seek out a potential mentor from among successful senior engineers or managers and *ask* if they would be willing to provide periodic coaching sessions.

4. Stage 4, sponsor, requires that the individual move up from influencing groups of individuals to affecting the direction of the organization or a major segment of it....These people can play one or more of three roles: manager, internal entrepreneur, or idea innovator [the senior technical person in a field].

COMMUNICATING YOUR IDEAS

Importance of Communication

Some Definitions.   Communication is the means by which information is made productive. The word stems from the Latin *communicare*, meaning *to impart* or *to make common*. Weihrich and Koontz define communication as "the transfer of information from the sender to the receiver, with the information being understood by the receiver." It is this mutual understanding that makes communication difficult when poorly done, but effective when done well.

Importance to the Engineer.   Engineering may be considered as a transformation process in which information is received, transformed in some way, and the results transmitted to others, as shown in Figure 17-1.

An Engineering Career in the "Gig" Economy

As noted in Chapter 6, as of 2018, Gallup estimated that 36 percent of U.S. workers were engaged in the "Gig" economy, a shocking 57 million people.[1] But what does this mean for a new engineering graduate as they enter their engineering career? First, we must understand what the gig economy is and is not. While the definition is still evolving, the gig economy is generally accepted to include all workers employed as independent contractors. That includes those working full time in this capacity and those who find part time work—either as their only source of employment or in addition to a regular full-time job (such as the consulting work of many engineering faculty). While those working full time in this manner gain the most attention, a far larger number are part of the latter group. In fact, Jolley disputes the oft repeated statement that the gig economy will continue to grow and there will be few regular positions left.[2]

So, while the gig economy might not replace the regular economy, the move to contract work will likely impact your engineering career and for that reason, it is important to understand how to succeed in these roles. Fortunately, engineering is a profession well suited to this type of contract work, given our focus on projects (discussed in Chapters 14 and 15), how commonly design work is put out to contract, and the defined nature of the engineering skillset.[3] Benefits of pursuing this path include an ability to focus on developing depth in a specific area, while enjoying the benefit of remote work to live anywhere of the engineer's choosing. Drawbacks include the lack of a predictable income stream and loss of employer paid benefits along with the risk of social isolation.

Engineers interested in this career path are advised to consider the actions that will make them successful in this environment and how well those actions fit with their skillset. In a recent *Harvard Business Review* article, Petriglieri, Ashford and Wrzesniewski distilled a study of 65 gig workers into a list of what separates those who are successful from those who are not:[4]

- *Produce or Perish*—in this environment, you are only as successful as your last deliverable. This tends to create a deep focus on productivity. In other words, the freedom provided by the gig economy is precarious.
- *Connection to Place*—successful workers in this environment have a defined workspace they can retreat to and avoid distractions. Even if that space is a favorite booth at a favorite coffee shop.
- *Connection to Routine*—by building a set routine, freelance workers remove the thoughts required to determine where to go or what to do, thus freeing mind space to focus on the truly important work.
- *Connection to Purpose*—over time, successful workers build a sense of why they work and why it matters to them and others. This provides resiliency when downturns happen.
- *Connection to People*—successful gig workers are aware of the risks of isolation and actively work to avoid it by cultivating networks and relationships with mentors.

Engineers seeking to strike out on their own are advised to consider how interested they are in developing these skills and how successful they think that development will be.

Sources

1. McCue, T.J., *57 Million U.S. Workers Are Part Of The Gig Economy*. 2019, @forbes.
2. Jolley, D. *Myths of the Gig Economy, Corrected*. Harvard Business Review, 2018.
3. Ordman, N. *How Will the Gig Economy Affect Engineering Careers?* Engineering 360, 2018.
4. Petriglieri, G. S. Ashford, and A. Wrzesniewski, Thriving in the Gig Economy. *Harvard Business Review,* March–April 2018.

Figure 17-1   Engineering as a transformation process.

The information *input* includes the statement of work, directives, methodologies learned in college or elsewhere, and company standards and practices. The *engineering transformation* involves a complex process of analysis and synthesis that requires substantial resources, time, and skill, but which is largely hidden to the outsider. The information *output* may be in the form of physical models, drawings, specifications, technical reports, and/or oral briefings.

Unless the information output is properly communicated, the meticulous engineering performed may be of little utility. Yet many engineers and scientists lose interest in a problem once they have solved it, and do not spend that extra effort in writing an effective report or documenting a computer program that would give their work real utility. Critics of engineering education have complained for generations, with no resolution, that *engineers can't communicate*. A problem that continues to persist today.

Importance to the Manager.   Communication is the principal business of the manager, consuming an estimated 90 percent of their time. Mintzberg estimates that managers spend 78 percent of their time just in oral communication (59 percent in scheduled and 10 percent in unscheduled meetings, 6 percent on the telephone, and 3 percent "managing by walking around" on tours); the remaining 22 percent of their time is deskwork, which consists in large measure of reading the communications of others and preparing written communications. Obviously, as engineers make the transition to manager, they must improve their communication skills further if they are to be effective.

Modeling the Communication Process.   Figure 17-2 diagrams the communications process concisely; let us discuss each step in turn:

- The thought of the sender must be *encoded* into English or some other language, a computer code, mathematical expression, or drawing with special consideration of the nature of the intended receiver.
- The code must then be *transmitted* via some selected *medium*; several are discussed in the next section.
- *Reception* of the message may be hindered because of distractions (*noise*) inhibiting the transmission or causing inattention in the reception.
- The message then must be *decoded*, which is effective only if sender and receiver both attach the same or similar meanings to the symbols used in the message.
- *Understanding* may be obstructed by prejudices, or by a desire not to hear or believe what is actually being said.
- *Feedback* that enables the sender to determine what the receiver actually understood of the message permits the correction of misunderstanding. Verbal feedback offers the same potential for misunderstanding as did the initial transmission, but face-to-face feedback is enhanced by nonverbal communication (discussed later).

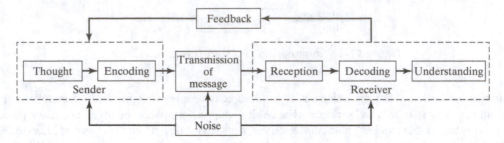

Figure 17-2   Communications process model. (From Harold Koontz and Heinz Weihrich, *Management*, 9th ed., McGraw-Hill Book Company, Inc., New York, 1988, p. 463.)

Communication Methods Compared

Characteristics.   Communications are transmitted in a variety of forms, each with its advantages and disadvantages.

The nature of the most common methods (until the mobile device) is indicated in Table 17-1.

Recently texting has become one of the **most common** forms of **communication.** It has also become a **business** and marketing tool, and a classroom helper. Today, even texting is being replaced with other forms of instant or structured communications in business, such as Slack, Instagram, and others.

Retention of Information.   Studies of learning and experience equate retention of information as follows:

We Tend to Remember: When Our Involvement Is:

10 percent of what we read	Passive reading
20 percent of what we hear	Passive verbal receiving
30 percent of what we see	Passive visual receiving
50 percent of what we hear and see	Passive verbal and visual receiving
70 percent of what we say	Receiving and participating
90 percent of what we say and do	Being

Table 17-1   Characteristics of Common Communication Methods

Communication Method	Speed	Feedback	Record Kept?	Formality	Complexity	Cost
Informal conversation	Fast	High	No	Informal	Simple	Low
Telephone conversation	Fast	Medium	No	Informal	Simple	Low-medium
Formal oral presentation	Medium	High	Varies	Formal	Medium	Medium
Informal note	Medium	Low	Maybe	Informal	Simple	Low
Memo	Medium	Low	Yes	Informal	Low	Low-medium
Letter	Slow	Low	Yes	Formal	Medium	Medium
Formal report	Very slow	Low	Yes	Very formal	Complex	High

Source: Paul R. Timm, *Managerial Communication: A Finger on the Pulse*, 2d ed., 1986, p. 59. Adapted by permission of Prentice-Hall, Inc., Englewood Cliffs, NJ.

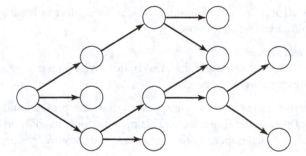

Figure 17-3 Grapevine.

Effectiveness. Timm describes a communications experiment in which the effectiveness of four media were compared. In order of *decreasing effectiveness* they were as follows:

- Oral plus written presentation
- Oral only
- Written only
- The grapevine

Timm's *oral plus written presentation* is strengthened further by (1) effective diagrams, illustrations, or demonstration; (2) feedback involving participation or *repeat back* presentation by the listener; and (3) where full comprehension is essential, simulation or on-the-job practice.

The *grapevine* of an organization is its informal communication system. It is a natural and inevitable occurrence in every organization that formal communication will be supplemented by informal transmission of information and rumors (both true and false) from employee to employee. The general pattern of a grapevine, shown in Figure 17-3, is not unlike a nuclear reaction. One person (fissioning atom) transmits a fact or rumor to several people (emits several neutrons). One or more (seldom all) of the recipients will pass the rumor on, often inaccurately, to one or several other people. When a rumor concerns something of great importance to most employees, it can spread at explosive speed (exceeding critical mass, as it were). Effective managers learn to monitor the grapevine (or persuade a trusted employee to do so), then make sure that factual information is published by official means to moderate transmission of misinformation.

Other Factors in Effective Communication

Active Listening. The art of effective listening is as important as effective communication. Listen positively and attentively, allowing the speaker to make his or her point. Analyze the speaker's attitude and frame of mind. Is the person an optimist or pessimist? Generally reliable or unpredictable? Try to reach beyond the speaker's words to his or her meaning. When in doubt, rephrase the speaker's words with a "Do I understand that … ?" Take notes of essential points unless that inhibits the communication. Finally, consider the speaker's nonverbal language (discussed next), as well.

Nonverbal Communication.   Albert Mehrabian divides the relative influence of the verbal, vocal, and facial aspects of oral communication as follows:

- 7 percent: Verbal (words used)
- 38 percent: Vocal (pitch, stress, tone, length, and frequency of pauses)
- 55 percent: Facial (expression, eye contact)

Obviously, the effective manager must learn, just as does the professional actor, that the way one makes a presentation is of paramount importance. This must include, in addition to the presentation from the neck up just described, the importance of *body language* (your posture, gestures, and body movement).

Communication Tools of Special Importance to the Engineer

The Written Report.   The results of engineering studies are often documented in formal written reports and executive summaries, and the usefulness of the study is determined by whether the report is (1) read and (2) understood. A few important considerations for effective report writing include the following:

- Be sure that you take sufficient time at the end of an assignment to report on it effectively.
- Begin your report with an **executive summary** of one page, whose content answers the following question: If the busy reader is only going to look at this page, what do they need to know?
- Consider putting your conclusions and recommendations at the front, followed by essential discussion, with peripheral material relegated to appendixes.
- Outline your report carefully, then write to your outline, and finally, take the time to review your work for clarity. Spelling and grammar will be judged as an indication of the value of the writer and the report.

Executive Summary

Executive summaries are much like other summaries in that their main goal is to provide a condensed version of a longer report's content. Executive summaries are also usually the first things read, and so it is a very important part of your report. The key difference from other summaries, however, is that executive summaries are written for someone who most likely does not have time to read the original. Your executive summary should be no more than a page (two max), and it should summarize all the other sections of your report.

The Oral Briefing or Presentation.   A briefing is an oral presentation of analyzed and synthesized information, presented to a person or group of people who have a need or desire for knowledge, but who do not have the time to become thoroughly familiar with all the details of a subject. It generally includes audiovisual aids such as PowerPoint slides, videos, models, or samples, and it may be supplemented by a written report. Usually, there will be an opportunity during or after the presentation for listeners to ask questions or make comments.

The essence of effective oral briefing is preparation and practice. Preparation includes steps of defining the objectives you wish to achieve, identifying your audience, outlining what you plan to cover, filling in the details, and preparing effective supporting materials. Practicing identifies weak areas in your presentation, helps ensure that the briefing flows smoothly, and assures that your presentation will fit within the allotted time. Effective briefing skills are essential for career success, for the busy executives who will make the final decision on whatever you are proposing often can be approached in no other way.

Visual Aids.   Visual aids used for the oral presentation commonly include PowerPoint slides. Today there are other visual aids like *Prezi* or Google Slides. Morse provides some guidelines for using visual aids.

- As in all lectures, there must be **a logical flow** to your presentation, not just a series of bulleted lists.
- Make your **presentation readable.** Good guidelines are: Title: 44 font bold: Text: 36 font bold Arial or Times New Roman or another clear font
- **Avoid sentences, paragraphs, or long blocks of text.** If you must use a paragraph, use an excerpt or a couple of sentences.
- **Avoid "Title Capitalization"** unless it is a title. Sentence capitalization is much easier to read.
- **Fancy slide transitions and fly-ins get old quickly.** Keep things simple. Avoid sound effects—they serve no other purpose than distracting the audience.
- **Expand one slide into two.** If your text does not fit well on one slide, split it into two.
- **Use the slide master for consistency.** Start with the slide master to set up the layout for your slides and create a coherent, consistent look.
- **Choose your background based on the room's lighting.** While conventional wisdom is to design with a dark background when you plan to project your slideshow onto a screen, in reality, the best background has to do with the presentation location.
- **Black or white out a screen.** If you stop to discuss a point and do not want people staring at the screen, black it out. Press Shift+B. In a light room, you can also white it out—press Shift+W. Press B or W again to continue your presentation.
- **Keep the format of the visual aids the same** throughout your presentation. Do not switch from horizontal to vertical layouts. Be consistent in your format, color, and style.
- **For optimum readability**, have the text of your visual aids set flush left, non justified.
- **Avoid reading your slides.** Slides supplement your presentation, not the other way around. Remember you deliver the content, not the slides.

STAYING TECHNICALLY COMPETENT

The Threat of Obsolescence

The Knowledge Explosion.   Well over 300 years ago, when the first scientific journals were begun, it was possible for a single scientist to be aware, through correspondence, of most of the scientific discoveries being made in the Western world. Once begun, scientific journals began multiplying rapidly, doubling in number about every 15 years. Putka estimated in 1987 that U.S. scientific journals then totaled about 5,000. (These include journals dealing with very narrow disciplines, such as the *International Review for the*

Sociology of Sport, the *Journal of Molluscan Studies*, and the *Fibonacci Quarterly*, which deals only with applications of the Fibonacci number series.) Dieter estimated in 1983 that current world output "amounts to 2 million technical papers a day, or a daily output that would fill seven sets of the *Encyclopedia Brittanica*." The rate of information creation has only accelerated in the last 25 years.

Kaufman reports that, when the number of journals reached the critical number of 300 in the mid-nineteenth century, abstract journals began to appear to provide access to the literature. By the mid-twentieth century, abstract journals themselves reached the critical 300 level. Computer abstracting and computer searches of massive national databases are current attempts to master the exponentially increasing flood of knowledge in science and technology, but today no single technologist can know, or even locate, all the relevant information even in a fairly narrow specialty. Electronic publishing is a new area of information dissemination. One definition of electronic publishing is in the context of the scientific journal. It is the presentation of scholarly scientific results in only an electronic (nonpaper) form.

Obsolescence Defined.   Obsolescence has been defined as the process of passing out of use or usefulness or even the process of being replaced by something newer or better (which, to the midcareer engineer, could be a new graduate equipped with the most modern education). Shannon in his text amplifies this:

> Persons are obsolescent technically if, when compared to other members of their profession, they are not familiar with, or are otherwise unfitted to apply, the knowledge, methods, and techniques that are generally considered important by members of their profession.

Organizational Obsolescence.   Thompson and Dalton believe that organizational obsolescence is a greater culprit, and they suggest three areas in which managers can make improvements and thus avoid having an obsolete organization: reward technical contribution, reduce barriers to movement, and focus on careers.

However, individual engineers cannot and should not depend on their employer to combat obsolescence; they must take personal responsibility for their own career progress.

Methods of Reducing Obsolescence

Mastering the Technical Literature.   If, as has been repeatedly stated, technical knowledge doubles in quantity every 10 years, the working engineer or scientist who quits learning upon leaving college will know only half what they need to know 10 years later, and a quarter 20 years later. Technical managers have an even greater problem, since they need to stay generally knowledgeable about a wide range of technology, including areas they did not touch in school. Yet, the average time engineers and scientists spend on keeping up to date with the professional literature is only *one hour a week*, a small fraction of the time required to stay competent.

Continuing Education.   One of the conclusions of the National Academy of Sciences in its 2004 report *The Engineer of 2020: Visions of Engineering in the New Century* highlights the national importance of continuing technical education:

> They will need this [continuing education] not only because technology will change quickly, but also because the career trajectories of engineers will take on many new directions—directions that include different parts of the world and different types of challenges and that engage different types of people and objectives. To be successful the engineer of 2020 will learn continuously throughout their career, not just about engineering, but also about history, politics, business, and so forth.

Certainly, the need for lifelong learning is nothing new. Hsun Tzu (298–238 b.c.) stressed the following lesson:

> Learning continues until death and only then does it cease....the objective of learning must never for an instant be given up. To pursue it is to be a [person] to give it up is to become a beast.

Many engineers try to keep themselves current by taking a series of educational courses; these may be formal graduate-level university courses, whether leading to a graduate degree or not, or in-house courses offered by their employer. Younger engineers tend to prefer formal graduate courses leading to a master's degree where a suitable program is available. Kaufman reports that "there is consistent evidence that professionals who have taken graduate courses are perceived as less obsolescent by themselves, their supervisors, and their colleagues," that they are less likely to be laid off by their employers, and that professionals with a graduate degree are unemployed a shorter period if they do lose their jobs. Kaufman compares graduate and in-house noncredit courses:

> A study of 2,500 technical professionals in six organizations found that not only were those with graduate training at the M.S. level better performers than those who had only a B.S. but their high level of performance was maintained ten years after the B.S. holders began to decline in performance. It would appear that a heavy dosage of graduate courses can push obsolescence back by ten years....

> Many organizations offer in-house noncredit courses for their professionals on a regular basis. The courses are offered primarily to supplement university-sponsored courses. They are usually more directly applicable to the work of the organization and are less demanding than are graduate courses, since they typically do not involve grades, examinations, or even homework.

On-the-Job Activity. The most important vehicle to reduce obsolescence is personal growth on the job itself. In a survey of 290 professionals, Margulies and Raia reported that 42 percent saw on-the-job problem solving and 20 percent saw colleague interaction as most important for professional growth, with 16 percent citing publishing and independent reading and only 14 percent citing formal coursework as most important. More recently, Farr reported as follows:

> Another group, one hundred engineers in several organizations, indicated that the best aids to updating their technical knowledge and skills were immersion in state-of-the-art technology in their work and having free time available to work on new ideas. The primary inhibitors to updating were non-challenging assignments and lots of nontechnical work.

Supervisors of professionals play an important role in providing challenging work assignments to the engineer and adequate technician and clerical assistance to perform routine activities; if the supervisor does not provide such growth opportunities, the engineer must find them, even if in another organization. The employer can assist professional development in a number of other ways: by providing time and a supportive atmosphere for technical reading, self-learning, and preparation of technical papers; by providing a truly professional information acquisition and distribution system; and by a professional approach to selecting the kinds of continuing education programs to support, and assessing results obtained from them.

PROFESSIONAL ACTIVITY

Professional Societies

Types and Purpose of Technical Societies.   There are a bewildering variety of professional societies seeking the membership and support of the engineer. Weinert divides the U.S. organizations roughly into four major groupings:

1. Those focused on established or emerging engineering disciplines. These include the five **founder societies**:
    *   American Society of Civil Engineers (ASCE, founded 1852, www.asce.org)
    *   American Institute of Mining, Metallurgical, and Petroleum Engineers (AIME, founded 1871, and now divided into four member societies)
    *   American Society of Mechanical Engineers (ASME, founded 1880, www.asme.org)
    *   Institute of Electrical and Electronic Engineers (IEEE, founded 1884 as AIEE, www.ieee.org)
    *   American Institute of Chemical Engineers (AIChE, founded 1908, www.AIChE.org)
2. Those focused on a broad occupational field, such as the Society of Automotive Engineers (SAE, www.sae.org) or the Society of American Military Engineers (SAME, www.same.org).
3. Weinert's fastest growing group are those focused on a specific technology, group of technologies, or one of the specific materials or forces of nature always referred to in classic definitions of engineering. These include such groups as the American Society of Heating, Refrigerating, and Air-Conditioning Engineers (ASHRAE, www.ashrae.org), the Society of Plastics Engineers (SPE, www.4spe.org), Project Management Institute (PMI, www.pmi.org), the Society of Manufacturing Engineers (SME, www.sme.org), and the American Society for Quality (ASQ, www.asq.org). (In universities these technologies are usually covered, if at all, as subdivisions of major engineering disciplines, such as mechanical, chemical, or industrial engineering.)
4. His "final group is composed of those associations and societies formed either by individual engineers or by groups of societies to accomplish a specific purpose." This includes ABET (formerly the Accreditation Board for Engineering and Technology, www.abet.org), formed by the "founder societies" and other disciplinary societies to accredit engineering college programs; the National Society of Professional Engineers (NSPE, www.nspe.org), with its interest in professional registration, engineering ethics, and public policy; and the National Council of Examiners for Engineering and Surveying (NCEE, www.ncees.org), formed to coordinate the state licensing process. Although most large disciplinary societies have divisions or committees on engineering education and on management or engineering management, the American Society for Engineering Education (ASEE, www.asee.org) provides an interdisciplinary forum for engineering education as an entity, and the American Society for Engineering Management (ASEM, www.asem.org) an interdisciplinary forum in the management of technological activities.

Reasons for Getting Involved.   Engineers owe it to their profession and to themselves to belong to at least one professional society and to support it with their dues and their effort. The larger disciplinary engineering societies maintain (through their activity in ABET) the quality of engineers entering the field. Societies of all four groupings provide a range of professional publications that you will need to keep professionally current in your field(s) of interest, hold annual (and often regional and local) conferences where you can keep abreast of new developments and share problems and solutions with leaders in your field of interest, and often sponsor educational programs important to your development. Many societies will have local sections in your city (or the metropolitan area nearest you). Local sections often meet for lunch or dinner monthly, giving you an opportunity to meet others in the area who share common problems and listen to speakers on topics of interest. Most local sections need additional volunteers for section activity, and service as a local section committee chair or officer can provide an early and satisfying opportunity to demonstrate your capacity for professional leadership while developing your professional network.

Many technical professionals find it important to maintain an active membership in a number of societies because of their varied interests and responsibilities, and find that their memberships change as they progress through their career. Most engineers will first join the society of their undergraduate discipline (ASCE, ASME, IEEE, IISE), often as a student, and may continue in that society for life or as long as they continue to view themselves as belonging to that discipline. Current memberships often include a society focused on the industry they work in, and one or more on the function they are currently practicing. Industrial employers are typically willing to pay dues of appropriate societies for their employees, easing the burden of multiple memberships.

Technical Papers and Publications.   Professions depend on the willingness of individuals to share their discoveries and observations with others for their progress. You can do this by offering to write a paper for presentation at a regional or national technical meeting; often you will find your employer willing to pay your expenses to attend a meeting at which you are speaking. Or you can offer an article to a professional magazine or journal. For a university professor, such publication is usually necessary for continued employment. As a result, professional society journals are often inundated with theoretical articles from faculty authors, but these journals are eager to receive more applied articles from practitioners. If you are working in industry or government, you may not have the same compelling reason to offer papers to meetings or publications to professional magazines, but you should look for opportunities to do so as a part of your personal growth and as a service to your profession.

Accreditation, Registration, and Certification

Three related topics are discussed in this section. **Accreditation** is a voluntary process where a designated agency grants recognition to an educational program that meets certain minimum standards. Engineering **registration** in the United States is granted by the several states to individuals meeting specified criteria. Professional **certification** of engineers is a voluntary process regulated by certain professional societies.

ABET Accreditation.   ABET (www.abet.org) has the responsibility for accrediting those U.S. engineering curricula that apply for such consideration. ABET is controlled by a board of directors consisting of members designated by the major engineering professional societies, and is funded by those societies and the universities seeking accreditation. ABET has an Engineering Accreditation Commission (EAC) and a Technology Accreditation Commission (TAC) responsible for establishing accreditation criteria, visiting universities, and recommending accrediting action to the ABET Board in engineering and engineering technology, respectively.

The ABET visiting team includes a visitor for each curriculum being examined, chosen from a list of practicing engineers and engineering educators established by the engineering society responsible for that discipline. Visitors analyze curriculum content, faculty, facilities, funding, student preparation, and other factors against ABET guidelines. For example, an undergraduate engineering curriculum must be shown to include 32 semester hours (or equivalent) of a combined mathematics and basic sciences), and 48 semester hours of engineering topics, consisting of engineering sciences and engineering design appropriate to the student's field of study. Accreditation visitors are reimbursed for their travel expenses, but not for their time. The engineer who volunteers for such activity provides a service that is the hallmark of serving their profession.

Professional Engineering Registration.   Engineers who are eligible should seriously consider becoming registered with their state as a Professional Engineer. This is usually a two-step process. The first step, leading to the Engineer in Training (EIT) designation, usually requires an approved engineering degree (which in some states can only be from a curriculum accredited by ABET as just described) and successful completion of the Fundamentals of Engineering (FE) test on topics such as calculus, physics, statics, thermodynamics, electrical circuits, and engineering economy (which may be taken while still an undergraduate). Not all topics are needed to pass the test. The EIT provides an additional credential for the young engineer, and engineering students are well advised to sit for the engineering fundamentals examination while the content is still fresh in their minds (and before it changes), rather than waiting until their career path might take them in a direction requiring registration.

After graduation and about four years of acceptable engineering practice, the candidate can then sit for an eight-hour test on the principles and practices of engineering; if successful, they will be registered as a Professional Engineer and can then append the designation PE following their name. Although initial registration is by the state giving the examination, other states may extend reciprocity and offer registration to engineers already registered with a state whose requirements are at least as rigid.

Requirements for the EIT and PE differ from state to state (www.nspe.org). Some states permit graduates with approved science, less-than-approved engineering, or (in some cases) engineering technology degrees to sit for the EIT and PE, commonly after a longer experience requirement; a number of states have a long-established practice (LEP) category requiring 12 to 20 years of responsible engineering practice and tests, but without a degree specification.

Corporations engaging in the practice of engineering legally do so only under the direction of individuals in the firm who are registered professional engineers, and in some local government activities, such as public works, an engineer cannot progress very far without becoming registered. Civil engineers are most likely to need registration for career success, but others may, too. Although the majority of engineers in industry are not required to be registered, there is continual pressure to increase the kinds of positions for which society demands registration.

Certification.   A number of professions in America have programs to examine individuals and grant those especially knowledgeable in their field recognition, often characterized by a designation they can place after their names. Common examples are Certified Professional Engineering Manager (CPEM), Project Management Professional (PMP), and Professional Engineer (PE). A number of engineering professional societies, especially those whose body of knowledge does not commonly lead to an undergraduate degree, have chosen to do this as well.

A typical engineering certification requires a combination of up to 10 years of education and experience, and passing of a day-long examination offered twice a year; several societies offer an associate certification to recognize earlier levels of preparation. Certification is not restricted to members of the sponsoring society, but nonmembers commonly pay a fee differential not greater than annual society dues. The sponsoring society may offer short courses and study guides to help prepare for examination and may require recertification every three to five years at a lesser fee to demonstrate continuing competence.

DIVERSITY IN ENGINEERING AND MANAGEMENT

Forty years ago, the ranks of American engineers, especially engineering managers, were almost completely white and male. Although this situation has improved and continues to change, there still exist attitudes and perceptions that can make career opportunities different for underrepresented groups (e.g., women or minorities) than for a white male, and they need to be understood. Also, employers who expect to be successful in the twenty-first century and beyond must learn how to employ the full range of its workers effectively, since it has been estimated that 80 percent of those being added to the work force in the near future will be groups currently underrepresented in engineering. In addition to underrepresented groups in the United States, there are a growing number of engineers educated elsewhere in the world, and those changes are discussed in Chapter 18.

Women as Engineering Students

Although four-year degree programs in engineering for men began in 1817 (at the U.S. Military Academy) and grew rapidly, it was not until 1892 that the first woman received a U.S. engineering degree (Elmina Wilson, in civil engineering from Iowa State). Careers of the few who followed in the next several decades often ended with marriage. By 1920, when women got the right to vote, only 90 women had received engineering degrees (from 20 U.S. schools). During World War II many engineering schools set up engineering training programs for women, but both before that war (during the Great Depression) and immediately following it women were pressured to become homemakers, so men could have the jobs. During the 62-year period from the first degree in 1892 to 1953, only one engineering school averaged over one degree to a woman per year (Purdue, with 103; the next highest was the University of Colorado with 47; the presence of Lillian Gilbreth [see Chapter 2] on the Purdue faculty between 1935 and 1950 and as advisor to women students there long after that surely contributed to Purdue's record). In addition, many engineering schools and professional engineering societies did not admit women (Tau Beta Pi awarded them a *Badge* rather than membership), and this led to the formation of the Society of Women Engineers about 1949.

In 1970 women made up less than 1 percent of engineering bachelor's degrees. The number then began to climb, reaching almost 10 percent by 1980 and 15 percent by 1985, but had reached only 21.3 percent in 2017. Graduate engineering degrees for women have increased and reached 25.7 percent of master's and 23.5 percent of doctoral engineering degrees in 2017 (Table 17-2).

Research shows that this discrepancy is not due to a lack of motivation or ability or to academic preparation of women students. Instead, it seems that environmental and societal factors are largely responsible for deterring women from entering or persisting in engineering. Competitive and unwelcoming classroom environments hinder women from persisting in their pursuit of engineering degrees. Because of the

Table 17-2   Bachelor Degrees for Underrepresented Groups

Engineering	1985	%	2008	%	2012	%	2017	%
Total of all bachelor degrees	77,571		74,170		88,176		124,477	
Women		14.5		18.1		18.4		21.3
African Americans		2.6		4.7		4.2		4.1
Asian Americans		5.8		13.0		12.2		14.6
Hispanic Americans		2.8		6.5		9.0		11.1
Native Americans		0.3						0.3

Source: http://www.asee.org/publications/college-profiles/2017-Engineering-by-Numbers-Engineering-Statistics.pdf, December 2017.

propensity of male-dominated stereotypes and examples in society and the college classroom, women in engineering may question their ability or commitment more than their male counterparts. Also, two highly important predictors of academic persistence and success—mentoring and research experiences—may be less readily available to women students.

Formalized women in engineering programs, which often include mentoring components, have become an important part of supporting and encouraging women students in engineering. Intervention and support programs during the first year of undergraduate education may be particularly important for underrepresented students, such as women in engineering. Unfortunately, from the beginning, external and internal deterrents may hinder women from their pursuit of engineering degrees. Much like graduate students and new faculty members during their first year, the first year of undergraduate education has important implications for socialization, achievement, and persistence. For undergraduate students, early academic and social experiences matter most during the first year of college and greatly influence persistence. So, if women enter these programs at a disadvantage, not due to motivation, academic preparation, and support among first-year students but due to lower confidence in their abilities, or if during the first year they show a significant drop in self-confidence, there could be negative consequences for retention.

Other Underrepresented Groups in Engineering

The number of bachelor's degrees obtained by underrepresented groups has increased slightly in the last few years, as shown in Table 17-2.

The low percentages of African Americans, Hispanics, and Native Americans entering engineering colleges and their low retention rates can be traced to a number of problems. Several engineering schools have been able to improve the retention rates for these groups through carefully conceived minority engineering programs (MEPs). The MEP model emphasizes three structural elements: (1) a formal orientation course for new freshmen featuring orientation to the university and its engineering program; (2) clustering MEP students in common sections of their classes to reduce ethnic isolation and encourage group study; and (3) providing a student study center for MEP student use.

A number of organizations exist to support members of under-represented groups in engineering. These include the following:

- American Indian Science and Engineering Society (AISES)
- League of United Latin American Citizens (LULAC)

- Mathematics, Engineering, Science Achievement (MESA)
- Mexican American Engineering Society (MAES)
- National Action Council for Minorities in Engineering (NACME)
- National Consortium for Graduate Degrees for Minorities in Engineering (GEM)
- National Society of Black Engineers (NSBE)
- SECME (Formerly Southeast Consortium for Minorities in Engineering)
- Society of Hispanic Professional Engineers (SHPE)
- National Association of Minority Engineering Program Administrators.

MANAGEMENT AND THE ENGINEER

Relation to the Engineering Career

As presented in Chapter 1, management responsibilities are part of the normal career progression of the engineer. In the early 1990s, however, American corporations began an organizational revolution (called downsizing, rightsizing, restructuring, or reengineering) that has led to massive reduction of middle management levels and positions and loss of staff specialist positions in favor of empowered teams, incorporating some specialist expertise. These structural changes appear to have become permanent and these teams are making decisions formerly reserved for managers. Today the engineer going into industry can expect a lower probability of being "in management" by a particular age, but they can also expect to need a broader understanding of "management" concerns, such as marketing and finance, than was needed of nonmanagers in earlier hierarchical organizations.

The probability and speed of transition into management positions and responsibilities will also depend on which of the career fields the engineer chooses to pursue. Careers beginning in operational areas such as manufacturing, customer service, and sales involve business considerations from the beginning, and in the past, they have commonly led to operations management positions. Careers in research or design (especially in advanced technology), in college teaching, or in technical writing can involve professional activity of increasing responsibility for much or all of a career without formally leading to management. Entrepreneurial and most consulting careers involve managerial concerns (if not the title) almost from the beginning. In government or military service, on the other hand, your power, prestige, and pay still seem to depend on the number of organizational levels. Here promotion into a managerial position still seems to come earlier, but often to involve less decision-making authority at a given level.

Need for Engineers in Top Management

Lawrence Grayson, former president of the American Society for Engineering Education, reported in 1989 that engineers were not reaching the top levels in American corporations as frequently as their counterparts in other countries:

> Engineers must be prepared for leadership—leadership in technical, corporate, and national affairs. More and more problems facing this country have strong technical components. Yet, engineers are not attaining the appropriate leadership positions and therefore have not been able to make the decisions that the nation requires.

In France, most of the leaders of business and government have graduated from the elite Grand Ecoles. The approximately 175 schools concentrate primarily on teaching engineering and technology. In Germany, a majority of the corporate leaders are alumni of the technical universities, whose graduate engineers have completed a period in industry and a thesis on an industrial problem. In Japan, more than 65 percent of the members of the boards of directors of the nation's leading companies have graduated from engineering and science programs, not graduate schools of business. In contrast, roughly two-thirds of the seats on boards of American companies are occupied by people trained in law, finance, or accounting.

Lester Thurow has explained the importance of a technical background in today's executives:

> These nontechnical managers may understand the technologies being employed by their firms, but they don't have enough background to develop intuitions on which of the possible technologies now on the horizon are apt to further develop and which are apt to be discarded.

> As a result, incumbent managers have no way to judge the merits of revolutionary changes in production technology. So, they procrastinate, waiting for it to become clear which technology is the best. By the time the answer is clear, foreign firms may have a two- to three-year lead in understanding and employing these new technologies.

> The problem is found not just among managers of manufacturing facilities. Those in the investment and marketing communities also don't know where to place their bets. The ignorance and resulting risk aversion of the industrial manager is reinforced by the ignorance and risk aversion of his investment banker, advertising manager and accountant.

On the other hand, the engineering mind-set can be a disadvantage in top-level politics (whether they be in government or a corporation). Columnist David Broder provides an example:

> [President Jimmy] Carter was a Naval Academy graduate and an engineer. His model of policy-making was rational, efficient, and introspective. Governing to him was an exercise in problem-solving. Come up with the right solution, check and recheck your calculations, and then act. Congress did not respond to that approach. And Carter did not react well when legislators tinkered with his solutions. Soon they were at odds.

Should You Choose Management?

Why Technologists Switch.   Badawy has found in career goal interviews he has conducted with engineers and scientists that about 80 percent have indicated their career goal was to become a supervisor or manager within five years. He classifies the reasons they have given into six categories:

1. *Financial advancement.* Technologists believe that the managerial ladder offers *better compensation* than does the technical path, even where a strong dual-ladder system (discussed later) exists.
2. *Authority, responsibility, and leadership.* Many believe that becoming a manager is the best way to make the right things happen.
3. *Power, influence, status, and prestige.* Engineers who seek managerial positions tend to rank higher in McClelland's need for power and to find satisfaction in influencing events.
4. *Advancement, achievement, and recognition.* Advancement in management positions provides a clear recognition for achievement.

5. *Fear of technological obsolescence.* Some engineers and scientists find it extremely difficult to keep up with the complexity and rate of technological change, and they may see management as the only alternative.
6. *Random circumstances.* An engineer may drift into management because a position has suddenly become open or for a variety of other almost accidental reasons.

Making the Right Choice for You. Many of these reasons can be the wrong reasons, and they can encourage an effective engineer to become an ineffective and unhappy manager. If the opportunity to move events and to achieve larger things through the leadership of other people gives you the most satisfaction, you will find this in a management career. If, instead, you find more satisfaction in what you accomplish and create personally, you may want to remain primarily a technologist.

Each professional needs to take personal charge of their career, and to determine its direction rather than letting events take their own course. By letting your managers know what your career goals are, preparing yourself for them, asking for transfers or reassignments that will enhance your chance of getting there, and seeking a position elsewhere if you cannot achieve it in your present organization, you can largely become the master of your own fate. If you really want to be a technical specialist, you should be aware that the movement into management for any period of time may be largely irreversible, since most engineers find it much more difficult to stay competitive in an advanced technology once they get into a managerial job.

The Dual Career Ladder

Many organizations attempt to provide a technical career ladder that is the equivalent of the management career ladder. Figure 17-4 provides an example of a dual career ladder; position titles will vary from organization to organization, but the typical dual ladder offers three or four levels of parallel position titles. In a survey of dual-ladder systems, companies gave the following reasons for implementing them (with percent responding):

- Retain the best professional and technical people (90 percent).
- Create a career path for those not interested in management (88 percent).
- Increase morale of the technical staff (67 percent).
- Create a more equitable nonmanagement compensation structure (61 percent).

Unfortunately, in many organizations the technical ladder does not live up to its promise of equality with the managerial alternative. Badawy identifies five criteria that must be met for a dual-ladder system to work efficiently:

1. The technical and administrative ladders must be equally attractive to technologists in terms of salary scales and status symbols and other noneconomic rewards.

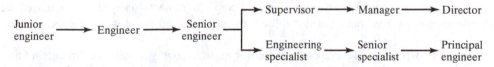

Figure 17-4 Example of dual-ladder system.

2. Neither ladder should be used as a dumping ground for individuals who are unsuccessful on the other ladder.
3. Criteria for promotion on the technical ladder must be rigorous and based on high technical competence and achievement.
4. Both ladders must have the full support of management.
5. The system must be fully accepted by the technical staff.

Preparing for Managerial Responsibilities

Requirements.   Effectiveness as a manager requires a combination of attributes, knowledge, and skills. Edwin Gee, while a senior vice president of DuPont, listed the following attributes of researchers that were indications that they would become good managers:

1. They are able to identify a problem, analyze it, and synthesize a solution.
2. They are willing to accept and even seek responsibility.
3. They view their current assignment as the most important thing they have to do rather than as a step toward promotion.
4. They have good work habits, set personal goals, and plan ahead.
5. They are able to get results without upsetting people.
6. They have integrity.
7. In addition to technical talent, they have demonstrated at least some skill in such fields as marketing, finance, and employee relations.
8. Finally, the ability to make sound judgments is the key to both long-term potential and current readiness for promotion.

Thamhain reports statistically significant correlation between personal desire to become a manager, actual promotion, and subsequent performance; but he adds that "personal desire alone is insufficient to gain a promotion. In the final analysis personal competence and organizational needs are the deciding factors. People who get promoted usually meet five key requirements":

1. Competence in current assignment and respect of colleagues
2. Demonstrated ability and willingness to take on greater responsibility (and good time management)
3. Preparation for new assignment through courses, seminars, on-the-job training, professional activities, and special assignments
4. Match of capabilities with current and long-range needs of the organization
5. Perceived aptitude for management

The skills that bring an engineer to prominence and result in promotion into a first management position are not necessarily the skills needed in the new management position. The engineer is a doer; an effective manager is a facilitator of other people's work. As we learned in Chapter 7, the new manager must learn to delegate, and the engineer's education and past experience have not prepared the new manager to do so. Indeed, the engineer and engineer-manager have quite different roles; these are contrasted by Amos and Sarchet in Table 17-3.

Formal Degree Programs.   To prepare yourself for management, you will need to obtain knowledge about management and to develop management skills, which are two different things. The most common

Table 17-3 Differences in Roles Between Engineers and Engineer Managers

Engineer's Roles	Engineer-Manager's Roles
Originates projects	Evaluates projects
Creates, seeks new ideas	Provides facilities to help engineers
Works on specific programs	Does overall planning
Has limited responsibilities	Has responsibility for a department or group of people
Is specialized, is technically oriented	Is people-oriented, is responsible for and responsive to people
Obtains facts themselves, is objective	Motivates others
Utilizes own skills	Utilizes skills of others to obtain goals
Has limited concern for finances	Has fiscal responsibilities

Source: John M. Amos and Bernard R. Sarchet, *Management for Engineers*, © Prentice-Hall, Inc., Englewood Cliffs, NJ, 1981, p. 16. Reprinted by permission.

programs for teaching knowledge about management are the Engineering Management degree, the Masters of Technology (MOT) degree, and the Master's degree in Business Administration (MBA).

The master's degree programs in engineering management typically require 30 to 36 semester hours of work beyond an acceptable engineering B.S.. This compares with requirements of from 42 to 60 semester hours for an engineer to obtain an MBA, largely because of the demand that engineers take remedial undergraduate courses in business subjects before tackling them at the master's level. The engineering management degree typically differs in that (1) courses are accelerated beyond the usual business first course, especially where quantitative capabilities are involved; (2) courses emphasize management of technical enterprises; and (3) faculty often have engineering backgrounds.

A commonly cited definition of Management of Technology comes from the 1987 National Research Council report *Management of Technology: The Hidden Competitive Advantage*:

> Management of technology links engineering, science, and management disciplines to plan, develop, and implement technological capabilities to shape and accomplish the strategic and operational objective of the organization.

Nondegree Coursework.   Of course, even formal courses in management do not have to be part of a credit program. Noncredit short courses are offered by universities, by professional societies, by independent consultant/entrepreneurs, and through the educational programs of large employers. Courses may be full-time for from one day to two weeks, or may be one or several times a week for some period, and may take place at your work site or elsewhere on your own time or your employer's time.

Experiential Training.   All the formal or informal coursework described previously still only provides *knowledge about* management. Role-playing, case studies, and management games attempt to develop understanding and enhance skills, but they are no substitute for real experience. Organizations that recognize the vital importance of developing future managers will have deliberate policies and programs to do so, and they will evaluate current managers partly on their success in developing future talent. Some of the methods used will be on-the-job training, coaching, selection of job assignments for their developmental value, job rotation, and temporary assignments to other areas. Often project assignments will give the new engineer an initial insight into the relation of his or her technical work to the total organization. McCall et al. give examples of the "veritable encyclopedia of executive education" that can be gleaned from appropriate assignments:

Assignments such as:	Can provide learning about:
Project/task forces	Giving up technical mastery; understanding other points of view
Line-to-staff switches	Coping with an ambiguous situation and understanding corporate strategies and cultures
Starting something from scratch	Identifying what is important and building a team
Turnaround jobs	Being tough, persuasive, and instrumental
Leaps in scope	Relying on other people and thinking like an executive

In the final analysis, each person must take responsibility for his or her own career. Let your immediate supervisor and others know, tactfully, what your interests and objectives are, and give them a chance to help you. Ask for an evaluation of what you need to do and what assignments you should look for to prepare yourself. Then, if your present organization cannot or will not offer what you are looking for, you owe it to yourself to consider if another organization can.

MANAGING YOUR TIME

Work Smarter, Not Longer

Time is a very democratic resource: The prince and the pauper both have exactly the same amount to spend in a day. Yet the engineering manager (and the engineer as well) never seems to have enough. Amos and Sarchet explain the problem:

> All new engineer[ing] managers constantly hear that to get ahead takes "hard work," which implies long hours, and that as [they are] promoted, [they] will assume greater responsibilities and have a greater span of management. However, each promotion brings more authority, which allows more delegation to get the work done through others. When the engineer puts in long hours at [their] office, does not take regular vacations, and spends little time with [their] family, [they] also [fail] to have time to develop the creative aspects of engineering management that are important to [their] success.

Amos and Sarchet provide some good ideas on planning time:

> Being busy is simple for the engineer manager, but being effective is difficult. Planning activities is a necessary requirement because [they do] not have time to do all the things that [their] conscience or imagination tells [them they] need to do, but [they] must decide what *to do* and what *not to do*.... The solution is not working long hours. Instead, it is setting priorities. Otherwise, [they] will constantly put second things first by default. Then [they] will be in trouble.

David Allen, a productivity consultant and the creator of the time management method known as *Getting Things Done*, suggests an organizational process for getting things done. The organizational process is a five-step approach: collect, process, organize, review, and do. Starting with the collect phase, the commitments must be collected into tangible elements and must be placed into a centralized *in* box. The next step is the process in which the *in* box is reviewed and each item is categorized as to what is the next action step for the item. This is critical to the organizational process, because understanding the next step can free the brain of what to do next. Tied to organizing is the process of reviewing, at regular intervals, the current commitments in your *in box*. The commitments must be reviewed and reorganized until a finite set of actions are required to be accomplished now. Next, without procrastinating any further, these actions must be completed until there are no more actions left.

Time Wasters and What to Do About Them

Every writer on managing time seems to have a laundry list of activities that waste time. Especially applicable is one by LeBoeuf, who surveyed 50 engineering managers in a number of countries and found their top 10 ranked as follows:

1. Inadequate, inaccurate, or delayed information
2. Ineffective delegation
3. Telephone interruptions
4. Meetings
5. Unclear communication
6. Crises
7. Leaving tasks unfinished
8. Indecision and procrastination
9. Drop-in visitors
10. Lack of self-discipline

No manager can completely avoid these problems, but the problems can be minimized through good time-management practices. These practices are:

1. *Information.* The effective manager thinks through the decisions that will require external information well ahead of time, sends out requests that define clearly what information will be needed, why, and when, and then gets on with other matters in the interim.
2. *Delegation.* As we have seen in Chapter 7, new engineering managers often find it difficult to trust others for matters they used to handle personally. Nonetheless, they will progress no further until they learn not only to assign jobs, but to delegate authority and still exact accountability for results.
3. *Telephone (and now email and instant message) interruptions.* Instant communications can be tyrants if you permit it. Make effective use of voicemail and electronic mail to avoid the *phone tag* of always catching the other person away from their phone. When you do get on the phone, give it your full attention and learn how to probe for the essentials of the call and bring it to a courteous yet prompt close.
4. *Meetings.* Think through what meetings you must attend, and which ones you can ask a subordinate to represent you in. For meetings that you initiate, ask first if the meeting is really necessary; if it is, use the techniques of preparation, meeting conduct, and follow-up discussed in Chapter 7 to make the meeting effective and efficient.

5. *Unclear communication.* Practice the techniques of oral and written communication discussed in Chapter 16.

6. *Crises.* Crises will occur and must be handled, and you must leave a degree of freedom in your schedule to handle the unexpected. Before charging off to fix the problem yourself, however, ask yourself who among your team or colleagues can shoulder some of the unexpected burden.

7. *Drop-in visitors.* Learn the difficult dividing line between being available to your team (and your colleagues, since you will need their help another time, and your manager, since helping them is your job) and spending excessive time on nonessentials. If you can make the contact in *their* office, it is much easier to get away!

Tools of Time Management

List Goals and Set Priorities.   The Swiss economist and sociologist Vilfredo Pareto is credited with the observation that in many collections of items, 80 percent of the value or importance is represented in only 20 percent of the items. Lakein emphasizes the need to (1) list possible long-, intermediate-, and short-term goals; (2) set priorities; (3) schedule the most important goals; and (4) follow through as scheduled.

Categorize your goals as either A (highest priority), B (lower priority), or C (desirable, but postponable), and the time that you estimate each will require. Schedule your most important A task first; if it is overwhelming and looks like it needs a large block of time that you do not have at the moment, ask yourself what part of it can be tackled *now* to prepare for the rest. Avoid C tasks: They are beguilingly easy, but they do not help solve the real problems.

Make a Daily Action List.   List things you plan to do that day, with assigned priorities (A or B). Maintain a tickler file of important deadlines.

Make a Time Log.   Periodically, list what you *actually* do with your time, minute by minute, for a week or two. Summarize your activities by category and by ABC priority. Most people who do this find that they are spending much more time than they expected on matters that are trivial or could be delegated to someone else, and much less on their A items; knowing this, they can make some needed changes in the way they work.

Handle Each Email Once if Possible.   Identify categories of items that you do not want to see, and get yourself off mailing lists. If a member of your team can help, mark the item "Please recommend a solution by [date]" or, better yet, "Please handle," and get it out of your inbox. Set aside a "C folder" in your email application for other less important matters; if they do not come up again, delete them. The remaining items are your As and Bs; schedule them and start on them.

Consider Your Energy Cycle and Your Environment.   You should be aware that your energy level is not a constant, but varies from hour to hour. When project managers were asked to rate their perceived energy level hour by hour, they reported a peak in the late morning, a low around lunchtime, and a second peak in the afternoon. But people differ: Some do their best work in the morning, some in the afternoon, and a wise manager learns when they function best and try to schedule difficult decisions or confrontations for that period. Days matter, too: The same project managers perceived their energy to be highest on Tuesday through Thursday and lowest on Sunday.

The environment in which work is done is also very important. The bullpen, or sea of desks in which some engineers work, helps foster communication, which can be either desirable or time wasting, depending on whether your work of the moment involves coordination or independent creativity. Noise, poor lighting, uncomfortable seating, and inadequate space also inhibit your best work. The open door of the manager's office almost rules out those blocks of time needed for the tough problems during the day, forcing them into evenings and weekends.

DISCUSSION QUESTIONS

17-1. How should Fayol's 1916 "Advice to Future Engineers" be modified to better suit the world of today?

17-2. Suggest some ways in which newly graduated engineers can "make their mark" in early assignments on the first job.

17-3. The text urges you to choose your manager carefully. How can you try to accomplish this practically?

17-4. Someone in an outside group declines to provide the support services you need, despite your most tactful requests and explanations. What do you do next?

17-5. Outline how the "gig" economy might impact your future career. What steps can you take now to be ready for those impacts?

17-7. You must meet with your staff of 20 engineers and technicians to introduce them to a demanding new program they will all be involved in. (**a**) Discuss how you will prepare for this oral presentation. (**b**) What precautions will you take during this talk? (**c**) Why might you prepare a written communication in addition to the oral presentation?

17-8. Outline how the use of email and social media impacts the communication aspects of your future career.

17-9. You have been promoted into managerial work. Outline what steps you can take to stay up to date in technology.

17-10. As the new vice president for R&D in a high-technology firm, you are concerned about the increasing technical obsolescence of your professional staff. Discuss the actions you might take and the programs you might support to reduce this problem.

17-11. Discuss the extent to which professional registration and/or certification is useful in your broad field of professional interest.

17-12. For a company with which you are familiar, identify any intentional programs to increase the opportunity for underrepresented groups.

17-13. The dual career ladder concept was developed to offer technical professionals an equal alternative to a management career, but it often proves less successful than hoped. Why do you think this is so?

17-14. You are an experienced engineering manager in a growing organization, have just selected several capable engineers for promotion to their first management positions, and are responsible for guiding their growth into their new responsibilities. What are some of the skills your new managers will need to acquire, and what advice will you give them about the new problems they will face?

17-15. In what ways are master's programs in business administration (MBA) or in engineering management deficient as preparation for a management position?

17-16. Identify the top several "time wasters" in your life (from the "top 10" in the chapter and any others you identify). What might you do to reduce their effect?

17-17. Estimate how your personal energy level varies throughout the day. How might you use this information in planning your activities?

SOURCES

Allen, David, *Getting Things Done: The Art of Stress-Free Productivity* (New York: Penguin Books, 2001).

At the Crossroads: Crisis and Opportunity for American Engineers in the 1990s (Washington, DC: American Association of Engineering Societies, 1994), reported in "Study Sees Structural Changes in Engineers' Work Environment," *Engineering Times*, March 1994.

Badawy, Michael K., *Developing Managerial Skills in Engineers and Scientists* (New York: Van Nostrand Reinhold Company, Inc., 1982), pp. 37–42.

Baum, Eleanor, "Women in Engineering, Engineering Education in the Twenty-first Century" Lecture Series, University of Missouri–Rolla, October 9, 1990.

Boice, R., "New Faculty Involvement for Women and Minorities," *Research in Higher Education*, 1993, 34, pp. 291–341.

Boyle, P. and Boice, R., "Best Practices for Enculturation: Collegiality, Mentoring, and Coaching," In M. S. Anderson (Ed.), *New Directions in Higher Education,* 101 (San Francisco: Jossey-Bass, 1998).

Broder, David, "Clinton's Performance Won't Mimic Carter's," *St. Louis Post-Dispatch*, November 17, 1992, editorial page.

Crawford, M. and MacLeod, M., "Gender in the College Classroom: An Assessment of the 'Chilly Climate' for Women," *Sex Roles*, 23, 1990, pp. 101–122.

Dalton, G. W. and Thompson, P. H., *Novations: Strategies for Career Management* (Glenview, IL: Scott, Foresman and Company, 1986).

Drew, T. L. and Work, G. G., "Gender-Based Differences in Perception of Experiences in Higher Education," *Journal of Higher Education*, 69, 1998, pp. 542–555.

Engineer of 2020: Visions of Engineering in the New Century, National Academy of Engineering (National Academies Press: Washington, 2005).

Farr, J. L. et al., "The Measurement of Organizational Factors Affecting the Technical Updating of Engineers," presentation to the annual meeting of the Academy of Management, Atlanta, GA, August 1979, as reported in Morrison and Vosburgh, *Career Development*, p. 26.

Fayol, Henri, *Administration Industrielle et Générale*, Constance Storrs, trans. (London: Sir Isaac Pitman & Sons Ltd., 1949), pp. 89–90.

Felder, R. M., Felder, G. N., Mauney, M., Hamrin, C. E., Jr., and Dietz, E. J., "A Longitudinal Study of Engineering Student Performance and Retention III: Gender Differences in Student Performance and Attitudes," *Journal of Engineering Education*, 84, 1995, pp. 151–163.

Gee, Edwin A. and Tyler, Chaplin, *Managing Innovation* (New York: John Wiley & Sons, Inc., 1976), p. 172.

Gouldner, Helen, "The Social Context of Minorities in Engineering," Appendix C in National Research Council, *Engineering Employment Characteristics* (Washington, DC: National Academy Press, 1985), pp. 59–62.

Grayson, Lawrence T., "Education for Leadership," *Engineering Education News*, February 1989, p. 3.

Hawks, B. K. and Spade, J. Z., "Women and Men Engineering and Science Students: Anticipation of Family and Work Roles," *Journal of Engineering and Science Education*, 87, 1998, pp. 249–256.

Hewitt Associates Survey reported in "Dual-Career Ladders: New Evidence That They Really Work," *Engineering Department Management & Administration Report*, March 1994, pp. 5–6.

Kaufman, H. G., *Obsolescence and Professional Career Development* (New York: AMACOM, A Division of American Management Associations, Inc., 1974), pp. 10–11.

Kerzner, Harold, *Project Management: A Systems Approach to Planning, Scheduling, and Controlling*, 4th ed. (New York: Van Nostrand Reinhold Company, Inc., 1992), pp. 367–368.

King, W. J., *The Unwritten Laws of Engineering* (New York: American Society of Mechanical Engineers, 1944).

Lakein, Alan, *How to Get Control of Your Time and Your Life* (New York: Peter H. Wyden, Inc., 1973).

Landis, Raymond B., "The Case for Minority Engineering Programs," *Engineering Education*, May 1988, pp. 756–761.

Lane, Melissa J., "The Current Status of Women and Minorities in Engineering and Science," *Engineering Education*, May 1988, p. 753 (from National Science Foundation data).

LeBoeuf, Michael, "Managing Time Means Managing Yourself," *Business Horizons*, February 1980, p. 41.

LeBold, William K. and LeBold, Dona J., "An Historical Perspective on Women Engineers and a Futuristic Look," *Annual Conference Proceedings, American Society for Engineering Education*, University of Illinois, Urbana-Champaign, June 20–24, 1993, pp. 1259–1263.

Lee, Denis M. S., "Job Challenge, Work Effort, and Job Performance of Young Engineers: A Causal Analysis," *IEEE Transactions on Engineering Management*, August 1992, pp. 214–226.

Manpower Comments, December 1992. Commission on Professionals in Science and Technology. (Vol. 29, No. 9), p. 21.

Margulies, N. and Raia, A., "Scientists, Engineers, and Technological Obsolescence," *California Management Review*, 10:2, 1967, pp. 43–48.

McCall, M. W., Lombardo, M. M., and Morrison, A. M., *The Lessons of Experience: How Successful Executives Develop on the Job* (Lexington, MA: Lexington Books, 1988), summarized in Alan W. Pearson, "Management Development for Scientists and Engineers" *Research Technology Management*, January–February 1993, p. 45.

Mehrabian, Albert, "Communication Without Words," *Psychology Today*, September 1968, pp. 53–55.

Meinholt, C. and Murray, S. L., "Why Aren't There More Women Engineers?" *Journal of Women and Minorities in Science and Engineering*, 5, 1999, 239–263.

Mintzberg, Henry, *The Nature of Managerial Work* (New York: Harper & Row, Publishers, Inc., 1973), p. 39.

Morrison, Robert F. and Vosburgh, Richard M., *Career Development for Engineers and Scientists: Organizational Programs and Individual Choices* (New York: Van Nostrand Reinhold Company, Inc., 1987), p. 169.

Morse, Lucy. "Distance Education Tools for Engineering." International Conference on Engineering Education—ICEE 2007, Portugal, 2007.

National Research Council, *Management of Technology: The Hidden Competitive Advantage* (Washington, DC: National Academy Press, 1987).

Proceedings, National Congress on Engineering Education, Washington, DC, November 20–22, 1986 (New York: Accrediting Board for Engineering and Technology, 1987), p. B-8.

Seymour, E. and Hewitt, N. M., *Talking About Leaving: Why Undergraduates Leave the Sciences* (Boulder, CO: Westview Press, 1997).

Shannon, Robert E., *Engineering Management* (New York: John Wiley & Sons, Inc., 1980), p. 126.

Smith, Leonard J., "Keeping Your Career on a Success Trajectory," in John E. Ullmann, ed., *Handbook of Engineering Management* (New York: © John Wiley & Sons, Inc., 1986), pp. 729–736. Reprinted by permission.

Steele, C., "A Threat in the Air: How Stereotypes Shape Intellectual Identity and Performance," *American Psychologist*, 52, 1997, pp. 613–629.

Super, Donald E., *The Psychology of Careers* (New York: Harper & Row, Publishers, Inc., 1957).

Super, Donald E., "A Life-Span, Life-Space Approach to Career Development," *Journal of Vocational Behavior*, 16, 1980, pp. 282–298.

Thamhain, Hans J., "Determining Aptitudes for Engineering Managers," *Proceedings of the International Engineering Management Conference, Institute of Electrical and Electronic Engineers*, Santa Clara, CA, October, 21–24, 1990, pp. 5–10.

Thompson, P. H. and Dalton, G. W., "Are R&D Organizations Obsolete?" *Harvard Business Review*, 54, November 1976, pp. 105–116.

Thurow, Lester C., "The Task at Hand," *Wall Street Journal*, June 12, 1987, p. 46D.

Timm, Paul, *Managerial Communication: A Finger on the Pulse* (Englewood Cliffs, NJ: Prentice-Hall, Inc., 1980).

Tobias, S., *They're Not Dumb, They're Different: Stalking the Second Tier* (Tucson, AZ: Research Corporation, 1990).

Vetter, Betty M.. "Women in Science III—Ferment: Yes...Progress: Maybe...Change: Slow," *Mosaic*, (National Science Foundation), Fall 1992 (final issue), pp. 34–41.

Wadsworth, E. M., "Women's Activities and Women Engineers: Expansions over Time," *Initiatives*, 55(2), 1992, pp. 59–65.

Weihrich, Heinz and Harold Koontz, *Management: A Global Perspective*, 10th ed. (New York: McGraw-Hill, Inc., 1993), p. 573.

Weinert, Donald W., "The Structure of Engineering," in Ullmann, *Handbook of Engineering Management* (New York: © John Wiley & Sons, Inc., 1986), pp. 28–30. Reprinted by permission.

Young, Edmund, "Management Thought for Today from the Ancient Chinese," *Management Bulletin*, October 1980, p. 31. *Occupational Outlook Handbook*, Bulletin 2300, 1988–1989 ed. (Washington, DC: Bureau of Labor Statistics, April 1988), p. 53.

Zimbler, Linda J., "Faculty and Instructional Staff: Who Are They and What Do They Do?" U.S. Dept. of Education National Center for Education Statistics Survey Report NCES 94–346, October 1994, p. 14.

18

Globalization and Challenges for the Future

PREVIEW

Discussions of globalization have grown at a fast pace over the last generation. This chapter begins by discussing globalization and our *flattened* world brought about by many factors, including multinational organizations and the web. A basic exploration of Japanese management styles and their utility for U.S. companies are outlined. The significance of the European Union, BREXIT, and changes to global trading norms are discussed. Next, problems and differences found in managing activities in developing countries are outlined.

The second major subject of this chapter are the challenges for future engineers proposed by The **National Academy of Engineering** (NAE). Through the engineering accomplishments of the past, the world has become flatter, more inclusive, and more connected. The challenges facing engineering today are not those of isolated locales, but of the world.

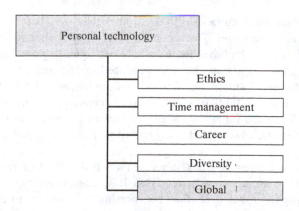

LEARNING OBJECTIVES

When you have finished studying this chapter, you should be able to do the following:

- Explain aspects of globalization.
- Discuss how the flattening of the world affects engineers.
- Discuss the challenges for engineers of the twenty-first century.

GLOBALIZATION

Two groups of topics are discussed in this section. First, trends in globalization, the nature of the multinational corporation, and their significance to the engineering career are considered. Second, the effect of cultural and economic differences on management is discussed.

At the top political and economic level, **globalization** is the process of denationalization of markets, politics, and legal systems. Globalization broadly refers to the expansion of global linkages, the organization of social life on a global scale, and the growth of a global consciousness. The term can be interpreted as either positive of negative and different definitions express different assessments of global change. The following definition from Wikipedia illustrates both sides:

> Globalization is the process of interaction and integration between people, companies, and governments worldwide. Globalization has grown due to advances in transportation and communication technology. With increased global interactions comes the growth of international trade, ideas, and culture. Globalization is primarily an economic process of interaction and integration that's associated with social and cultural aspects. However, conflicts and diplomacy are also large parts of the history of globalization, and modern globalization.

People around the globe are more connected to each other than ever before. Information and money flow more quickly and easily than ever. The economic effect of globalization was felt during the 2008 to 2012 global financial crisis, which has led to a backlash against globalization in certain political movements around the globe. Goods and services produced in one part of the world are increasingly available in all parts of the world. International travel is more frequent. International communication is commonplace. While some people think of globalization as primarily a synonym for global business, it is much more than that. The same forces that allow businesses to operate as if national borders do not exist also allow social activists, labor organizers, journalists, academics, and many others to work on a global stage.

A key aspect of globalization is global trade, either in the form of free trade or through systems of tariffs. These paths are not new, as early as 1988 Japanese firms began exporting parts to the United States, assembling them here, then reexporting the product to Europe in an attempt to avoid European tariffs and quotas on Japanese-made goods! Human societies across the globe have established progressively closer contacts over many centuries, but recently the pace has dramatically increased. Jet airplanes, cheap (or free) communication options via the web, instant capital flows—all these have made the world more interdependent than ever. Multinational corporations manufacture products in many countries and sell to consumers around the world. Along with products and finances, ideas and cultures circulate more freely. As a result, laws, economies, and social movements are forming at the international level. Many politicians, academics, and journalists treat these trends as both inevitable and (on the whole) welcomed. But for billions of the world's people, business-driven globalization means uprooting old ways of life and threatening livelihoods and cultures. The global social justice movement, itself a product of globalization, proposes an alternative path, more responsive to public needs. Intense political disputes will continue over globalization's meaning and its future direction.

However, the age of globalization has brought significant changes to the relationship between rich and poor states. In this new age, where the transfer of information to and from remote parts of the world is nearly instantaneous, U.S. corporations outsource significant portions of their manufacturing base to poorer countries around the globe. U.S. dominance of information technology serves as a tool with which it may exploit poor countries' low-wage workers, weaker environmental laws, and other factors to perpetuate its dominance of global markets. No longer are states in the periphery viewed as merely potential consumers of high-priced U.S. goods. They now serve as the inexpensive labor for the production of goods, including software, to be sold by U.S. manufacturers at higher prices in global markets.

As knowledge-based globalization expands, there are many reasons for these trends including the growing number of nations with increased technical skills, and a willingness of global corporations to locate facilities where profit opportunities exist. These offshore opportunities include: lower cost of labor, relaxed environmental standards, and nonexistent corporate tax structures. Thomas Friedman in his book, *The World is Flat,* has described this next phase of the global world as a flat world. Friedman describes 10 "flatteners" that he sees as leveling the global playing field:

1. **The new age of creativity (the fall of the Berlin Wall)**, November, 1989. This event "tipped the balance of power across the world toward those advocating democratic, consensual, free-market-oriented governance, and away from those advocating authoritarian rule with centrally planned economies." Friedman uses the fall of the Berlin Wall as a symbol for a general global shift toward democratic governments and free-market economies. This shift is currently being tested as politics in many regions (including the United States) lurch away from free trade.

2. **The new age of connectivity.** The web emerged as a tool of low-cost global connectivity. This event "enabled more people to communicate and interact with other people anywhere on the planet than ever before."

3. **Workflow software** is the ability of machines to talk to other machines with no humans involved. Friedman believes these first three flatteners have become a "crude foundation of a whole new global platform for collaboration."

4. **Open sourcing** is the act of releasing previously proprietary software under an open source/free software license. Notable software packages that have been open sourced include: Netscape Navigator, whose code became the basis of Firefox; StarOffice, which became the base of the OpenOffice.org office suite; and Wikipedia, which is written collaboratively by volunteers from

all around the world. Instead of people just downloading music or news, they are increasingly likely to contribute information—writing a review of a product they bought on amazon.com, rating their professor at ratemyprofessor.com, or editing an encyclopedia entry on their favorite trivia topic on wikipedia.com.

5. **Outsourcing** allows companies to split service activities into components that can be subcontracted and performed in the most efficient, cost-effective way and then reintegrating their work back into the overall operation. Outsourcing includes services such as book production, accounting, and customer service.

6. **Offshoring** is manufacturing's version of outsourcing. Offshoring is when a company moves its production from its home country to another country, where it can be done with "cheaper labor, lower taxes, subsidized energy, and lower health-care costs."

7. **Supply chaining** allows horizontal collaboration among suppliers, retailers, and customers— to create value, resulting in "the adoption of common standards between companies" and more efficient "global collaboration."

8. **Insourcing** is distinct from supply chain management because it goes beyond supply chain management since it is third-party-managed logistics and requires more extensive types of collaboration. Insourcing has become more common over the last decade as a backlash to negative experiences with outsourcing (including customer support).

9. **In-forming.** Google and other search engines are prime examples. "Never before in the history of the planet have so many people—on their own—had the ability to find so much information about so many things and about so many other people," writes Friedman.

10. **"The Steroids"** are made up of specific technologies, supercharged by all the other flatteners, the most notable of which is the ubiquitous smart phone and various forms of social networking.

European Union

The United States has long been the world's most attractive single market, with over 300 million affluent people in a single market free of internal barriers; it has therefore been a prime target of European and Pacific Rim manufacturers of every sort. Over 25 years ago, Western Europe was transformed into a market free of internal barriers that equals or exceeds that of the United States, with profound implications for international trade. The *Wall Street Journal* summarized these implications in early 1988:

> The EC [European Community] may never become the United States of Europe its founders dreamed of in the 1950s. The Europe of 1992 will continue to differ from the United States of 1988. The 12 nations will still speak nine different languages, spend twelve different currencies in different ways, and be governed by different laws.

Today the European Union (EU), which has as its foundation the European Community, is an economic and political partnership between 28 European countries. The results are frontier-free travel and trade for its 513 million citizens, the euro (the single European currency), safer food and a greener environment, better living standards in poorer regions, joint action on crime and terror, cheaper communications and travel, and millions of opportunities to study abroad.

The euro (€) is probably the EU's most tangible achievement. The single currency is shared by 19 countries (2018).

The State of Globalization Today

The flattening forces, along with further changes to technology and communications since Friedman released his book in 2005, continued to accelerate changes in many areas of the global economy. However, political changes around the world—notably the United Kingdom vote to leave the European Union, the refugee crisis in Europe created by those fleeing North Africa, and changes in trade policy by the Trump Administration in the United States—have challenged the previously prevailing wisdom that globalization is accelerating and unstoppable. At its most basic level, the forces of globalization disrupt the availability of labor and lead to the migration of jobs. This migration has a tendency to impact the lowest earning members of developed nations and these negative impacts have snowballed into a movement that resists open borders and free trade. Only time will show which of these views will become the dominant force in a future engineering career.

Multinational Organizations

A multinational corporation is one with significant operations in more than one country. Organizations of this type are the very foundation of global commerce. Jacoby identifies six stages of multinationalization:

1. Exporting products to foreign countries
2. Establishing sales organizations abroad
3. Licensing patents and know-how to permit foreign firms to make and sell its products
4. Establishing foreign manufacturing facilities
5. Multinationalizing management from top to bottom
6. Multinationalizing ownership of corporate stock

A wide variety of factors makes international management more complex than single-country management. Robinson categorizes the important variables in the international system into six groups, based on differences in (1) national sovereignty, (2) national economic conditions, (3) national values and institutions, (4) timing of national industrial revolutions, (5) geographical distances, and (6) areas and population.

Pricing policies are a strategic tool of multinationals. One example is **transfer pricing.** For example, an automobile firm may produce an engine in country A, a transmission in country B, and assemble these into a car in country C. The price at which components are transferred to the assembly process determines where profits are accumulated and taxed and the level of import duties, yet often there is no competitive price for such components. Obviously, the countries losing revenue in this process may press for "more realistic" transfer prices. Multinationals also have engaged in *penetration pricing* to gain market share in a new market by unusually low prices. Japan and other Pacific Rim countries have been accused of *dumping* goods in the United States and Europe at prices far below their own internal markets, leading to legislation and quotas limiting such imports. In mid-1994, North Dakota farmers were accusing Canada of dumping durum wheat in the United States; Canadians replied that U.S. subsidies of wheat shipments to Europe left a shortage of wheat needed by U.S. spaghetti makers, which they were filling.

Understanding Globalization's Impact on a U.S.-Based Engineering Career

When I began my work as an engineer with a large multinational company in the early 2000s, I experienced a world where U.S.-based engineers generally did the design work and those designs were implemented by other employees around the world. However, within only five years, that view began to change dramatically. This experience is in line with statistics that estimate 40 percent of the world's engineering work in 1995 was based in the United States and dropped to 10 percent by 2010. My own work was part of that change as I hired more engineers in our non-U.S. locations to continually increase the capacity of our engineering group. While this trend might be grounded in many of the same forces flattening the world (the McKinsey Global Institute has calculated 59 percent of engineering work could be off-shored), it is also driven by a need for engineers who understand the needs of global customers. In other words, engineers who can design for the Indian or Chinese market, because that is their native culture. It is also driven by the increasing availability of engineers trained outside U.S. universities. Studies estimate that fewer than 5 percent of U.S. college students go into engineering, a much lower rate than the 12 percent typical in Europe and over 40 percent in China.[1]

What does this mean for current engineering undergraduates? First it indicates that you will be competing in a global talent market. This change is clearly highlighted by the controversy created in 2018 when Foxconn indicated plans to staff their new Wisconsin plant with Chinese trained engineers.[2] Second, it highlights that new engineers will be working with an ever broader mix of engineers from around the world. To be successful in this environment, students need to develop an understanding of other cultures and customs.

Sources

1. Bugliarell, G., *Globalization and Engineering (editorial)*, in *The Bridge*. 2005, National Academy of Engineering. p. 1.
2. Anderson, S., *Why Foxconn Is Considering Engineers From China*, in *Forbes*. 2018.

Japanese Management Styles

Ever since the Japanese began their ascendancy in the 1970s, an ocean of printing ink has been consumed in books and articles about Japanese management and cultural practices alleged to be at the root of their success. Among the practices most frequently mentioned, with some discussion of their implications, are the following:

1. Recruitment of employees directly out of secondary school or college, which (**a**) makes acceptance into a prestigious university critical, and (**b**) limits the opportunity for midcareer transfer to another company although the recent success of the *heddo-hantaa* (from the American "headhunter") in arranging midcareer movement of people with needed skills is moderating this.
2. The fabled lifetime employment of Japanese firms only applied to full-time male employees of the larger firms; at its peak in the mid-1970s it applied to only half of Japan's workers, and

is estimated to be only half that now; nearly a fifth of the workforce is part-time (threefold that of 20 years earlier) with no security (a practice that has recently increased in the United States); further cutbacks are accommodated by bringing work in-house from subcontractors or foreign plants.

3. Promotions are infrequent and based (along with salary) on seniority and teamwork, although Honda and other firms are now beginning to base bonuses on performance rather than seniority.

4. Retirement or restricted employment for older (55-year-old) workers, except for the few rising to top management; with a large aging population, this practice will give Japan serious problems.

5. *Michi,* or "the way" (to heaven or to corporate success), which emphasizes mastery of even a small task—a Zen philosophy of self-improvement to do the present job better, contrasted with the Western (and Confucian) emphasis on the next promotion.

6. *Giri,* a sense of duty of honor or obligation to observe community (read, company) customs; for the modern-day Japanese worker this may involve compulsory parties and unwanted assignments overseas or away from family, yet require loyalty to the company. Respected Japanese managers and technicians are *moretsu shain* (fanatical workers) and *yoi kigyo senshi* (good corporate soldiers), and thousands become victims of *karoshi* (death from overwork), leaving *karoshi* widows behind.

7. From the company in return, an almost womblike concern for the (full-time) worker, including family support, housing, recreational activities, social gatherings and cultural events, and festivities to commemorate promotion, internal transfer, and retirement, a security that has diminished over the last generation.

8. Ceremonies and rituals (such as group calisthenics and the company song) to foster love of company and group identification.

9. Rotation between functional departments to ensure broad exposure to the company (at the cost of the skills honed in specialization).

10. *Omikoshi* management, "the practice of having the middle [and lower] levels in the corporate organization... plan new projects on their own initiative, obtain top management approval for it, and then carry it out on their own... so that they carry the company like a group of people carrying the small portable shrine (*omikoshi*) that is a traditional feature of Shinto festivals."

11. The *ringi* system of having people sign on the bottom-up proposals, after which responsibility for implementation is borne by the group rather than a specific individual.

These practices—although many now apply only to a minority of Japanese workers—are not without their problems. Odaka cites four disadvantages whose drawbacks have become increasingly obvious in the postwar years:

1. Encouraging employee dependency and suppressing individual creativity
2. Discriminatory employment and impediments to the formation of a free horizontal labor market
3. Harmful effects of the escalator system [in which those first on the employment escalator rise before later arrivals] and middle management promotion gridlock
4. Work that has no joy and seemingly has no meaning

Japanese management is moderating some of the sharpest differences from methods of Western management—the gradual, but slow, improvement in opportunities for women is a case in point. The Japanese postwar baby boomers, or *dankai sedai,* are more affluent and better educated (especially in Western culture); as they gain ascendancy they are insisting on a balance between Western ideas and Japanese values. On the other side, leading American corporations have selectively adopted the most useful and transferable of the lessons from Japanese management.

Japanese production plants have several characteristics that set them aside from American competitors. Some of these have already been mentioned in earlier chapters: the almost fanatic emphasis on quality, the just-in-time approach that almost eliminates inventory on the plant floor, and the quality control circles and other devices through which the production employee directly works to improve the product. Others are a high priority on plant cleanliness, regarded as a prerequisite to quality; operating equipment below its ultimate capacity to extend its life and precision; automation and warning systems that require fewer workers; and a mutual trust (and unified language and ethnicity) that requires fewer management review levels.

Any of the management methods being successfully employed in Japan or other countries can be adapted to U.S. needs, and many of them were originally developed in the United States. As Drucker puts it, "What we can learn from foreign management is not what to do. What we can learn is to do it."

International Trade Agreements

In the past several decades, ending in 2016, international trade and multinational management organizations have increased dramatically. From 1950 to about 1988, real U.S. gross domestic product (GDP) tripled, but real-world GDP quadrupled, and world trade grew sevenfold; NAFTA, WTO, World Bank, International Monetary Fund, and other trade agreements should assure that this continues. More and more American companies are finding that, even if they are not heavily involved in international markets themselves, they cannot escape the impact of international competition in their home markets.

North American Free Trade Agreement and the USMCA.   Implementation of the North American Free Trade Agreement (NAFTA) began on January 1, 1994. This agreement removed most barriers to trade and investment among the United States, Canada, and Mexico. The United States and Canada have long been touted as having the world's longest undefended border. What few U.S. citizens realize is that Canada is also our largest trading partner and Mexico is second. Under the Trump administration, the very existence of NAFTA was challenged. Months of negotiations led to the development of a new provisional agreement among the United States, Mexico, and Canada, labeled the USMCA (United States–Mexico–Canada Agreement) by the U.S. administration. While consensus says the core of the agreement is little changed from NAFTA, there are a number of new provisions that address points of trade contention between the parties, including trade in dairy and automotive, and wage minimums in certain areas for all partners. Time will tell the impact of these negotiations and the potential of a new agreement to change the balance of trade between the three countries.

World Trade Organization.   The World Trade Organization (WTO) is an international organization designed to supervise and liberalize international trade. The WTO came into being on January 1, 1995, and is the successor to the General Agreement on Tariffs and Trade (GATT), which was created in 1947. The WTO has over 150 members, accounting for over 97 percent of world trade. The WTO is the only global international organization dealing with the rules of trade between nations. At its heart are the WTO agreements, negotiated and signed by the bulk of the world's trading nations and ratified in their parliaments. The goal is to help producers of goods and services, exporters, and importers conduct their business. The past 50 years have seen growth in world trade. Merchandise exports grew on average by 6 percent annually. Total trade in 2000 was 22 times the level of 1950. GATT and the WTO have helped to create a strong and prosperous trading system contributing to unprecedented growth.

Management in Developing Countries

Background.   There is certainly a *world of variety* in the problems of managing activities in developing countries, which may be defined for our purposes as the entire world except for the United States, Canada, Europe, Japan, Australia, New Zealand, Russia, and a few other countries such as Israel and the most industrialized parts of Brazil and South Africa. The Pacific Rim countries of South Korea, Taiwan, Hong Kong, and Singapore are rapidly developing in a manner similar to Japan's development and will also not be considered further. Next are presented a few general concerns with operations in developing countries.

Characteristics of Developing Countries.   Developing countries differ widely, but the following characteristics will be found in many of them:

1. A shortage of capital, often with financial resources controlled by a few families
2. Government planning of the economy, and frequently, government operation of utilities and major industries
3. A shortage of skilled workers, professionals, and support services
4. A high level of government control of foreign subsidiaries in terms of approvals to establish or modify business activities; import controls; currency exchange rates; control of repatriation of profits and capital; requirement of partial or majority local ownership of foreign subsidiaries; and taxation policies, including tax incentives
5. Different preferences regarding leadership style. Hofstede measured the "power distance" between the superordinate and subordinate (a measure of the extent of autocratic leadership style). For example, consider the following facts:
   - Five South American countries and Mexico had power distance indexes from 63 to 81, indicating an authoritarian style (Argentina was 49).
   - Five Romance-language European countries (Italy, France, Belgium, Spain, and Portugal) had indexes from 50 to 68.
   - Eight South and Southeast Asian countries ranged from 54 (Japan) to 77 (India).
   - The Philippines, both Asian and Spanish in influence, had the highest index (94).
   - The United States had an index of 40.
   - Nine Western and North European countries had indexes of 18 to 39, indicating a less authoritative style in these developed countries.

6. In many Asian and South American countries the family is paramount, and enterprises tend to be small, with management confined to family members.

7. While engineers from developed countries tend to specify the advanced technology they are familiar with for use in developing countries (and the host countries often encourage this), there is often limits on local ability to operate and, especially, repair such technology. Often a more *appropriate technology* is possible, which may be more labor intensive (not necessarily a penalty where labor is cheap and unemployment high), that can be used and maintained more easily.

8. Without long experience in an industrial economy, nationals (including government officials) may take a much more relaxed attitude about getting things done on time.

Ethical Considerations.   Because cultures vary so much, engineering and managerial work in developing countries may involve ethical decisions that would not present themselves in the United States. For example, bribery of public officials may be a way of life and may be necessary to get permits issued or spare parts released from customs. Minor officials may be paid so little that to maintain a decent livelihood, they count on supplements that would clearly be illegal in the United States. Should the American plant manager participate? On a higher level, should a U.S. sales manager pay millions of dollars in bribes to officials in a foreign government to facilitate sales of jet aircraft to the national airline, an illegal act under U.S. law due to the Foreign Corrupt Practices Act, or abandon the sale to less fastidious competitors?

Plant operation presents its own questions. Should a U.S.-owned plant just maintain levels of pollution control and plant safety consistent with local requirements, or should the higher levels required in U.S. plants be maintained, at higher cost? To what extent should an American plant manager insist on equal opportunity regardless of race, religion, and genders in a society that distinctly discourages such equality? Like many of the issues in professional ethics discussed in Chapter 16, actual cases of such problems often do not have simple solutions.

Significance for the Engineering Career

American engineers have been in demand internationally for a century and a half, especially since World War II. Many engineers have found they could earn up to twice as much abroad in places like high growth Middle Eastern countries, returning after as little as 15 years with a comfortable retirement. Engineers in truly global concerns like the large design and construction firms previously cited or manufacturers with half their sales overseas often find at least one international tour necessary to rise very far in the firm. Many engineers find an international tour satisfies their desire for adventure, or returns them to the culture their family originated from, or adds an invaluable experience in the education and growth of their children.

There are negative factors as well. Companies who are not fully committed to global operations still send engineers and managers abroad, but when they return, the expatriates find they have been *out of sight—out of mind* and have lost momentum in their career. Two-career families may find that the career of the expatriate's spouse is "on hold" during the overseas period, and families with children or with elderly parents may find being overseas disruptive. Today the young engineer needs to consider carefully what an international assignment will do for (or the harm it might do to) their overall career or progress within a specific firm, and the impact it might have on personal goals and family life.

ENGINEERING GRAND CHALLENGES

A diverse committee of experts in engineering and science from around the world was convened by the U.S. National Academy of Engineering and announced in February 2008, 14 challenges for engineers in the twenty-first century. As was pointed out in the previous section of the text the world has become flatter, more inclusive, and more connected. The Challenges belong to four realms of human concern—sustainability, health, vulnerability, and joy of living. The Challenges are summarized as follows:

Sustainability

- **Make solar energy economical.** Sunshine has long offered a tantalizing source of **environmentally friendly power**, bathing the earth with more energy each hour than the planet's population consumes in a year. But capturing that power, converting it into useful forms, and especially storing it for a rainy day, poses provocative engineering challenges.
- **Provide energy from fusion.** Another popular proposal for long-term energy supplies is **nuclear fusion**, the artificial recreation of the sun's source of power on earth. The quest for fusion has stretched the limits of engineering ingenuity, but hopeful developments suggest the goal of practical fusion power may yet be attainable.
- **Develop carbon sequestration methods.** It remains unlikely that fossil fuels will be eliminated from the planet's energy-source budget anytime soon, leaving their environment-associated issues for engineers to address. Most notoriously, evidence is clear that the carbon dioxide pumped into the air by the burning of fossil fuels is increasing the planet's temperature and threatens disruptive effects on climate. Anticipating the continued use of fossil fuels, engineers have explored technological methods of **capturing the carbon dioxide** produced from fuel burning and sequestering it underground.
- **Manage the nitrogen cycle.** A further but less publicized environmental concern involves the atmosphere's dominant component, the element nitrogen. The biogeochemical cycle that extracts nitrogen from the air for its incorporation into plants—and hence food—has become altered by human activity. With widespread use of fertilizers and high-temperature industrial combustion, humans have doubled the rate at which nitrogen is removed from the air relative to preindustrial times, contributing to smog and acid rain, polluting drinking water, and even worsening global warming. Engineers must design **countermeasures for nitrogen cycle problems**, while maintaining the ability of agriculture to produce adequate food supplies.
- **Provide access to clean water.** Chief among concerns in this regard is the **quality and quantity of water**, which is in seriously short supply in many regions of the world. Both for personal use—drinking, cleaning, cooking, and removal of waste—and large-scale use such as irrigation for agriculture, water must be available and sustainably provided to maintain quality of life. New technologies for desalinating seawater may be helpful, but small-scale technologies for local water purification may be even more effective for personal needs.

Health

- Advanced health informatics and **computerized catalogs of health information** should enhance the medical system's ability to track the spread of disease and analyze the comparative effectiveness of different approaches to prevention and therapy.

- *Engineer better medicines.* Another reason to **develop new medicines** is the growing danger of attacks from novel disease-causing agents. Certain deadly bacteria, for instance, have repeatedly evolved new properties, conferring resistance against even the most powerful antibiotics. New viruses arise with the power to kill and spread more rapidly than disease-prevention systems are designed to counteract.

Vulnerability

- **Restore and improve urban infrastructure.** Even as terrorist attacks, medical epidemics, and natural disasters represent acute threats to the quality of life, more general concerns pose challenges for the continued enhancement of living. Engineers face the grand challenge of renewing and **sustaining the aging infrastructures of cities and services**, while preserving ecological balances and enhancing the aesthetic appeal of living spaces.
- **Prevent nuclear terror.** As a consequence, vulnerability to biological disaster ranks high on the list of unmet challenges for biomedical engineers—just as engineering solutions are badly needed to **counter the violence of terrorists** and the destructiveness of earthquakes, hurricanes, and other natural dangers. Technologies for early detection of such threats and rapid deployment of countermeasures (such as vaccines and antiviral drugs) rank among the most urgent of today's engineering challenges.
- **Secure cyberspace.** A prime example where such a barrier exists is in the challenge of **reducing vulnerability to assaults on cyberspace**, such as identity theft and computer viruses designed to disrupt Internet traffic. Systems for keeping cyberspace secure must be designed to be compatible with human users—cumbersome methods that have to be rigorously observed do not work, because people find them inconvenient. Part of the engineering task will be discovering which approaches work best at ensuring user cooperation with new technologies.

Joy of Living

- **Reverse-engineer the brain.** An important way of exploiting such information would be the development of methods that allow doctors to forecast the benefits and side effects of potential treatments. **"Reverse-engineering" the brain,** to determine how it performs its magic, should offer the dual benefits of helping treat diseases while providing clues for new approaches to computerized artificial intelligence. Advanced computer intelligence, in turn, should enable automated diagnosis and prescriptions for treatment.
- **Enhance virtual reality.** Some new methods of instruction, such as **computer-created virtual realities,** will no doubt also be adopted for entertainment and leisure, furthering engineering's contributions to the joy of living. Advanced personalized learning and the external world are not the only places where engineering matters; the inner world of the mind should benefit from **improved methods of instruction and learning,** including ways to tailor the mind's growth to its owner's propensities and abilities.
- **Engineer the tools of scientific discovery.** The spirit of curiosity in individual minds and in society as a whole can be further promoted through engineering endeavors **enhancing exploration** at the

frontiers of reality and knowledge, by providing new tools for investigating the vastness of the cosmos or the inner intricacy of life and atoms.

FUTURE CONSIDERATIONS IN ENGINEERING AND MANAGEMENT

Future directions in engineering and management are, of course, inseparable from future trends of society as a whole. Some of the driving forces affecting the beginning of the twenty-first century include the following:

- Continuing computer-based information revolution
- Increasing technological sophistication of society
- International and political considerations in an interdependent world
- Demographic shifts
- Interactions of food, energy, materials, and the environment

The Information Revolution

We are in the early phases of the knowledge management revolution, which is just as significant to world history as was the industrial revolution that preceded it.

- **The virtual organization.** The availability of integrated management information and decision-making systems, together with user-friendly software and increased computer literacy among managers, has decreased the number and levels of middle managers and staff professionals. Yet the empowerment and responsibility of nonmanagers has increased. As members throughout an organization more and more share common purposes and a common information system with trusted suppliers and customers, boundaries between them will become more fluid as they participate in virtual teams, and they will become interdependent members of the new *virtual organization*. Ultimately, the lessons learned in reducing hierarchy, breaking down functional boundaries, and empowering individuals in profit-making industry will find real (but perhaps more limited) application in government, military organizations, and academia.
- **The virtual product.** The traditional practice of manufacturing standard goods for inventory and ultimate sale will increasingly disappear. Designs and specifications for families of *virtual products* will continue to exist in computer memory, and they will be transmitted to computer-integrated manufacturing systems and 3D printers when needed and as specified by the customer. The need for repetitive labor in manufacturing will continue to decline, and the availability of inexpensive labor will become less critical than the availability of knowledge workers in determining industrial locations.
- **The home.** The computer and information revolution has already begun to affect the way we live, and we can expect that it will do so increasingly. Computer systems in the home can be expected to handle home security, climate control, and optimization of energy use. Computer, mobile devices, television, and telephone will blend into a total system that connects us with the rest of the world for communication (oral, visual, and written), education, and interactive banking, shopping, and entertainment.

Education and the Technological Society

To function effectively in this technological society will require a more general understanding of technology. Education to prepare for this world will require a good understanding of mathematics and science and a widespread computer literacy. Such a change can be only gradual, since elementary and secondary school teachers are, often poorly prepared in such areas (at least in the United States). However, we can speculate on the age when something close to an engineering education will be the standard for an educated person—especially the policymakers who control society.

The Flattened World

The world has become flat and is becoming flatter. At the same time, political changes are slowing, if not reversing, the movement to global free trade. There will be a continuing increase in the size and ubiquity of the great multinational companies, to the point that the determination of "what country they belong to" will become more and more uncertain. Clearly, the engineering manager of the future will need a much greater world view in designing, manufacturing, and marketing products.

Demographics

The increasing demand for knowledgeable workers will hasten the integration of traditionally under-represented groups in all levels of our economy, especially in engineering, and in all levels of management and government. New approaches in childcare, part-time work, and work at home may accompany this. The increase in elderly citizens will begin to blur the age of retirement, with some workers retiring early and beginning a second career, while others continue with their employer in at least a limited way into their 70s. While medical science (and engineering) continues to find cures that keep us alive longer, this increases the burden on the working population and changes the ethical decisions that medicine must face.

Food, Energy, Materials, and the Environment

While birthrates can be expected to decrease as countries develop, world population will place increasing demands on static or decreasing farmlands and especially on diminishing natural resources of fossil energy and minerals. As industrialization continues to spread and the world's largest forests shrink, world problems of air, water, and thermal pollution can only increase. We can expect continuing pressure for energy conservation, and seeking alternate sources of energy production to avoid the air and water pollution from fossil fuels, unless fusion energy is proven feasible, economical, and safe. All of these problems present difficult decisions in public policy at local, national, and world levels. They also offer endless challenges to be met by the engineer and the engineering manager.

Changes in the Engineering Career

As a result of the decline of middle management in the industrial organization, engineering (and other) specialists might expect to spend a longer portion of their career working as professionals before being considered for management positions when working in traditional organizations. At the same time, when working in smaller companies or pursuing entrepreneurial ventures, engineers might expect to need managerial skills much earlier in their careers than previous generations.

Engineers will find their careers less stable than in previous generations. Bahrami explains:

> Many firms have reexamined their employment policies—initiating early retirement programs and other incentives to reduce the size of their workforce. As pointed out in other studies, the critical tradeoff in this context is between corporate flexibility and individual security. Many corporations rely on temporary workers, specialized vendors, and consultants in order to flexibly deal with unique contingencies. Additionally, this trend points to a fundamental shift in the foundation of employer–employee relationship, away from the traditional patriarchal orientation toward what may be characterized as a peer-to-peer relationship. This sentiment is echoed in the following comment which encapsulates the implicit relationship between Apple Computer and its employees: "You own your own careers; we provide you with the opportunity."

Engineers will therefore have to manage their career progress more than ever before. Because of rapidly changing technology, they will need to emphasize continual learning and development, not just through coursework but also by learning to follow changes in their specialty, their industry, and general business trends. Design engineers will find that the evolution of computer-aided design makes them more efficient and productive; as long as they continue to master new tools—but more vulnerable if they do not, since fewer designers may be needed. And, like production workers, they will find that they are competing in a global market, and that modern communication permits engineering to be integrated across the globe. In conclusion, engineers are a key element today in the continued success of the society of the United States. Even with political changes in the United States and Europe, much of the world continues to become flatter, more inclusive, and more connected. The challenges facing engineering today are not those of isolated locales, but of the world and the people of the world. To fit into this world, the engineer needs to remain alert to changing products, processes, technologies, and opportunities and to manage the progress of his or her career. These engineers will find the twenty-first century exciting and rewarding.

DISCUSSION QUESTIONS

18-1. Describe some of the effects of globalization.

18-2. Congratulations! You have served as an R&D manager so effectively that you have been appointed Director of R&D for your company's affiliate in Tokyo. What differences will you expect to find in your new assignment (other than language), and how might this affect your life and your management style?

18-3. What is the significance of current political shifts in the United States and Europe on a future engineering career.

18-4. China was not explicitly considered as either a developed or developing nation in the text. Why is this?

18-5. The discussion of globalization on an engineering career notes the higher number of engineers being trained outside the United States. How should engineering programs prepare their students for this future of a "blended" engineering workforce with engineers trained to different standards?

18-6. Give an example of an "appropriate technology" that might be preferable in a specific developing country situation instead of the technology used to solve the same problem in the United States.

SOURCES

Bahrami, Homa, "The Emerging Flexible Organization: Perspectives from Silicon Valley," *California Management Review,* Summer 1992, pp. 33–52.

Drucker, Peter F., "Learning from Foreign Management," *Wall Street Journal,* June 4, 1980, editorial page.

Friedman, Thomas L., *The World Is Flat: A Brief History of the Twenty-first Century.* (New York: Farrar, Straus and Giroux, 2007).

Hofstede, Geert, "Hierarchical Power Distance in Forty Countries," in C. T. Lammers and D. J. Hickson, eds., *Organizations Alike and Unlike* (London: Routledge & Kegan Paul Ltd., 1975), p. 105, quoted in Anant R. Negandhi, *International Management* (Newton, MA: Allyn and Bacon, 1987), pp. 322–323.

"Is Japan Using the U.S. as a Back Door to Europe?" *BusinessWeek,* November 14, 1988, p. 57.

Jacoby, Neil H., "The Multinational Corporation," *The Center Magazine,* 3, May 1970, pp. 37–55.

"Many Japanese Find Their 'Lifetime' Jobs Can Be Short-Lived," *Wall Street Journal,* October 8, 1992, pp. A1, A8.

National Academy of Engineering, *The Technological Dimensions of International Competitiveness* (Washington, DC: National Academy Press, 1988), p. 53.

Odaka, Kunio, *Japanese Management: A Forward-Looking Analysis* (Tokyo: Asian Productivity Organization, 1986), p. 68.

Robinson, Richard D., *Internationalization of Business: An Introduction* (Hinsdale, IL: CBS College Publishing, 1984).

Squires, Frank H., "Karoshi Widows," *Quality,* November 1990, p. 66.

GLOBAL WEBSITES

The following are useful source websites and government publications.

http://www.globalsherpa.org, December 2018 Global Sherpa is dedicated to promoting awareness and knowledge about important issues and ideas in international development, sustainability, globalization and world cities.

http://www.globalization101.org/what-is-globalization/, December 2018. Globalization101.org is dedicated to providing students with information and interdisciplinary learning opportunities on this complex phenomenon.

http://www.engineeringchallenges.org/, December 2018

http://www.wto.org/

https://ustr.gov/about-us/policy-offices/press-office/fact-sheets/2018/october/united-states%E2%80%93mexico%E2%80%93canada-trade-fa-2, December 2018. Office of the United States Trade Representative, **United States – Mexico – Canada Trade Fact Sheet**.

http://www.imf.org, December 2018. The International Monetary Fund (IMF) is an organization of 188 countries, working to foster global monetary cooperation, secure financial stability, facilitate international trade, promote high employment and sustainable economic growth, and reduce poverty around the world.

http://globalization.icaap.org/, December 2018. Globalization is a peer-reviewed journal devoted to the examination of social, political, economic, and technological globalization.

Index

ABET. *See* Accrediting Board for Engineering and Technology (ABET)
Abilene paradox, 39
Absolute liability, 210
Academic careers, 366
Academy of Management, 13
Acceptable quality level (AQL), 257, 262
Acceptance theory, 147
Account executives, 321
Accountability, 148–149
Accounts receivable turnover, 166
Accreditation, 377
Accrediting Board for Engineering and Technology (ABET), 5, 254, 377–378
Acid test ratio, 165
Acquired need theory, 48
Acquisitions, 289
 project planning, 289–308
 value engineering (VE), 220
Active listening, 371
Activities, 120, 276–285
 for organizing efficiency, 120–121
 span of control, 125
Activity-based costing, 160
 to drive improvements, 161
Activity-on-arrow method, 300
Activity-on-node (AON) diagram, 300
Activity ratios, 165–166
Actual cost of work performed (ACWP), 308
ACWP. *See* Actual cost of work performed (ACWP)
Adams, J. Stacey, 64

Adam's equity theory, 48
Ad-hoc committees, 150–151
Administration Industrielle et Générale, 33
Administrative management, 33–35
 Fayol's role in, 33–34
 Robb's role in, 35
 Urwick's role in, 35
 Weber and bureaucracy, 34–35
Administrative skills, project matrix, 323
Advanced therapeutic and surgical devices, 285
'Advice to Future Engineers' (Fayol), 34, 362
Aerospace engineers, 6
 employment, 7
Affiliation needs, 60
African Americans, as engineering students, 380
After-sales service, 276
Agenda, 323
Agile project, 309–310, 329
AGVs. *See* Automatic guided vehicles (AGVs)
AI. *See* Artificial intelligence (AI)
AIChE. *See* American Institute of Chemical Engineers (AIChE)
AIME. *See* American Institute of Mining, Metallurgical, and Petroleum Engineers (AIME)
AISES. *See* American Indian Science and Engineering Society (AISES)
Albanese, Robert, 9
Alexander the Great, 21, 37

Allen, David, 387
Allen, Thomas J., 67, 178, 193, 300, 328
Amazon, 7, 108, 227
American Association of Engineers, 339
American Cyanamid Company, 146
American Indian Science and Engineering Society (AISES), 380
American Institute of Chemical Engineers (AIChE), 376
American Institute of Mining Engineers, 28
American Institute of Mining, Metallurgical, and Petroleum Engineers (AIME), 376
American Machinist, 28
American Management Association, 13
American National Standards Institute (ANSI), 219
American Revolution, 25
American Society for Engineering Education (ASEE), 27, 37, 376, 381
American Society for Engineering Management (ASEM), 13, 323, 340, 353, 376
American Society for Quality (ASQ), 216, 252, 376
American Society of Civil Engineers (ASCE), 28, 376
American Society of Heating, Refrigerating, and Air-Conditioning Engineers (ASHRAE), 376

409

American Society of Mechanical
  Engineers (ASME), 28–29,
  339, 376
Amos, John M., 75, 97, 384, 385, 386
Amrine, Harold T., 23, 228
Analytic potential, 329
Ancient civilizations, 21–22
'And-also' method, brainstorming, 191
Andrew Hawkins v. Town of Shaw, 343
ANOVA (analysis of variance)
  methods, 262
ANSI. *See* American National
  Standards Institute (ANSI)
Ansoff, H. Igor, 177
Apollo program, 20, 40, 219, 365
  standardization in, 219
Apple, 16, 177, 278, 293
Application engineering, 177
Applied research, defined, 176
Apportionment, 215
Appraisal costs, 252
Apprentice, 367
AQL. *See* Acceptable quality level
  (AQL)
Architect/Engineering (A/E),
  225, 229
Arkwright, Richard, 23
Armstrong, G., 277
Arsenal of Venice, 20, 22–23, 37
  management practices of, 22–23
  tasks of, 22
Art of War, The (Tzu), 75
Arthashastra, 22
Artificial intelligence (AI), 115
Artificial organs, 285
ASEE. *See* American Society for
  Engineering Education (ASEE)
ASEM. *See* American Society for
  Engineering Management
  (ASEM)
ASQ. *See* American Society for
  Quality (ASQ)
Assets, 161
Asset turnover, 166
Assignment of duties, 148
Assumptions, 294
A. T. Kearney, Inc., 227
Attinger, E. O., 284
Attribute-listing approach, 192

Attributes methods, 254
Audits, 166
Augustine, Norman R., 89, 218
Augustine-Morrison Law of
  Unidirectional Flight, 218
Augustine's Laws, 89
Authority
  acceptance theory, 147
  delegation, 148–150
  formal, 147
  nature of, 58
Autocratic style of leadership, 55
Automated patient monitoring, 285
Automated storage and retrieval
  systems (AS/RS), 244
Automated version control, 207
Automatic control, 155–156
Automatic guided vehicles
  (AGVs), 244
Availability, 218
Avoidance, 66
Avolio, B. J., 57

Babbage, Charles, 27
Babcock, Daniel L., 4
Babcock, George D., 22, 32
Babcock, Jean, 267
Babylon, 21
Badawy, Michael K., 382, 383
Bahrami, Homa, 407
Baker, Bruce N., 327–328
Balance sheet, 161
  budget, 158–159
Balanced matrix, 322
Balderston, Jack, 180, 182, 329
Bank Wiring Observation Room
  Experiment, Hawthorn
  studies, 36
Bar charts, 297–300, 305
Barnard, Chester I., 41, 97, 98, 147
Barth, Carl, 32–33
Base technologies, 182
Basic research, defined, 176
Bass, B. M., 56–57
Bathtub curve model, 215
BCWP. *See* Budgeted cost of work
  performed (BCWP)
BCWS. *See* Budgeted cost of work
  scheduled (BCWS)

Bechtel, 282
Bedeian, Arthur G., 9, 11
Begley, Sharon, 113
Behavior modification, 48, 64, 66
Behavioral management, 35–39
  Abilene paradox, 39
  Hawthorne studies, 35–39
Bell system, 77, 262
Benchmarking, 40, 144, 163
Berelson, B., 58
Berthollet, Claude Louis, 23
Bessemer, Henry, 4, 25
Beta distribution, 301
Betz, Frederick, 91, 175, 193
Bid, 278
Bill of materials (BOM), 239
Bioengineering, 284
Biomaterials design, 285
Biomechanics of injury and wound
  healing, 285
Biomedical engineering, 284–285
Biomedical engineers, 284, 404
Biomedical technology, 285
Birnbaum, Philip, 329
BIT. *See* Built-in test (BIT)
Black engineers, 381
Blake, Robert R., 52–53
Blake, Stewart P., 182
Blanchard, Benjamin S., 217
Blanchard, Kenneth, 42
Blood chemistry sensors, 285
Body language, 372
Boeing (case study), 350–351
Boilerplate, 292
Boisjoly, Roger, 351–355
Boisjoly, Russell J., 352–353
BOM. *See* Bill of materials (BOM)
Boone, Louis E., 54
Boulton, Matthew, 24
Boulton and Watt (manufacturers), 24
Bounded rationality, 98–99
Bowen, Donald D., 54
Brainstorming, 191–192, 259
Brandeis, Louis D., 32
Breadboard-level testing, 216
Break-even charts, 223, 232, 233–236
  effect of automation, 235
Breakpoint leadership, 69
Briefings, 233, 369

Broder, David, 382
Brown v. Board of Education, 343
Budgeted cost of work performed (BCWP), 308
Budgeted cost of work scheduled (BCWS), 308
Budgets, 158–160
  balance sheet, 159
  capital expenditure, 159
  cash, 159
  financial, 159
  operating, 159
    expense, 159
    process, 159–160
    profit, 159
    revenue, 159
  projects, 308
Built-in test (BIT), 217
Bureaucracy, 34–35, 68
Bureaucrats, 27
Burgess, Susan, 241
Burn, Jerry, 354
Burns, J. M., 56
Bush, Vannevar, 36
Business strategy, 181–182
  base technologies, 182
  key technologies, 182
  pacing technologies, 182
Buzan, Tony, 192

CAC. *See* Cost at completion (CAC)
CAD/CAM, 41, 227
CAM. *See* Computer-aided manufacturing (CAM)
Camden and Amboy Railroad, 25
Campbell, John P., 58
Campus interview, 136, 138, 140–141
Canals, 4–5, 21, 25–26, 37
Canons of Ethics, Engineers' Council for Professional Development, 339
Capital expenditure budgets, 159
Carbon dioxide, 403
Carbon sequestration methods, 403
Career fields, 365–366
  academic, 366
  consulting, 366
  engineering management, 365
  entrepreneurial, 366

operational, 365
other careers, 366
research and design, 365
writing, 366
Career stages, 366–367
  disengagement, 367
  establishment, 366
  exploration, 366
  growth, 366
  maintenance, 366
Carlson, Chester, 91–92
Carnegie, Andrew, 25
Carnegie, Dale, 53, 58
Cartwright, Edmund, 23
Cash budgets, 159
Cash flow statement, 163
Categorical imperatives, 377
Caveat emptor, 209
C-chart, 257
CCB. *See* Configuration control board (CCB)
CEES. *See* Center for Engineering Ethics and Society (CEES)
Center for Engineering Ethics and Society (CEES), 339
Central limit theorem, 302
Centralized control, 150
Certainty, 65
  decision making under, 96–99
Certification, 378
Certification marks, 186
Certified Professional Engineering Manager (CPEM), 378
Certified Reliability Engineer (CRE), 216
  designation, American Society for Quality (ASQ), 216
Certo, Samuel C., 9
Cetron, Marvin, 89
Challenger disaster (case study), 347, 351–355
Chance node, 107
Chandler, M. K., 69
Character attributes, leaders, 50
Charisma, 147
Charter, 323
Charter, project matrix, 325
Chemical engineers, employment, 6
Chief engineer, and policies, 84

Chlorine bleach, 23
Choate, Pat, 65
CIM. *See* Computer integrated manufacturing (CIM)
Civil engineers employment, 7
Cleland, David I., 130, 320, 328
Client communications, 328
Clinical engineering, 285
Closed-loop control, 155–156
Closing, 308
Closing the project phase, project management, 293
CNC. *See* Computer numerically controlled (CNC) machines
Coca-Cola Company, trade secrets, 187
Codes of conduct, 341
Coercive/punishment power, 147
Cohen, D. I., 193
Cohen, Stephen S., 224
Colleague, 340
Collective marks, 186
College teaching and research, 284
Collins, Jim, 42, 59
Commercial validation, 179, 204
Commitment principle, 83
Committees, 150–151
  executive, 151
  making effective, 151
    purpose and chair, 151
    size and membership, 151
  reasons for using, 151
    administration, 151
    policy making, 151
    problems with, 151
    representation, 151
    securing cooperation in execution, 151
    sharing knowledge and expertise, 151
    training of participants, 151
Communication, 367
  active listening, 371
  characteristics of, 370
  comparison of communications method, 370–371
  defined, 367
  effectiveness of, 371
  encoding, 370

Communication (*continued*)
  and engineers, 367
  executive summaries, 372
  feedback, 370
  importance of, 367–369
  and managers, 369
  nonverbal, 372
  oral briefing/presentation, 372–373
  process, modeling, 369
  reception, 369
  retention of information, 370
  transmission, 369
  understanding, 369
  visual aids, 373
  written report, 372
Communications process model, 369
Competition, 15, 74
Compilation of values, 337–338
Completed staff work, 364
Computer-aided manufacturing
  (CAM), 244
Computer applications, 282–283
Computer-based automation, factory
  workers skills required
  for, 132
Computer-based information systems,
  114–115
  decision support systems, 114
  expert systems, 115
  integrated databases and the
    cloud, 114
  management information, 114
Computer hardware engineers, 283
Computer integrated manufacturing
  (CIM), 244, 245, 405
Computer modeling, physiologic
  systems, 285
Computer numerically controlled
  (CNC) machines, 238
Computer revolution, 244
Conceptual design review, 208
Conceptual model, 100
Conceptual skills, 11
Conceptual stage, 207
Concordia Publishing House, 259
Concurrent control, 157
Concurrent engineering, 203–205
Concurrent (simultaneous)
  engineering, 198, 203–205

Configuration control board (CCB),
  151, 206
Configuration management, 206
  and automated version control,
    benefits of, 207
Conflict
  management methods, 326
  sources, 326
Connolly, Terry, 52
Constraints, 103
Construction (Hyatt Regency disaster),
  344–345
Consultative style of leadership, 55
Consulting careers, 366, 381
Consumer Reports, 212
Content theories, 60–64
  affiliation needs, 60
  esteem needs, 60
  Herzberg's two-factor theory, 62–63
  Maslow's hierarchy of needs, 60
  McClelland's trio of needs, 63
  physiological needs, 60
  vs. process theories, 60
  security/safety needs, 60
  self-actualization needs, 61
Contingency plan, 83
Contingency theory, 54
Contractor, 328
Contracts, 300–331
  classification, 330–331
    cost type, 330
    fixed-price, 330
Contributory negligence, 209
Control, 238–240, 305–309
Control spans, 124–127
Control systems, 239–243
  in engineering design, 205–209
    configuration management,
      206–208
    design reviews, 208–209
    drawing/design release, 205
Controlled freedom, 69
Controlling, 154–168
  cost and schedule, 327–331
    crashing the project, 306–308
    earned value system, 308
    project control systems, 305–306
    reducing project duration, 306
  financial controls, 158–166

  nonfinancial controls, 168
  process of, 155–157
    characteristics of effective, 158
    mechanical process, 155–157
    steps in, 155
    three perspectives on the timing
      of, 157–158
Conventional rating scale, 145
Cook, Tim, 16
Cooke, Morris L., 32
Cooperatives, 120
Coordination, management
  function, 168
Copyrights, 184, 186
Corporate culture, 148, 167
  learning, 364
Corporate research organizations,
  177–178
Corporate restructuring, and
  size of specialist staff
  organizations, 128
Corporate values, 341
Corporations, 119–120
Corps des Ponts et Chaussées, 4
Corrective maintenance, 217, 269
Cost accounting, 160–161
Cost at completion (CAC), 308
Cost centers, 159
Cost plus fixed-fee contracts, 331
Cost plus incentive fee contracts, 330
Cost type contracts, 330–331
  cost plus fixed-fee, 331
  cost plus incentive fee, 330
  letter contract, 331
  time and materials, 331
Cost variance, 308
Costs
  actual cost of work performed
    (ACWP), 308
  budgeted cost of work scheduled
    (BCWS), 308
  controlling, 327–329
  crash, 306
Cost/schedule control system criteria
  (C/SCSC), 308
Cottage industry, end, 23–24
Courage, 115
Cover letter, 140–141
  do's and don'ts of, 140–141

Covey, Stephen, 42
CPM. *See* Critical path method (CPM)
CPEM. *See* Certified Professional
   Engineering Manager
   (CPEM)
Crash cost, 306
Crash time, 306
Crashing, 306
Crashing projects, 306
CRE. *See* Certified Reliability
   Engineer (CRE)
Creative environment, providing, 193
Creative people, characteristics of,
   192–193
Creative process, 190–191
  frustration, 190
  inspiration, 190
  preparation, 190
  verification, 190
Creativity, 190–194
  brainstorming, 191–192
  creative environment, providing a,
   193
  creative people, characteristics of,
   192–193
   approach to problems, 192
   curiosity, 192
   independence, 192
   personal attributes, 193
   self-confidence, 192
  creative process, 190–191
  innovation and, 190–191
  nature of, 190
  technological gatekeepers, 193–195
Crises, 12, 387–388
Critical design review, 208–209
Critical path, 300
Critical path method (CPM), 289, 300
Crompton, Samuel, 23
Crosby, Phil, 258
C/SCSC. *See* Cost/schedule control
   system criteria (C/SCSC)
Current assets, 161–163
Current liabilities, 163
Current ratio, 165
Curtis, Ellen Foster, 352–353
Customer, 40
Customer communications, 328–329
Customer focus, 40, 76, 259

Customer satisfaction, 42, 80, 203,
   208, 231, 252, 260
Custom-made items, 278

Daily action list, 388
Daimler Chrysler, 210
Dalton, G. W., 367, 374
Dankai sedai, 400
Dannenbring, David G., 89
Danzig, George, 105
Deadlines, meeting, 363
Decentralized management, 150
Decision analysis, 120
Decision making, 96–115
  categories of, 102
  computer-based information
   systems, 114–115
   expert systems, 115
   integrated databases, 114
   management information/
    decision support systems, 114
  implementation, 115
  level of certainty, 99
  management science, 99–102
   models, 100–101
   origins, 99–100
  nature of, 97–98
   occasions for decision, 97–98
   relation to planning, 97
   types, 98–99
  nonroutine, 98
  objective *vs.* bounded rationality,
   98–99
  occasions for decision, 97–98
  payoff table (decision matrix), 102
  rational, 98–99
  relation to planning, 97
  routine, 98
  tools for, 102–114
   categories, 102
   under certainty, 102–104
   under risk, 105–110
   under uncertainty, 110–113
  types of decisions, 98–99
  under certainty, 102–104
  under risk, 105–110
   examples, 105–107, 111
   expected value, 111
   nature of, 105

   queuing (waiting-line) theory,
    107–110
   risk as variance, 110
  under uncertainty, 110–113
Decision matrix, 102
Decision node, 107
Decision Support System (DSS), 96, 114
Decision trees, 96, 107
Decision variables, 103
Decisional roles, managers, 11–12
Decisions, 96–97
  effect of management level, 114
  level of certainty, 99
  nonroutine, 98
  routine, 98
Defects per sample, 257
DeGeorge, Richard T., 349
Dekker, Don L., 338
Delegation, 148–150
  of authority, 148
   barriers to, 149–150
   decentralization, 150
   reasons for, 149
  control systems, 157
Delphi Method, 84–85, 90
Delta Airlines, 282
Deming, W. Edwards, 146, 252, 254,
   261–263
  14 points, 262–263
Demographics, 406
Deontological ethics, 337
Department of Defense (DOD), 122, 200
Departmentation, 118, 121–124
  functional, 121
  geographic divisions, 122
  methods of, 122, 123
  patterns of, 121–124
   type of customer, 122
Dependent (unknown) variable, 87
Depletion, 162–163
Depreciation, 162–163
Derating, 216
Design criteria, others, 217–220
  availability, 218
  human factors, 218–219
  maintainability, 217
  producibility, 219
  standardization, 219
  value engineering/analysis, 220

Design engineers, 121, 134
Design patent, 185, 189
Design release, 205
Design review, 208–209
    conceptual, 208
    critical, 209
    software, 209
    system, 208
Deutsch, James B., 345
Developing countries
    characteristics of, 401–402
    management in, 401–407
Development, defined, 176
Development stage, 200
Dhillon, B. S., 180, 191–192
Dieter, George E., 374
Differential piecework method, 29
Digital economy, 41
Diplomatic style of leadership, 55
Directing, 22
Disengagement stage, careers, 367
Disney Productions, 282
Dispatching, 239
Disposable organization, 130
Disposal stage, 203
Disseminator role, managers, 11
Disturbance handler role,
        managers, 11
DOD. *See* Department of Defense
        (DOD)
Double-entry bookkeeping, 23
Double-sampling, 257
Drawing release, 205, 297
Drop-in visitors, 387, 388
Drucker, Peter, 42, 57, 60, 70, 77,
        79–80, 120, 132, 400
Drum-buffer-rope, 241
DSS. *See* Decision Support System
        (DSS)
Dual career ladder, 383–384
Dual-ladder system, 383
Dummy activities, 300
Duncan, Daniel M., 345
Duty-based ethics, 337

EAC. *See* Engineering Accreditation
        Commission (EAC)
Early learning path, 329
Earned value methods, 157

Earned value system, 308, 323
Eastman Chemical Company, 259
Ecole des Ponts et Chaussées, 4, 26
Ecole Polytechnic, 26
Economic order quantity (EOQ),
        223–224, 232–233
ECPD. *See* Engineers' Council for
        Professional Development
        (ECPD)
Effective control systems,
        characteristics of, 154, 158
*The Effective Executive: The Definitive
        Guide to Getting the Right
        Things Done* (Drucker), 42
Effectiveness, defined, 264
Efficiency, defined, 264
Effort-to-performance expectancy, 64
EFQM (European Foundation for
        Quality Management), 261
Electrical engineers, 6
    employment, 7
Elements of Administration, The
        (Urwick), 35
Emergent leader, 49, 52
Emerson, Harrington, 32
Employment application, 136–137
Employment, engineering, 7
Empowered, use of term, 121, 128, 134
Engineer, achieving effectiveness as,
        360–389
    career, 365–367
        fields, 365–366
        stages, 366–367
        success definition, 365
    communication methods
        comparison, 370–371
        characteristics, 370–371
        effectiveness, 371
        retention of information, 370
    communication tools of special
        importance, 372–373
        oral briefing/presentation,
            372–373
        visual aids, 373
        written report, 372
    ideas communication, 367–370
        definitions, 367
        engineer, importance to, 367–369
        manager, importance to, 369

    modeling the communication
        process, 369
    management, 381–386
        choice of, 383
        dual career ladder, 383–384
        relation to the engineering
            career, 381
        responsibilities, preparing for,
            384–386
    other factors in effective
        communication, 371
        active listening, 371
        nonverbal communication, 372
    professional activity, 376–378
        ABET accreditation, 377–378
        certification, 378–379
        professional engineering
            registration, 378
        professional societies, 376–377
        registration, 378
    right start, getting off to, 361–367
        regarding associates and
            outsiders, 364
        regarding boss, 364
        regarding work, 362–363
    running marketing, 281
    staying technically competent,
        373–375
        threat of obsolescence, 373–374
    time management, 387–389
        time wasters, 387–388
        tools of, 388–389
        work smarter, 386
Engineer in Training (EIT)
        designation, 378
Engineer manager, and forecasting, 84
*Engineer of 2020: Visions of Engineering
        in the New Century, The*, 374
Engineered time standards, 266–267
Engineering, 4–7
    biomedical engineering, 284–285
    career in "Gig" economy, 367–368
    career significance, 402
    careers, changes in, 406–407
    concurrent, 203–205
    definition by ECPD, 6
    education beginnings, 4–5
    education development, 26–27
    employment, 7

engineering jobs in a company, 7
ethics, 337–358
future considerations, 405–407
    changes in career, 406–407
    demographics, 406
    education and the technological
        society, 406
    energy, 406
    environment, 406
    flattened world, 406
    food, 406
    information revolution, 405
    materials, 406
grand challenges, 403–405
    health, 403–404
    joy of living, 404–405
    sustainability, 403
    vulnerability, 404
groups in, 380–381
jobs in a company, 7
management, synthesis with, 13–17
    careers, 15–17
origins of, 4–5
as a profession, 5–6
roles in organization (Amazon), 7–8
systems, 200–203
as a transformation process, 367–368
value, 220
women and minorities in, 379–381
Engineering Accreditation
    Commission (EAC), 377
Engineering codes of ethics, 337,
    339–347
Engineering design
availability, 218
concurrent (simultaneous)
    engineering, 203–205
control systems in design, 205–209
    configuration management,
        206–208
    design review, 208–209
    drawing/design release, 205
human factors engineering, 218–219
maintainability, 217
managing, 198–220
nature of, 199
new product development, 200–205
producibility, 219
product liability

development of, 209–210
    reducing liability, 210–212
reliability, 212–216
standardization, 219
systems engineering, 200–203
value engineering/analysis
    (VE/A), 220
Engineering design, managing,
    198–220
concurrent engineering, 203–206
control systems, 205–209
nature of, 199
new product development, 200–203
    tasks/stages in, 200–203
other design criteria, 217–219
    availability, 218
    human factors, 218–219
    maintainability, 217
    producibility, 219
    standardization, 219
    value engineering/analysis, 220
product liability, 209–212
    development of, 209–210
    reducing, 210–212
    safety, 210–211
reliability, 215–216
Engineering design review, 151
Engineering education, 26–27
beginnings of, 4–5
development of, 26–27
and technology, 406
Engineering employment, 7
Engineering ethics, 339–358
codes, 339–340
concepts in, 341
guidelines for ethical dilemmas,
    342–343
postemployment limitations,
    348–349
problems in consulting and
    construction, 342–347
    construction safety, 344
    distribution of public
        services, 343
    political contributions, 342
problems in industrial practice,
    342–347
    conflict of interest, 348
    environmental responsibilities, 348

professional ethics, 336
    definitions, 336–337
    significance, 342
    whistleblowing, 348
Engineering Grand Challenges,
    403–405
health, 403–404
joy of living, 404–405
sustainability, 403
vulnerability, 404
Engineering management careers, 365
Engineering management, historical
    development of, 13–17, 20–42
administrative management, 33–35
    Fayol, Henri, 33–34
    Robb, Russell, 35
    Urwick, Lyndall, 35
    Weber, Max and bureaucracy,
        34–35
behavioral management, 35–39
    Abilene paradox, 39
    Hawthorne studies, 35–39
contemporary contributions, 39–42
    customer focus, 40
    globalization, 40
    information technology, 41
    management theory and
        leadership, 41–42
    project management, 40
    quality management, 39–40
industrial revolution, 23–27
    America, industrial development
        in, 25
    cottage industry, end of, 23–24
    engineering education
        development, 26–27
    factory system, problems of, 24–25
management philosophies, 27
origins, 21–23
    ancient civilizations, 21–22
    Arsenal of Venice, 22–23
scientific management, 27–33
    Babbage, Charles, 27–28
    Gilbreths, 30–31
    growth and implications, 32–33
    Taylor, Frederick W., 29–30
    Towne, Henry and ASME, 28–29
Engineering manager, compared to
    other managers, 13

Engineering problem solving, 96–97, 101
Engineering stages of new product development, 174
Engineering technology, 6
Engineers
  and communication, 367–373
  defined, 3, 6
  delegation, barriers to, 149–150
  dual career ladder, 383–384
  future demands on manufacturing, 227
  heritage of, 4
  intensity of use of, 225–227
  key attribute for, 6
  and management, 381–386
  in management, need for, 14, 381–382
  marketing, 276–284 (*see also* Marketing)
  process, 225
  in purchasing, 272
  service activities, 276–285
    characteristics of the service sector, 282
    importance of Service-producing industries, 281–282
  in service economy, 281–285
  specific service industry examples, 282–285
    biomedical engineering and the healthcare, 284–285
    college teaching and research, 284
    computer applications, 282–283
    government service, 283
    green engineering, 283
    sustainability environment, 284
  in top management, 282–283
  responsibility and bridge failure, 346–347
  types, 6–7
  use of motivational theories by, 69
  vs. scientists, 67–68
Engineers' Council for Professional Development (ECPD), 5
  engineering definition, 5–6
English Iron Act (1750), 25
Enterprise resource planning (ERP), 240–241

just-in-time (JIT), 241–243
  model changeover, 243
  synchronized manufacturing, 241
Enterprise self-audit, 167
Entrepreneur, 193
Entrepreneurial careers, 366
Entrepreneurial mindset, 194
Entrepreneurship, 190
  defined, 193
  intrapreneurship and, 195
Environment
  creative, providing, 193
  and industrial practice, 347
  sustainable energy and, 284
Environmental ethics, 337
EOQ. *See* Economic order quantity (EOQ)
Equity, 161
Equity theory, 64
Ergonomics, 218
Erickson, Tamara J., 182
Erie Canal, 4, 25
ERP. *See* Enterprise resource planning
Establishment stage, careers, 366
Esteem needs, 60
Ethical egoism, 337
Ethical leadership and motivation, 63
Ethics, 335–358. *See also* Professional ethics
  corporate codes of, 341–342
  defined, 336
  engineering, core concepts in, 341
  in industrial practice, 347–351
    conflict of interest, 348–349
    environmental responsibilities, 348
    gifts, 348
    inside information, 348
    moonlighting, 348
    postemployment limitations, 348–349
    significance, 347
    whistle-blowing, 349
European Union, 396–397
Events, 300
Executing phase, project management, 293
Executive committee, 151
Executive summaries, 372

Expectancy theory, 64–65
Expected value, 105
Expense budget, 159
Expense centers, 159
Experience level, 329
Experiential training, 385–386
Expert power, 147
Expert systems, 115
Explanatory forecasting models, 87, 89
Exploration stage, careers, 366
Exploratory technological forecast, 90–91
Exponential smoothing, 86–87
External audits, 166
External failure costs, 253
Extinction, 66

Fabricated items, 278
Face-to-face feedback, 372
Facilitative activities, 127
Facilities and Plant Engineering Handbook, 268
Factor of safety, and reliability, 216
'Factory of the future,' 224
Factory system, problems, 24–25
FAF. *See* Fully automated factory (FAF)
Fail-safe design, 216
Failure CDF (cumulative distribution function), 213
Failure/hazard rate, 213
Failure PDF (probability density function), 213
Fair day's work, 28, 29
'Fair use exception,' to copyright law, 187
Farr, J. L., 375
Fasal, John H., 220
Fast tracking, 306
Fayol, Henri, 12, 20, 33–34, 37, 41, 47, 83, 168, 318, 362
  and administrative management, 33–35
Feedback, 369
Feedback control, 157–158
Feedforward control, 157–158, 168
Feigenbaum, Armand, 258
Financial budgets, 158–160

Financial controls, 158–160
  audits of financial data, 166
  budgeting process, 158–160
  budgets, 158–160
  cost accounting, 160–161
  financial statements, 161–163
  human resource controls,
      166–168
  ratio analysis, 163–166
Financial data, audits of, 166
Financial ratios, 163–166
Financial statements, 161–163
  assets, 161
  balance sheet, 161
  cash flow statement, 163
  income statement, 163
  liabilities, 161
Finished goods, 132, 162, 232
Firm fixed-price contracts, 330
First-line managers, 10, 138
First-line production management
      positions, 225
First-to-market strategy, 177
Fisher, Dalmar, 317
Fixed assets, 159, 161
Fixed costs, 233
Fixed matrix, 321
Fixed-position layout, 229
Fixed-price contracts, 330
Fixed-price incentive contracts, 330
Fixed-price, redeterminable
      contracts, 330
Fixed-price with escalation
      contracts, 330
Flexible manufacturing cell (FMC),
      232, 244
Flexible manufacturing systems
      (FMS), 239, 244–248
  advantages of, 245
  definitions, 245
  flexible manufacturing cell
      (FMC), 244
  fully automated factory
      (FAF), 245
  Lean Manufacturing, 246–247
  stand-alone machines, 244
Float, 300, 303
Florman, Samuel C., 4–5, 354
Fluor, 282

FMC. *See* Flexible manufacturing cell
      (FMC)
FMS. *See* Flexible manufacturing
      systems (FMS)
Follett, Mary Parker, 35
Follow-the-leader strategy, 177
Forced ranking, 145–146
Forecasting, 74–94
  entrepreneurship, 77
  exploratory technological
      forecast, 90
  and the Internet, 91
  intrapreneurship, 195
  invention and innovation, 91–92
  normative technological
      forecasting, 89
  qualitative methods, 84–85
    choice of method, 85
    Delphi method, 84–85
    jury of executive opinion, 84
    sales force composite, 85
    users' expectation, 85
  quantitative methods, 85–86
    exponential smoothing,
      86–87
    simple moving average, 85
    weighted moving average, 86
  regression models, 87–89
    multiple regression, 89
  technological forecasting, 89–101
  technological innovation, 92
  technology S-curve, 90
Formal authority, 147
Formal degree programs, 384–385
Formal leaders, 56, 68
Forrester, Jay W., 100
Fraction defective, 254–257
Frank, Barney, 49
Frederick, W. Taylor, 29–30,
      80, 218
French, John R. P. Jr., 147
Friedman, Thomas, 40, 133, 395–396
Full-scale production stage, 204
Fully automated factory (FAF),
      244–245
Fully projectized organization,
      316–317, 323
Functional analysis and allocation, 208
Functional departmentation, 121–122

Functional matrix, 322
Functional organization, 321
  advantages and disadvantages
      of, 321
  conducting projects within, 316
  matrix management, 318–319
  vs. projectized, 318
Functional (specialized)
      authority, 127
Functional status, 148
Functions of managers, 12–13
Functions of the Executive, The
      (Barnard), 42, 97
Future, 83, 121

Game theory, 113
  political application of, 113
Gantt
  charts, 297
  modified diagram, 305
Gantt, Henry L., 31–33, 37, 297
Gatekeeper, 367
GDP. *See* Gross domestic product
      (GDP)
Gee, Edwin, 384
Geiger, Gordon H., 254
GEM. *See* National Consortium
      for Graduate Degrees, for
      Minorities in Engineering
      (GEM)
General Electric, 42, 91, 177–178,
      227, 232
General Motors, 33, 35, 37, 42, 57,
      146, 150, 178, 210, 242, 291
  and product liability, 209
General Motors Research
      Laboratories, 178
General partner, 119
George, Claude S. Jr., 20
GIDEP. *See* Government–Industry
      Data Exchange Program
      (GIDEP)
Gifts, 348
"Gig" economy, 367–368
Gilbreth, Frank B., 30–32, 37, 218
Gilbreth, Lillian Moller, 31–32
Gillum, Jack D., 345
Giri, 399
Global Positioning Satellites (GPS), 41

Globalization, 393–407
  engineering career, significance for, 398
  Engineering Grand Challenges, 403–405
    health, 403–404
    joy of living, 404–405
    sustainability, 407–408
    vulnerability, 404
  European Union, 396–397
  flatteners, 395
    in-forming, 396
    insourcing, 396
    the new age of connectivity, 395
    offshoring, 396
    open sourcing, 395–396
    outsourcing, 396
    the steroids, 396
    supply chaining, 396
    workflow software, 395
  future considerations, 405–407
  impact on a U.S.-based engineering career, 398
  international trade agreements, 400–401
    NAFTA and the USMCA, 400
    WTO, 400–401
  Japanese management styles, 398–400
  management in developing countries, 401–402
    background, 401
    characteristics of developing countries, 401–402
    ethical considerations, 402
  multinational organizations, 397
  today's trends, 397
Gluck, Samuel E., 336
Goals, 52, 294–295
Goal statement, 78
Gobeli, David H., 321–322
Goetz, B. E., 155
Goldratt, Eliyahu M., 241
Gompers, Samuel, 32
Good to Great: Why Some Companies Make the Leap ... and Others Don't (Collins), 42
Goodman, Richard, 329
Government employment, 7

Government–Industry Data Exchange Program (GIDEP), 216
Government service, 7, 281, 283
GPS. *See* Global Positioning Satellites (GPS)
Graduate Degrees for Minorities in Engineering (GEM), 381
Graicunas, 125
Grant, David, 113
Granularity, 259
Grapevine, 371
Grayson, Lawrence T., 381
Great pyramid of Cheops, 21
Great Wall of China, 21
Greenleaf, Robert K., 56
Greenman v. Yuba Power Products, 210
Griffin, C. Steven, 350–351
Griffin, Ricky W., 9
Gross domestic product (GDP), 400
Ground support equipment (GSE), 295
Group technology, 229, 232, 241
Growth stage, careers, 366
GSE. *See* Ground support equipment (GSE)

Haloid Corporation, 92
Hamilton, Alexander, 224
Hammer, Michael, 42
Hammer, Willie, 271
Hammurabi of Babylon, 21
Hansen, John A., 132
Hardy, George, 353–354
Hargreaves, James, 23
Harris, E. Douglas, 50–52
Hartley, John, 245
Hart Scientific, 93
Harvey, Jerry, 39
Hawthorne effect, 37
Hawthorne studies, 35–37
Hawthorne Works (Western Electric Company), 20, 35
Health, 403–404
Healthcare, 284–285
Health care systems engineering, 285
Heating, ventilation, and air conditioning equipment (HVAC), 82
Heddo-hanta, 398
Heilmeier, George H., 15

Hennington v. Bloomfield Motors, 209
Henry Ford Health System, 259
Hersey, Paul, 54, 55
Hersey and Blanchard life cycle theory, 54
Herzberg, Frederick, 62, 69
Herzberg's two-factor theory, 48, 62–63, 69
Hewlett, Bill, 42
Hierarchy of needs, 39, 48, 60
High-technology enterprise, 13–14
Hispanics, as engineering students, 380
Hofstede, Geert, 401
Home computer systems, 405
Hoover, Herbert, 14, 26
Human factors engineering, 198, 218–219
Human resource accounting, 166–167
Human resource controls, 154, 166–168
  coordination, 168
  human resource accounting, 167
  management audits, 167
  non-financial controls, 168
  social controls, 167
Human resource planning, 136, 137–138
  job requisition/description, 138–139
  managers, hiring, 138
  technical professionals, hiring, 137–138
Human resources (personnel) management, 271–272
Humphrey, Watts S., 148
Hurwicz approach, 110, 112
HVAC. *See* Heating, ventilation, and air conditioning equipment (HVAC)
The Hyatt Regency disaster (construction case study), 344–347
Hygiene factors, 62

IBM, 37, 42, 183, 259
Identification of need, 174–175
IE. *See* Industrial engineering (IE)
IEEE Engineering Management Society, 14

IEEE Transactions on Reliability, 216
Illumination, 190
Illumination Experiments, Hawthorn
    studies, 36
Imperial Rome, 22
Implied warranty, 210
In Search of Excellence (Peters/
    Waterman), 42, 69–70, 167
Income statement, 158, 163
INCOSE. *See* International Council
    on Systems Engineering
    (INCOSE)
Incubation, 190
Independent (known) variables, 87
Indirect manager, 10
Individualism ethics, 337
Industrial engineering (IE), 225
Industrial Engineering magazine, 31
Industrial engineers, employment
    opportunities, 31
Industrial products
  engineering involvement in
      marketing of, 277–279
  types, 278
Industrial Revolution, 4, 20, 23–27
  cottage industry, end of, 24–25
  factory system, problems of, 24–25
  industrial development in
      America, 25
Inference engine, 115
Informal leaders, 49, 52
Informal time standards, 266
Information, 387
Information-based organization,
    special problems for
    management in, 133
Information Revolution, 126,
    132–134, 405–406
  impact of, 132–134
Information technology (IT), 41, 127,
    132–134, 205, 281–285, 309,
    360, 399
Informational roles, managers, 11–12
In-forming, 396
Ingenious, 4
Ingenium, 4
Inherent availability, 218
Initiating phase, project
    management, 294

Innovation, 79, 89, 91–92
Input, 264, 369
Inside information, 350
Insourcing, 400
Inspection, 257
Installations, 278
Institute of Electrical and Electronic
    Engineers (IEEE), 216, 376
  Reliability Division, 216
Instrumentality, 64
Integrated databases, 114
Integrated product teams, 205
Intellectual qualities, leaders, 50
Interchangeable manufacture, 23
Interface management, 202
Internal auditing, 166
Internal failure costs, 252
International Council on Systems
    Engineering (INCOSE), 100,
    200, 202
International Engineering
    Congress, 27
International management, 397
International Monetary Fund, 400
International trade agreements,
    400–401
Internet, 91
  and forecasting, 91
  usage, 92
  world usage and population
      statistics, 93
Internet of Things (IoT), 6
  smart retail and, 108
Interpersonal roles, managers, 11
Interpersonal skills, 11
  project matrix, 322
Interviewing well, 142
Intrapreneurship, 195
Intrauterine device (IUD), 210
Invention, 91–92
Inventory control, 232–233, 272
  break-even charts, 233–236
  economic order quantity (EOQ),
      232–233
    problems with, 232–233
  learning curves, 236–238
  types of inventory, 232
Inventory turnover, 99, 165
Iraq, 21

ISO, 200
ISO 9000, 258, 261
ISO 14000, 258
Isoprofit line, 103
IT. *See* Information technology (IT)
IUD. *See* Intrauterine device (IUD)

Jacoby, Neil H., 397
Japanese management styles, 398–400
  ceremonies and rituals, 399
  giri, 399
  lifetime employment, 398–399
  michi, 399
  omikoshi management, 399
  promotions, infrequent, 399
  recruitment of employees directly
      out of secondary school/
      college, 398
  retirement/restricted employment
      for older workers, 399
  ringi system, 399
  rotation between functional
      departments, 399
  worker security, 399
Japanese production plants,
    characteristics of, 404
Jay, John, 24
Jefferson, Thomas, 5, 24, 26
Jennings, Daniel F., 9
Jericho, 21
JIT. *See* Just-in-time (JIT)
Job enrichment, 62–63
Job requisition/description, 138–139
  example of, 138
Jobs, Steve, 16, 278
Johnson & Johnson, product liability
    and, 210
Johnson, Spencer, 42
Juran, J. M., 212, 254
Jury of executive opinion, 84
Just-in-time (JIT), 233, 241–243,
    270, 400

Kaizen teams, 264
Kanban, 233, 241–243
  cards, 241, 242
  cycle, 243
Kant, Immanuel, 337
Kantian ethics, 337

Karoshi, 399
Katzenbach, Jon R., 130
Katz, Ralph, 328
Katz, Robert L., 11
Kaufman, H. G., 374–375
Keller, Charles W., 291
Kennedy, Marilyn Moats, 89
Kerr, S., 66
Kerzner, Harold, 130, 315, 328
Key activities, 120–121
   identification of, 120–121
   organizing by, 120–121
Key technologies, 182
Kickoff meeting, proposal
   team, 292
Kilminster, Joe, 353–354
King, William R., 321
King, W. J., 367
Knowledge base, 115
Knowledge explosion, 373–374
Knowledge level, 329
Knowledge workers, 60, 405
Koestenbaum, Peter, 115, 336
Kohlberg's moral development,
   337–339
Koontz, Harold, 12, 81, 83, 120,
   367, 370
Kotler, P., 277
Kotter, J. P., 49

Lakein, Alan, 388–389
Lane, Frederick C., 22
Lanham Act, 186
Larson, Erik W., 321–322, 327
Laufer, Arthur, 230
Launch Escape System, Apollo
   command module, 219
Lawler, Edward E., III, 64
LCL. *See* Lower control limit (LCL)
Leadership, 47–70, 314–331
   contingency approaches, 54
     leadership continuum, 54–55
   failures, 63
   from good to great, 59
   full range model and transforma-
     tional (*see* Transformational
     leadership)
   and management, 48–49
     Myers–Briggs preferences, 52

nature of, 49
   traits, 50–52
   people/task matrix approaches,
     52–54
     Hersey and Blanchard life-cycle
      theory, 54
     leadership grid, 52
     Michigan State studies, 54
     Ohio State studies, 54
   secrets, 50
   servant, 56–57
     other viewpoints, 57
Leadership continuum, 54–55, 69
   controlled freedom, 69
   leader as metronome, 69
   technical competence, 68
   work challenge, 69
Leadership Grid, The, 52, 69
*Leadership 101: What Every Leader
   Needs to Know* (Maxwell), 42
*Leadership Secrets of the World's
   Most Successful CEOs*
   (Yaverbaum), 50
Leading, 12, 47, 66
League of United Latin American
   Citizens (LULAC), 380
Lean manufacturing, 246–247
   principles of, 246–247
Lean Six Sigma (LSS), 263
Learning curves, 236–238
Learning paths, 329
LeBoeuf, Michael, 387
Lee, Denis M. S., 362
Legitimate/position power
   (authority), 147
Leshner, Alan, 176
Letter contract, 331
Leverage ratios, 165
Liability, 163, 209–212
   product development of, 209–210
   product reducing, 210–212
Library of Congress, and copyrights,
   187
Lighthall, Frederick F., 354
Limited liability company (LLC), 119
Limited partners, 119
Limited partnership, 119
Line and staff relationships, 119
Line functions, 127

Line management, 81
Line relationships, 127
Linear programming, 103–105
Linear relationship, 87
Liquidity ratios, 165–166
LLC. *See* Limited liability company
   (LLC)
Loading, 239
Locke, John, 337
Logical planning, 83
Longnecker, Justin G., 9
Lower control limit (LCL), 256
Luddites, 24
Luegenbiehl, Heinz C., 338
LULAC. *See* League of United Latin
   American Citizens (LULAC)
Lumsdaine, Edward, 190, 192
Lund, Robert K., 352–354
Lund, Robert T., 132

MacArthur, Douglas, 254
Machiavelli, Niccolo, 22
MAES. *See* Mexican American
   Engineering Society
   (MAES)
Maintainability, 217
   components
     active maintenance time, 217
     administrative and preparation
      time, 217
     logistics time, 217
   types, 217
Maintenance
   items, 278
   predictive, 269
   preventive, 269, 270
   scope of, 268–269
   stage, 366–367
   types of, 269
Maintenance and facilities (plant)
   engineering, 268–271
Maintenance engineering, 225
*Maintenance Engineering
   Handbook*, 268
Maintenance management, 269–270
   facilities and plant engineering
     functions, 270–271
   maintenance staff, size of, 269
   repair parts inventory, 270

total productive maintenance
(TPM), 270
work scheduling, 270
Maintenance seekers, 62
Maintenance stage, careers, 366
Malcolm Baldrige National Quality
Award, 260
Management
administrative, 33–35
behavioral, 35–39
definition, 8, 10
engineering, synthesis with, 13–17
careers, 15
definitions, 13–14
engineers in, need for, 14
and engineering career, 15–17
future considerations, 405–407
demographics, 406
education and the technological
society, 406
energy, 406
environment, 406
flattened world, 406
food, 406
information revolution, 405–406
materials, 406
leadership, 48–49
levels, 10–11
first-line managers, 10
middle managers, 10
top managers, 10–11
managerial roles, 11–12
managerial skills, 11
conceptual, 12
interpersonal, 11
technical, 11
philosophies, 27
philosophies outside western
culture, 38
profession, 13
scientific, 27–32
growth and implications of,
32–33
vs. skills required, 11
women and minorities in,
379–381
Management audits, 167
Management by objectives (MBO),
69, 74, 80, 166, 183

advantages, 81
disadvantages, 81
Management information systems
(MIS), 114
Management of technology (MOT), 385
Management science, 13, 99–102
analyst, 101–102
manager, 68–69
models, 100–101
origins, 99–100
process, 101–102
Management skills, development of,
384–386
Management theory, 41
Manager
communication, 368
first-line, 10–11
functions of, 12–13
controlling, 12
leading, 12
organizing, 12
planning, 12
staffing, 12
middle, 10
top, 10–11
Managerial decision making, defined, 97
Managerial Grid, 52
Managerial responsibilities, preparing
for, 384–386
Managerial roles, 11–12
decisional, 12
disturbance handler, 12
negotiator, 12
resource allocator, 12
informational, 12
disseminator, 12
monitor, 12
interpersonal, 11
figurehead, 11
leader, 11
liaison, 11
Managerial skills
conceptual, 11
interpersonal, 11
technical, 11
Managing engineering and technology
(advance organizer), 17
Managing in the Next Society
(Drucker), 42

Maneggiare, 9
Manufacturing, 282
Manufacturing engineers, 41
future demands of, 227
Manufacturing facilities, planning,
227–232
Manufacturing resource planning
(MRP II), 240
Manville Corporation, and product
liability, 210
Margulies, N., 375
Market share, 79
Market research, 278
Marketing
and engineer, 277
definition, 277
electronic, 280
4PS of, 279
industrial products, 277
process of, 277–279
projected employment change,
281–282
SEO process, 280
Marketing industrial products, engi-
neering involvement in, 278.
See also Industrial products
Marriott, 282
Maslow, Abraham, 37, 39, 60, 61–63, 70
Maslow's hierarchy of needs, 39, 48, 60
Mass, 128
Master production schedule (MPS), 239
Master's degree programs in engineer-
ing management, 385
Materials management, 272
Materials requirements planning
(MRP), 239–240
Mathematical models, 100
Mathematics, Engineering, Science
Achievement (MESA), 381
Mathews, Marilynn C., 342
Matrix
management, 318–321
organization, 314, 319–321
applications for, 321
fixed, 321
relationships, 320
shifting, 321
relationships, 320
Matsushita, Konosuke, 33

Matsushita Electric Industrial
    Company, 33
Maudslay, Henry, 23
'Maximax' solution, 110
'Maximin' solution, 110
Maxwell, John C., 42
Mayan temples (South America), 21, 37
Mayo, Elton, 36
MBO. *See* Management by objectives
    (MBO)
MBWA ('management by walking
    around'), 42
McCall, Morgan W. Jr., 68, 386
McClelland, David, 48, 63, 70, 382
McClelland's trio of needs, 63–64
McConnell, John H., 62
McDonald's, 80, 157, 282
    and product liability, 209
McFarland, Dalton E., 9
McGregor, Douglas, 39, 48, 70
    McRee, William, 26
    theory X, 58–60
    theory Y, 58–60
MDT. *See* Mean downtime (MDT)
Mean downtime (MDT), 217
Mean time between failures (MTBF),
    215, 217
Mean time between maintenance
    (MTBM), 217
Mean time to repair (MTTR), 217
Mean value, 255
Measure of central tendency, 254
Measure of dispersion, 254–255
Mechanic arts, 5, 29
Mechanical engineers, employment, 6
Mechanical process control, 155–157
    closed-loop, 155–156
    open-loop, 156
Mechanistic management systems, 129
Medical engineering, 284–285
Medical imaging systems, 285
Mehrabian, Albert, 372
Mentor, 367
MEPs. *See* Minority engineering pro-
    grams (MEPs)
Merino, Donald N., 253
MESA. *See* Mathematics,
    Engineering, Science
    Achievement (MESA)

Metcalf, Henry, 29
Methods-Time Measurement
    (MTM-1), 267
Me-too strategy, 177
Mexican American Engineering
    Society (MAES), 381
Michi, 399
Michigan and Ohio State studies of
    leadership styles, 54
Microsoft Corporation, mission state, 77
Middle managers, 10
Middlesex Canal Company, 25
Mihalasky, John, 26–27
Milestones, 297
Mill towns (England), 24
Milliken & Company, 259
MIL-STD-105E, 257
MIL-STD-499B, 200
MIL-STD-973 (EIA/IS649), 207
Mindmapping, 192
Minimax regret solution, 110
Minority engineering programs
    (MEPs), 380
Mintzberg, Henry, 11, 49, 369
MIS. *See* Management information
    systems (MIS)
Mission statement, 77
MNCs. *See* Multinational corporations
    (MNCs)
Mockups, 217
Model
    changeover, 246–247
    conceptual, 100
    mathematical, 100
    physical, 100
Modified ranking, 146
Moen, Ronald D., 146
Monitor role, managers, 12
Monitoring, controlling, 293
Monoliths on Easter Island, 21
Monte Carlo method, 303
Mooney, James, 35
Moonlighting, 348
Moral philosophy, 336–337
Morality, defined, 336
Morals, 337
Moretsu shain, 399
Morrill Land Grant Act (1862), 26, 37
Morrison, Robert F., 365, 367

Morse, Lucy, 373
Morse, Robert, 25
Most likely time, PERT, 301
MOT. *See* Management of technology
    (MOT)
Motivation, 58–70
    content theories, 60–66
        affiliation needs, 60
        esteem needs, 60
        Herzberg's two-factor theory,
            62–63
        Maslow's hierarchy of needs, 60
        McClelland's trio of needs, 63–64
        physiological needs, 60
        security/safety needs, 60
        self-actualization needs, 61
    and direction of an individual's
        behavior, 58
    McGregor's theory X and theory Y,
        58–60
    and persistence of an individual's
        behavior, 58
    process theories, 64–66
        behavior modification, 66
        equity theory, 64
        expectancy theory, 64–65
        Porter–Lawler extension, 65–66
    project performance, 324–331
        customer communications,
            328–329
        early learning path, 329
        keys to project success, 327–328
        managing conflict, 326–327
        and strength of an individual's
            behavior, 58
        team building, 324–325
Motivation seekers, 62
    process theories, 64–66
Motive, 58
Motorola, Inc., 259
    quality control at, 258
    Six Sigma process, 259
Mouton, Jane S., 52, 326
MPS. *See* Master production schedule
    (MPS)
MRP. *See* Materials requirements
    planning (MRP)
MTBF. *See* Mean time between
    failures (MTBF)

MTBM. *See* Mean time between maintenance (MTBM)
MTTR. *See* Mean time to repair (MTTR)
Muda, 246
Mule, 23
Mulloy, Lawrence, 353–355
Multinational corporations (MNCs), 25, 395
  pricing policies, 397
Multinational organizations, 397
Multiple regression, 89
Multiple sampling, 257
Multistory, 229
Murphy, David C., 327
Murray, Phillip, 32
My Years with General Motors (Sloan), 42
Myers–Briggs Type Indicator (MBTI), 52
Myers, Donald D., 184
Myers, M. Scott, 62

NACME. *See* National Action Council for Minorities in Engineering (NACME)
NAFTA. *See* North American Free Trade Agreement (NAFTA)
NASA. *See* National Aeronautics and Space Administration (NASA)
National Action Council for Minorities in Engineering (NACME), 381
National Aeronautics and Space Administration (NASA), 10, 109, 200, 353–355, 365
National Association of Minority Engineering Program Administrators, 381
National Consortium for Graduate Degrees, for Minorities in Engineering (GEM), 381
National Council of Engineering Examiners (NCEE), 376
National Institute of Standards and Technology (NIST), 203, 219
National Society of Black Engineers (NSBE), 381
National Society of Professional Engineers (NSPE), 204, 340, 376

National Transportation Safety Board (NTSB), 346
Native Americans, as engineering students, 380
NCEE. *See* National Council of Engineering Examiners (NCEE)
Need for achievement, 63
Need for affiliation, 64
Need for power, 64
Negative reinforcement, 66
Negligence, 209
Negotiator role, managers, 12
Net worth, 163
Network-based project scheduling, 300
Newman, William H., 321
New product development, 200–203
  commercial validation and production preparation stage, 200
  conceptual stage, 207
  development stage, 204
  disposal stage, 203
  full-scale production stage, 204
  phases/stages in, 200–203
  product support stage, 204
  technical feasibility stage, 204
NIST. *See* National Institute of Standards and Technology (NIST)
Nitrogen cycle, 403
Nominal Group Technique, 191
Nondegree coursework, 385
Nonfinancial controls, 168
Non-project-driven organization, 315
Nonroutine decisions, 98
Nonverbal communication, 370
Normal cost, 306
Normal distribution, 302
Normal time, 306
Normative ethics, 337
Normative technological forecasting, 89
North American Free Trade Agreement (NAFTA), 400–401
NSBE. *See* National Society of Black Engineers (NSBE)
NSPE. *See* National Society of Professional Engineers (NSPE)

NTSB. *See* National Transportation Safety Board (NTSB)
Nuclear fusion, 403

Objective function, 103
Objectives, 79–80, 325
  vs. bounded rationality, 98–99
Obsolescence, 373–374
  definition, 374
  knowledge explosion, 373–374
  methods of reducing, 374–375
    continuing education, 374–375
    on-the-job activity, 375
    technical literature, mastering, 374
  organizational, 374
  threat of, 373–374
Obstacles, 294
Occupational Safety and Health Act (OSHA), 271
Odaka, Kunio, 399
Odds and evens (children's game), 113
Official goals, 79
Offshoring, 396
Oldenquist, Andrew G., 339–340
Omega, 93
Omikoshi, 399
On the Economy of Machinery and Manufactures (Babbage), 27
Oncken, William Jr., 150
One Minute Manager, The (Blanchard/Johnson), 42
Open-loop control, 156
Open-sourcing, 395
Operating budgets, 159
Operating plans, 83
Operating profit, 164
Operating ratios, 165
Operational availability, 218
Operational careers, 365
Operations management (case study), 231
Operations research, 99
Operative goals, 79
Opportunity statement, 294
Optimal clinical laboratories, design of, 285
Optimistic time, PERT, 301
Oral briefing/presentation, 372–373

Organic management systems, 129
Organization, legal forms of, 119–120
Organization man, 54
Organization structure, 321–322
Organization theory, traditional,
    121–124
  line functions, 127–128
  patterns of departmentation,
    121–124
  span of control, 124–127
  staff functions, 127–128
Organizational development, 52
Organizational obsolescence, 374
Organizations
  bad-mouth, 364
  grapevine, 371
  types of, 315–316
Organizing, 12, 74, 136–152
  authority and power, 146–148
    acceptance theory of
      authority, 147
    culture, 148
    formal authority, 147
    sources of power, 147–148
    status, 148
  committees, 150–151
  cooperatives, 120
  corporations, 119–120
  creation of, 119–120
    executive, 151
    policy making and
      administration, 151
    pooling of authority, 119
    reasons for using, 151
    representation, 151
    securing cooperation in
      execution, 151
    sharing knowledge and
      expertise, 151
    training of participants, 119
  decentralization, 150
  defined, 120
  delegation, 148–150
    barriers to, for engineers,
      149–150
    reasons for, 149
  departmentalization, 118
    patterns of, 121–124
  human aspects of, 136–152

human resource planning, 137–138
by key activities, 120–121
legal forms of, 119–120
line and staff relationships, 127–128
nature of, 119–120
orientation and training, 143–145
partnerships, 119
performance, appraising, 145–146
personnel selection, 50
  campus interview, 140–141
  cover letter, 140
  reference checks, 141
  résumé, 140
  site visits, 141
  starting salary, 141–142
sole proprietorship, 119
span of control, 124–127
staffing technical organizations,
  137–139
teams, 118, 130–131
traditional organization theory,
  121–124
Orientation, 143–145
Osborne, Adam, 92
Osborne, Alex, 191
OSHA. *See* Occupational Safety and
  Health Act (OSHA)
Ouchi, William G., 70
Out of the Crisis (Deming), 262
Outcome, 102
Output control, 157
Outsourcing, 396
Overhead (burden), 160
Owen, Robert, 24

Pacing technologies, 182
Pacioli, Luca, 23
Packard, Dave, 42
Pal's
  Baldrige award winner, 260
  mission state, 78
  and quality, 260
Parker, Jim, 35
Partially projectized organization, 318
Participative decision making, 328
Partnerships, 119
Partridge, Captain, 26
Patent rights, establishing, 184–185
Patent scorecard, 183

Patents, 183
  granted by the U.S. patent
    office, 183
Payoff table (decision matrix), 102
P-chart, 257
PE. *See* Professional engineer (PE)
Percentile, 146
Performance, appraising, 145–146
Performance-to-outcome expectancy,
  64–65
Personal ethical decision-making, 343
Personal staff, 127
Personnel selection, 136
  campus interview, 140–141
  cover letter, 140
  reference checks, 141
  résumé, 140
  site visits, 141
  starting salary, 141–142
PERT. *See* Program Evaluation
  Review Technique (PERT)
Pessimistic time, PERT, 301
Peters, Thomas J., 42, 69, 80, 132,
  167, 282
Peterson, Paul, 50
Physical and financial resources, 80
Physical model, 100, 369
Physical qualities, leaders, 50
Physiological needs, 60
Piece rate system, 29
Pilot processes, 138
P-kanban, 243
Planning, 12, 74–94, 293–294
  concepts, 81–84
    horizon, 83
    policies, 84
    premises, 81, 83
    procedures, 84
    responsibility, 81
    systems of plans, 83
  contingency plan, 83
  and decision making, 97
  foundation for, 76–81
    management by objectives, 80–81
    strategic planning, 76–80
  importance of, 75
  management by objectives (MBO),
    80–81
  nature of, 75–76

and decision-making process, 75–76
importance of, 75
operating plans, 83
planning horizon, 83
planning/decision-making process, 75–76
policies and procedures, 84
premises, 81, 83
production activity, 223–248
responsibility for, 81
staff, 79
strategic, 76–80
systems of plans, 83
technology at AAON, 82
Planning horizon, 83
Planning phase, project management, 293
Planning premises, 81
Plant design, 229
Plant layout, 229–230
Plant location, 227–228
Plant manager, 225
Plant patents, 185
Plowman, E. G., 50
Poisson probability distribution, 257
Policies, 84
Porter, Lyman W., 61
Porter–Lawler extension, 64, 65
Position power, 147
Positive reinforcement, 66
Post-action, 157
Postemployment limitations, 348–349
Power
loom, 23
position, 147
sources, 147–148
Prai-Barshana technique, 191
Precedence diagrams, 305–306
Predetermined time systems, 266–267
Predictive, 269
Predictive maintenance, 269
Preproduction prototypes, 202
Preproposal effort, 291–292
Prevention costs, 252
Preventive maintenance, 157, 217, 269–270
Price, Willard, 343
Prince, The (Machiavelli), 22

'Principle of insufficient reason,' 110
Principles of Scientific Management (Taylor), 30
Pringle, Charles D., 9, 98, 145, 148
Problem statement, 191
Procedures, 84, 98
Process control charts, 256–257
Process engineers, 225
Process layout, 229
Process planning, 239
Process theories, 60, 64–66
behavior modification
extinction, 66
negative reinforcement, 66
positive reinforcement, 66
punishment, 66
vs. content, 60
equity theory, 64
expectancy theory, 64–65
Porter–Lawler extension, 65
Producibility, 219
Product design function, 174
Product evaluation function, 174
Product layout, 229
Product liability
defined, 209
development of, 209–210
reducing liability, 210–212
Product life cycle, 132, 174–175
developing reliability over, 212
Product manager, 321
Product quality. See Quality
Product support stage, 204
Product use and logistic support function, 174
Production activity, planning, 223–148
and control systems, 238–243
manufacturing resource, 240
materials requirements, 239–240
steps in, 238–239
engineer in, 225–226
future demands on manufacturing engineers, 227
intensity of use of engineers, 225, 227
types of positions, 225
enterprise resource planning, 240–241
just-in-time, 241–243

model changeover, 243
synchronized manufacturing, 241
manufacturing facilities, 227–232
plant design, 229
plant layout, 229–230
plant location, 227–228
process layout, 229
product layout, 229
manufacturing systems, 244–248
advantages of an FMS, 245
definitions, 244–245
flexible, 244
lean manufacturing, 246–247
supply chain management, 247–248
quantitative tools in, 232–238
break-even charts, 233–236
inventory control, 232–233
learning curves, 236–238
vital nature of, 224
Production control, 238–239
Production function (and/or construction), 174
Production management, 168
Production operation
human resources (personnel) management, 271–272
materials management, 272
purchasing, 272
Production planning, 238–243
and control systems, 239–240
bill of materials (BOM), 239
just-in-time (JIT), 241–243
manufacturing resource planning (MRP II),240
materials requirements planning (MRP), 239–240
model changeover, 243
synchronized manufacturing, 241
steps in, 238–239
tools, 232–233
Production preparation stage, 204
Production technology scale, 129
Production, vital nature of, 224
Productivity, 79–80, 264–267
defined, 264
service industry, 282–283
Profession, definition, 5

Professional activity, 376–377
  technical societies, types/purpose
    of, 376
Professional engineer (PE), 342
Professional engineering
  registration, 378
Professional ethics, 336–344
  in consulting and construction,
    342–344
    construction safety, 344–345
    distribution of public
      services, 343
    guidelines for facilitating
      solutions to ethical dilemmas
      in professional practice,
      342–343
    personal ethical decision
      making, 343
    political contributions, 342
    significance, 347
  engineering code of ethics, 342
Professional societies, 376–377
  getting involved, reasons for,
    376–377
  purpose of technical societies, 376
  technical papers and
    publications, 377
  types of technical societies, 376
Profit and loss statement, 163
Profit budget, 159
Profit centers, 159
Profit margin, 166
Profitability, 80
Profitability ratios, 166
Program Evaluation Review Technique
  (PERT), 157, 289, 300, 301,
  303, 305
Programs, 291
  defined, 291
  and projects, 291
Project
  characteristics of a project, 290–291
  contracts, 330–331
  crashing, 306–308
  duration reduction, 305–306
  motivation, 324–325
  programs, 291
  project planning tools, 293–304
    costs, 306

schedule, 295–304
scope, 294–295
proposal process, 291–293
  contents, 292–293
  effort, 291–292
  internally driven, 293
  preparation, 292
Project-driven organization, 315–316
  defined, 315
  elements of, 315–316
  future business, 316
  key functional support, 316
  manufacturing and routine
    administration, 316
  project engineer, 315
  project office, 315–316
  projectized *vs.* functional
    organizations, 316–318
Project duration estimates, 305
Project management, 40, 168,
  289–310
  closing the project phase, 308–309
  common path to engineering
    management, 324
  essential considerations in, 290
  executing phase, 293
  initiating phase, 293
  methods, 290
  monitoring and controlling
    phase, 293
  planning phase, 293
  three-legged stool, 290
  under uncertainty, 309–310
Project Management Institute (PMI),
  293, 321, 322, 323, 376
Project Management Professional
  (PMP), 378
Project manager, 322–324
  characteristics of effective, 322–323
    administrative skills, 323
    interpersonal skills, 323
    technical skills, 322–323
    manager's charter, 323–324
    skills development, 323
Project Manager's charter, 323–324
Project matrix, 322
  administrative skills, 323
  charter, 323
  interpersonal skills, 323

project performance, motivating,
  324–329
skill development, 323
team building, 324–325
technical skills, 322–323
Project organization, 315–322
  elements of, 315–316
  matrix management, 318–321
  project-driven organization, 315
  projectized *vs.* functional, 316–318
  structure, 321–322
Project performance, motivating,
  324–329
Project planning
  and acquisition, 289–310
  schedule, 295–304
  scope, 294–295
  tools, 293–304
Project planning tools, 293–304
Project proposal process, 291–293
  internally driven projects, 293
  preproposal effort, 291–292
  proposal contents, 292–293
  proposal preparation, 292
  RFP (request for proposal),
    292–293
  statement of work, 294
Project team, 322
  building, 324–325
  compromising, 326
  conflict management, 326–327
    methods of, 326–327
  conflict, sources of, 326
  confrontation, 326
  forcing, 326
  problem solving, 326
  smoothing, 326
Projectized organizations,
  316–318, 322
  advantages and disadvantages
    of, 318
  vs. functional, 318
  matrix management, 318–321
Projects, 178–179, 289–310. *See also*
    Budgets
  characteristics of, 290–291
  conducting within the functional
    organization, 316
  contracts, 330–331

cost type contracts, 330–331
   fixed-price contracts, 330
costs, 330–331
crashing, 306–308
customer communications, 328–329
early learning path, 329
keys to success, 327–328
organization structure and project
      success, 321–322
and programs, 291
resources, 303
schedule, 295–304
Proposal contents, 292–293
Proposal preparation, 292
Proposal process, 291–293
Protection of ideas, 184
   comparison of means of
         protection, 188
   copyrights, 186–187
   patents, 184–185
   trade secrets, 187–188
   trademarks, 186
Prototypes, 201
Public works improvements, 343
Punishment, 66
Punishment power, 147
Purchased parts, 232
Pure rating, 146
Pyramids of Egypt, 21

QM. *See* Quality Management
Quality, 252–255
   assuring, 252–253
   better quality, importance of,
         254–255
   costs, 252–254
      appraisal, 253
      external failure, 253
      internal failure, 252
      prevention, 252
   defined, 252
   Deming's 14 points, 262–263
   inspection and sampling, 257
   process control charts, 256–257
   revolution, 244, 258
   statistics of, 254–255
   Taguchi methods, 261–262
Quality and production operations,
      managing, 251–272

assuring product quality, 252–257
   better quality, importance of, 252
   definitions, 252
   Deming's 14 points, 262–263
   inspection, 257
   process control charts, 256–257
   quality, 261–262
   quality costs, 252–254
   quality teams, 264
   sampling, 257
   statistics of quality, 254–255
   Taguchi methods, 261–262
facilities (plant) engineering, 268
maintenance engineering, 225–226
   repair parts inventory, 270
   scope, 268–269
   size of maintenance staff, 269
   total productive maintenance, 270
   types, 269
   work orders, 270
   work scheduling, 270
other manufacturing functions,
      271–272
   human resources (personnel)
         management, 271–272
   purchasing and materials
         management, 272
productivity, 264–267
   definition, 264
   manufacturing, 264–265
Quality circles, 264
Quality control, 254–255
Quality Management (QM), 39–40
   customer focus and, 40
   globalization, effect on, 40–41
   information technology, role in, 41
   performance capabilities and, 40
Quality of conformance, 252
Quality of design, 252
Quality of production, 252
Quality of use, 252
Quality teams, 264
Queuing (waiting-line) theory,
      107–110
Quick assets ratio, 165

Raia, A., 375
Railroads, 25, 32, 37
Randolph, Isham, 339

Range, 255
Ranky, Paul, 245
Rating scale, 145
Ratio analysis, 163–166
Rational decision-making, 98
Raven, Bertram, 147
Raw material, 228, 232
R-chart, 256
Recentralization, 150
Recreational Equipment Inc.
      (REI), 120
Reda, Hussein M., 241, 243
Redundancy, 214, 216
Reengineering the Corporation
      (Hammer/Champy), 42
Reference checks, 136, 141
Referent power, 147
Registered trademarks, 186
Registration, 377–379
Regression analysis, 89
Regression models, 87–89
   multiple regression, 89
   regression analysis, 89
Reinforcement, 66
Relations analysis, 121
Relay Assembly Test Room
      Experiments, Hawthorn
      studies, 36
Reliability, 212–216
   definition, 212
   measures, 213
   product life cycle, developing
         over, 215
      apportionment, 215
      designing, 215–216
      flattening the bathtub curve, 216
      growth, 216
      planning, 215
   profession, 216
   significance of, 212
   simple model, 212–216
      bathtub curve model, 215
      series–parallel models, 214–215
      simple parallel model, 214
      simple series model, 213
Reliability Division, American Society
      for Quality (ASQ), 216
Reliability engineering, 216
Reliability Review, 216

Rensselaer (New York) Polytechnic Institute, 26
Repair parts inventory, 270
Request for proposal (RFP), 130, 291–293
Research, defined, 176
Research and design careers, 365
Research and development (R&D)
  and business strategy, 181–182
  creativity, 190–193
  evaluating R&D effectiveness, 182–183
  individual effectiveness, 183
  nature of, 176–177
    definition, 174–175
    expenditure and performance, distribution by, 176–177
  organizational effectiveness, 182–183
  product life cycle, 174–175
  project selection, 178
    initial screening, 178–179
    need for, 178
    quantitative approaches, 179–181
  protection of ideas, 184–189
    comparison of means of, 188–189
    copyrights, 186–187
    patents, 184–185
    trade secrets, 187–188
    trademarks, 186
  strategy and organization, 177–178
    corporate research organizations, 177–178
    new product strategies, 177
  success of, 181–183
    business strategy, 181–182
    evaluating R&D effectiveness, 182–183
  support for, 183–184
  technology life cycle, 174–175
Research function
  creativity, 190–191
    brainstorming, 191–192
    characteristics of creative people, 192–193
    and innovation, 193
    nature of, 176–177
    process, 181–182

protection of ideas, 184–189
  providing creative environment, 193
  techniques for, 183–184
  technological gatekeepers in R&D organizations, 193
  mindmapping, 192
  product life cycle, 174–175
  research and development (R&D), 173
  technology life cycle, 174–175
Resistance to time standards, 267
Resource allocation, 303–304
Responsibility centers, 159
Résumé, 140
Reswick, J. B., 199
Retained earnings, 162–164
Retention of information, 370
Revenue, 159
Revenue and expense statement, 163
Revenue budget, 159
Revenue centers, 159
Revolutionary War, 4
Reward power, 147
Reynolds, Terry, 26
RFP. See Request for proposal (RFP)
Rhode Island, 25
Rights-based ethics, 337
Ringi system, 399
Risk, 294
  decision making under, 105–110
    examples, 105–107
    expected value, 105–106
    nature of, 105
    queuing (waiting-line) theory, 107–110
    risk as variance, 110
  definition, 212
  expected value, 105–107
  nature of, 105
  project planning, 294
  queuing (waiting-line) theory, 107–110
  as variance, 110
Ritz Carlton, 259
Robb, Russell, and administrative management, 35
Robbins, Stephen P., 58
Roberts, E. B., 193
Robinson, Richard D., 397

Roethlisberger, Fritz, 36
Root mean square, 302
Rosenbaum, Bernard L., 66, 68
Rosenwald, Julius, 83
Rost, J. C., 49
Routine decisions, 98
Rules, 84
Rural Electrification Administration, 32

SAE. See Society of Automotive Engineers (SAE)
Safety, 216
Saint Monday, 24
Salaries, 142
Sales force composite, 85
SAME. See Society of American Military Engineers (SAME)
Sample standard deviation, 255
Sample variance, 255
Sampling, 257
Sarchet, Bernard R., 75, 97, 384, 386
Sayles, L. R., 69
Scalar status, 148
Schedule/scheduling, 239, 295–304
  activity-on-node (AON) diagram, 300
  building, 298
  controlling, 326–329
  critical path method (CPM), 300
  Gantt charts, 297–305
  milestones, 297
  program evaluation review technique (PERT), 289, 300
  work breakdown structure (WBS), 295
Schedule C, 119
Schedule variance, 308
Schmidt, Warren H., 54–55, 69
Scientific management, 27–33, 218
  Gilbreths' legacy, 30–32
  growth and implications of, 32–33
  negative impact, 33
  spread of, 32
  Taylor's piece rate system, 29
  Towne and the ASME, 28–29
Scientific method (scientific process), 101
Scientists, vs. engineers, 67–68

SCM. *See* Software configuration management (SCM); Supply chain Management (SCM)
Scope, 294–295
Scoring model, 179–180
Screening (concurrent) control, 157
Screw-cutting lathe, 23
Search engine optimization (SEO), 280
SECME. *See* Southeast Consortium for Minorities in Engineering (SECME)
Security/safety needs, 60
See, James Waring, 28
Seely, Bruce, 26
Seiler, Robert E., 178
Self-actualization needs, 61
Semivariable costs, 233
SEO. *See* Search engine optimization (SEO)
Series–parallel models, 214–215
Servant As Leader (Greenleaf), 56
Servant leadership, 48, 56
Service economy, engineers in, 281–283
Service industry, 282–284
  biomedical engineering and the healthcare, 284–285
  college teaching and research, 284
  computer applications, 282–283
  government service, 283
  green engineering, 283
  sustainable environments, 284
Service marks, 186
Service-producing industries, importance of, 281–282
Service relationships, 118, 127
Service sector, 282–285
  characteristics, 282
  examples, 282–285
    biomedical engineering and the healthcare, 284–285
    college teaching research, 284
    computer applications, 282–283
    government service, 283
    green engineering, 283
    sustainable environments, 284
Setup cost, 232
Seven Habits of Highly Effective People, The (Covey), 42

1750 English Iron Act, 25
Shackleton, Sir Ernest, 55
Shannon, Robert E., 89, 155, 180, 190, 374
Sheldon, Oliver, 48
Shewhart, Walter, 263
Shifting matrix, 321
Shop-order system of accounts, 29
SHPE. *See* Society of Hispanic Professional Engineers (SHPE)
Sigma-chart, 256
Simon, Herbert A., 98–99
Simple moving average, 85
Simple parallel model, 214
Simple payback time, 180
Simple series model, 213
Simplex method, 105
Simulated sampling, 303
Simulation, 108–109
Simultaneous engineering, 203–205. *See also* Concurrent engineering
Single-story, 229
Site visits, 141
Six Sigma, 40, 259
Skinner, B. F., 48, 66
Slack, 300, 303
Slater, Samuel, 25
Slim, William, 49
Sloan, Alfred P. Jr., 42, 150
Slope, 306
Slowter, Edward E., 339, 340, 341
SME. *See* Society of Manufacturing Engineers (SME)
Smith, Adam, 337
Smith, Jack, 57
Smith, Leonard, 365
Social controls, 167
Social responsibility, organizations, 77, 80
Socialization, 145
Society of American Military Engineers (SAME), 376
Society of Automotive Engineers (SAE), 376
Society of Hispanic Professional Engineers (SHPE), 381
Society of Manufacturing Engineers (SME), 227, 376

Society of Plastics Engineers (SPE), 376
Society of Reliability Engineers (SRE), 216
Society of Women Engineers (SWE), 379
Software configuration management (SCM), 207
  benefits of, 207–208
Software design review, 208–209
Software engineers, 6
Soho Engineering Foundry, 24
Solar energy, 403
Solar Turbines Incorporated, 260
Sole proprietorship, 119–120
Sony, 183
SOPs. *See* Standard operating procedures (SOPs)
Sources and uses of funds statement, 163
Southeast Consortium for Minorities in Engineering (SECME), 381
Space Division, North American Aviation, 318
Span of control (span of management), 124–127
  factors determining effective spans, 125–126
  recent trends in spans, 126–127
Span of management, 124–127
SPE. *See* Society of Plastics Engineers (SPE)
Specialized knowledge, 13
Specialized staff, 127
Spinning jenny, 23
Sponsor, 367
Sports medicine, 285
Spread, 254
SRE. *See* Society of Reliability Engineers (SRE)
Srikanth, Mokshagundam L., 241
Staff functions, 127–128
Staff relationships, 127–128
Staff specialists, 127
Staffing, 12
Staffing technical organizations, 137–146
Stahl, Michael, 329
Stand-alone machine, 244

Standard, defined, 219
Standard operating procedures
    (SOPs), 84
Standardization, 219
Starr, Martin K., 89
Starting salary, 141–142
Statement of work, 292, 294, 369
Statistical methods, 254–255
Status, 148
    functional, 148
    scalar, 148
Steam engine, 23, 37
Steel mills, 37
Steiner, George A., 58
STEM (Science, Technology,
    Engineering and Math), 225
Steroids, the, 396
Stevens Institute of Technology, 28, 32
Stevens, John, 25
Stewart, John M., 177
Stonehenge (England), 21, 37
Stopwatch time study, 266
Strategic planning, 74, 76–80, 184
    goals, 79–80
    mission state, 78
    strategies, 76–77
    SWOT analysis (Strengths,
        Weaknesses, Opportunities,
        and Threats), 78–79
    vision statement, 77
Strategies, 50
    business, 231
    layout, 231
    location, 231
    for managing technology, 91–94
        changes in, 92–94
        innovations, 91–92
        inventions, 91–92
    new product, 177–178
        application engineering, 177
        first-to-market, 177
        follow-the-leader, 177
        me-too, 177
Strengths, Weaknesses, Opportunities,
    and Threats (SWOT) analysis,
    78
Strictly liability, 210
Subdivisions, 284
Success criteria, 294

Summa de arithmetica, geometria,
    proportioni et proportionalia
    (Pacioli), 23
Super, Donald E., 366
Supply chain management (SCM), 247
Supply chaining, 396
SWE. *See* Society of Women
    Engineers (SWE)
Swiss watch industry, and
    technological change, 93
SWOT. *See* Strengths, Weaknesses,
    Opportunities, and Threats
    (SWOT) analysis
Synchronized manufacturing/drum-
    buffer-rope, 241
Synthesis, 13
System I, 147
System II, 148
System design review, 208–209
System life cycle, 200, 202, 216
System Safety Society, 212
System/software design review, 209
Systems engineering, 100, 200–202
    functional analysis and allocation,
        208
    phases/stages in, 203–205
        commercial validation, 204
        conceptual stage, 204
        development stage, 204
        disposal stage, 203
        full-scale production stage, 204
        product support stage, 204
        production preparation
            stage, 204
        technical feasibility stage, 204
    requirements analysis, 200
    tasks
        alternative investigation, 201
        integration, 202
        performance assessment, 202
        problem statement, 201
        re-evaluation, 203
        system modeling, 201
Systems engineering phases, 174

TAC. *See* Technology Accreditation
    Commission (TAC)
Tafur, Pero, 22
Taguchi, Genichi, 252, 261

Taguchi methods, 261–262
    loss factor, 262
Tannenbaum, Robert, 54–55
Tasks, 200–203
    alternative investigation, 201
    integration, 202
    performance assessment, 202
    problem statement, 201
    re-evaluation, 203
    system modeling, 201
Taylor, Frederick Winslow, 29–33, 80,
    218, 246
TCP/IP, 41
Team building, 324–325
Team(s), 40, 130–134. *See also* Project
    team
    definition, 130
    information revolution, impact of,
        132–134
    organizational structures, 130
    quality, 264, 270
    virtual, 130
    workplace, 41
'Tear-down' approach,
    brainstorming, 191
Technical competence, 68
Technical functions, 13
Technical organizations, staffing
    human resource planning, 137–139
        hiring managers and technical
            professionals, 138
        job requisition/description, 138
    job application process, 140–143
        appraising performance,
            145–146
        campus interview, 140–141
        cover letter, 140
    job offer, 142–143
        orientation and training, 143–145
    reference checks, 141
    résumé, 140
    site visits, 141
    starting salary, 141–142
Technical people, 47–70
    leadership, 48–57
    motivating and leading, 66–68
    motivation, 58, 59
        content *vs.* process theories,
            60–66

McGregor's Theory X and
Theory Y, 58–60
Technical professionals
differences among, 67–68
general nature of, 66–67
leading, 68–69
breakpoint leadership, 69
dimensions of, 68
as orchestration, 68–69
scientists *vs.* engineers, 67–68
Technical skills, 11
project matrix, 322
Technical societies, 376
getting involved in, 376–377
technical papers and
publications, 377
types/purpose of, 376
Technological change, and top management, 93
Technological forecasting, 89–91
Technological innovation, 92
Technology Accreditation Commission
(TAC), 377
Technology, experiences connecting
with, 131
Technology life cycle, 174–175
Technology management, 385
Technology S-curve, 90
Telephone interruptions, 387
Tennant, Charles, 23
Thamhain, Hans J., 147–148,
324–326, 384
Thayer, Sylvanus, 5, 26
Theory X and Theory Y, 37, 39, 58, 69
Therbligs, 31
Thompson, Arnie, 353–354
Thompson, P. H., 367, 374
3M, 79
Thurow, Lester, 254, 382
Time and materials contracts, 331
Time management, 388–389
daily action list, 388
Email, handling, 388
energy cycle/environment, 388
need for, 386
time log, 388
time wasters, 387–388
tools of, 388–389
Time standards

defined, 265
engineered, 266–267
informal, 266
Time value of money, 180
Timing of control, 157–158
concurrent control, 157
feedback control, 157
feedforward control, 157–158
screening, 157
Timm, Paul, 371
Titanic (construction case study),
349–350
Titular leader, 49
*Tool and Manufacturing Engineers
Handbook*, 240
Top management
need for engineers in, 381–382
and technological change, 93
Top managers, 10
Toshiba, 183
Total Productive Maintenance
(TPM), 270
Total Quality Management (TQM),
258–261
criteria, 259–260
Towne, Henry R., 28–29, 32
Toyota and Takata, product liability,
211
Toyota Motor Corporation, 212, 233
and just-in-time (JIT), 241–243
Toyota Production System (TPS), 246
main goal of, 246
TPM. *See* Total Productive
Maintenance (TPM)
TPS. *See* Toyota production system
(TPS)
TQM. *See* Total Quality Management
(TQM)
Trade secrets, 187
Trademark Law Revision Act of
1988, 186
Trademarks, 186
Trade-off (trade) studies, 208
Training, 143–145
Tran, Kimberly, 350
Transfer pricing, 397
Transformational leadership
full range model, 56–57
contingent reward, 57

idealized influence (Charisma), 56
individualized consideration/attention, 57
inspirational motivation, 56
intellectual stimulation, 57
management by exception, 57
Treece, James B., 243
Tribus, Myron, 92
Trio of needs (McClelland),
63–64
Truman, Harry S., 47
Tsu, Lao, 57
Two-factor theory, 48, 62, 69
Tzu, Hsun, 75, 375

UCL. *See* Upper control limit (UCL)
Uncertainty
decision making under, 110–114
game theory, 113–114
Unclear communication, 388
Uniform Partnership Act, 119
Unifying agent, 315
Units, 128
University of Aston (Birmingham,
England), 129
University of Missouri–Rolla, 345
Unstructured brainstorming, 191–192
Unternehmer (German tradition), 77
Upper control limit (UCL), 256
Urban Dynamics model, 100
Urwick, Lyndall, 31, 33, 35
U.S. Department of Defense (DOD),
122, 308
U.S. Military Academy (West Point),
5, 26, 379
U.S. patent and trademark office
(USPTO), 183, 186, 189
patents granted by, 186
Users' expectation, 85
Utilitarian ethics, 337
Utility, 184
Utility patent, 185
Utley, Dawn R., 69

Vail, Theodore N., 77
Valence, 65
Value analysis, 220
Value engineering/analysis (VE/A), 220
Value proposition, 278, 356

Values, 337–338
Variable budgets, 160
Variable costs, 233
Variables methods, 254
VE/A. *See* Value engineering/analysis (VE/A)
Venice, 22
Verbal feedback, 370
Virtual organization, 405
Virtual product, 405
Virtual teams, 130–131, 405
Visibility, looking for, 363
Vision statement, 77
Visual aids, 373
Vosburgh, Richard M., 365, 367
Vroom, Victor, 48, 64
Vulnerability, 404

Wainer, H., 193
Waiting-line situations, 108
Wall Street Journal, 396
Wal Mart, 282
War of 1812, 5, 25–26
Wass, Donald L., 150
Water frame, 23
Waterman, Robert H. Jr., 42, 69, 80, 167, 282
Watson, Tom Jr., 42
Watt, James, 23–24
WBS. *See* Work breakdown structure (WBS)
Weber, Max, 20, 33–35
Weighted checklist (weighted scoring model), 179

Weighted moving average, 86–87
disadvantage, 86
Weihrich, Heinz, 12, 81, 83, 120, 368
Weinert, Donald W., 376
Weiss, Gorden E., 220
Welch, Jack, 42, 227
Westbrook, Jerry D., 69, 70
Weston, William, 25
Whistleblowing, 349, 350–355
case study
Boeing, 350–351
Challenger disaster, the, 351–355
Whitney, Willis R., 178
Wickenden, William, 4
Wiggins, Calvin, 354
Wilcox, John R., 347
Wilson, Elmina, 379
Wilson, Joseph, 92
Winterbottom v. Wright, 209
W-kanban, 243
Womack, James P., 246
Women as engineering students, 379
Women in engineering, 379–380
Woodward, Joan, and Aston studies, 128–129
Work breakdown structure (WBS), 295
Work challenge, 69
Work measurement, 265–267
engineered time standards, 266–267
informal time standards, 266
resistance to time standards, 267

Work packages, 295
Work sampling, 266
Worker performance and attitude, 80
Workflow software, 395
Work-in-process inventory, 232
Workplace teams, 41
World Bank, 400
World Is Flat, The (Friedman), 133, 395
World Trade Organization (WTO), 184, 400–401
Worthington Industries, 62
Wren, Daniel A., 20, 21, 24, 27–28, 124
Writing careers, 366
Written report, 372
WTO. *See* World Trade Organization (WTO)
Wysocki, Robert K., 308

X-bar control chart, 256
Xerox Corporation
Baldrige award winner, 259
technological innovation, 92

Yankelovich, Dan, 65
Yaverbaum, Eric, 50
Yoi kigyo senshi, 399
Young, Edmund, 7

Zenz, Gary J., 272
Zero-base budgeting, 159
Zumwalt, Admiral, 42
Zysman, John, 224